United States Nuclear Regulatory Commission

Protecting People and the Environment

NUREG-1829
Vol. 1

Estimating Loss-of-Coolant Accident (LOCA) Frequencies Through the Elicitation Process

Main Report

Office of Nuclear Regulatory Research

AVAILABILITY OF REFERENCE MATERIALS
IN NRC PUBLICATIONS

United States Nuclear Regulatory Commission

Protecting People and the Environment

NUREG-1829
Vol. 1

Estimating Loss-of-Coolant Accident (LOCA) Frequencies Through the Elicitation Process

Main Report

Manuscript Completed: March 2008
Date Published: April 2008

Prepared by
R. Tregoning (NRC), L. Abramson (NRC)
P. Scott (Battelle-Columbus)

A. Csontos, NRC Project Manager

Office of Nuclear Regulatory Research

ABSTRACT

The NRC is establishing a risk-informed revision of the design-basis pipe break size requirements in 10 CFR 50.46, Appendix K to Part 50, and GDC 35 which requires estimates of LOCA frequencies as a function of break size. Separate BWR and PWR piping and non-piping passive system LOCA frequency estimates were developed as a function of effective break size and operating time through the end of the plant license-renewal period. The estimates were based on an expert elicitation process which consolidated operating experience and insights from probabilistic fracture mechanics studies with knowledge of plant design, operation, and material performance. The elicitation required each member of an expert panel to qualitatively and quantitatively assess important LOCA contributing factors and quantify their uncertainty. The quantitative responses were combined to develop BWR and PWR total LOCA frequency estimates for each contributing panelist. The distributions for the six LOCA size categories and three time periods evaluated are represented by four parameters (mean, median, 5th and 95th percentiles). Finally, the individual estimates were aggregated to obtain group estimates, along with measures of panel diversity.

There is general qualitative agreement among the panelists about important technical issues and LOCA contributing factors, but the individual quantitative estimates are much more variable. Sensitivity studies were conducted to examine the effects on the estimated parameters of distribution shape, correlation structure, panelist overconfidence, panel diversity measure, and aggregation method. The group estimates are most sensitive to the method used to aggregate the individual estimates. Geometric-mean aggregation produces frequency estimates that approximate the medians of the panelists' estimates and also are generally consistent with both operating experience and prior LOCA frequency estimates, except where increases are supported by specific material aging-related concerns. However, arithmetic-mean and mixture-distribution aggregation are alternative methods that lead to significantly higher mean and 95th percentile group estimates. Because the results are sensitive to the aggregation method, a particular set of LOCA frequency estimates is not generically recommended for all risk-informed applications.

FOREWORD

Estimated frequencies of loss-of-coolant accidents (LOCAs; i.e., pipe ruptures as a function of break size) are used in a variety of regulatory applications, including probabilistic risk assessment (PRA) of nuclear power plants. Currently, the U.S. Nuclear Regulatory Commission (NRC) is using such information to establish a risk-informed alternative to the emergency core cooling system (ECCS) requirements in Title 10, Section 50.46, of the *Code of Federal Regulations* (10 CFR 50.46). Current requirements consider pipe breaks in the reactor coolant pressure boundary, up to and including breaks equivalent in size to the double-ended rupture of the largest pipe in the reactor coolant system. One aspect of this risk-informing activity is to evaluate the technical adequacy of redefining the design-basis break size (the largest pipe break to which 10 CFR 50.46 applies) to a smaller size that is consistent with updated estimates of pipe break frequencies.

To provide the technical basis for a risk-informed definition of the design-basis break size, this study developed LOCA frequency estimates using an expert elicitation process. This process consolidated operating experience and insights from probabilistic fracture mechanics studies with knowledge of plant design, operation, and material performance. Expert elicitation is a well-recognized technique for quantifying phenomenological knowledge when modeling approaches or data are insufficient.

The results from the expert elicitation provide LOCA frequency estimates for piping and non-piping passive systems, as a function of effective break size and operating time through the end of the plant license-renewal period, for both boiling- and pressurized-water reactors (BWRs and PWRs, respectively). The panelists generally agreed on the important technical issues and LOCA-contributing factors. However, as expected, the panelists' estimates exhibit both significant uncertainty and diversity. The uncertainty is reflected in the estimated parameters (mean, median, 5th and 95th percentiles) of the individual LOCA frequency distributions, and the diversity is captured by the confidence bounds on the group estimates. In addition, this study considered the sensitivity of the results to various analysis approaches. The results are most sensitive to the method used to aggregate the individual panelists' estimates to obtain group estimates. In this study, geometric-mean aggregation produces group frequency estimates that approximate the medians of the panelists' estimates and are also generally consistent with both operating experience and prior LOCA frequency estimates except where increases are supported by specific material aging-related concerns. However, arithmetic-mean and mixture-distribution aggregation are alternative methods that lead to significantly higher mean and 95th percentile group estimates.

Because the alternative aggregation methods can lead to significantly different results, a particular set of LOCA frequency estimates is not recommended for all risk-informed applications. The purposes and context of the application must be considered when determining the appropriateness of any set of elicitation results. In particular, during the selection of the BWR and PWR transition break sizes for the proposed 10 CFR50.46a rulemaking, the NRC staff considered the totality of the results from the sensitivity studies, rather than only the summary frequency estimates from this study. The NRC anticipates that a similar approach will be used in selecting appropriate replacement frequencies for the estimates provided in NUREG/CR-5750, "Rates of Initiating Events at U.S. Nuclear Power Plants: 1987 – 1995," and for other applications that require frequencies for break sizes other than those in NUREG/CR-5750.

Jennifer L. Uhle, Director
Division of Engineering
Office of Nuclear Regulatory Research
U.S. Nuclear Regulatory Commission

TABLE OF CONTENTS

LIST OF APPENDICES

LIST OF FIGURES

LIST OF TABLES

EXECUTIVE SUMMARY

The emergency core cooling system (ECCS) requirements are contained in 10 CFR 50.46, Appendix K to Part 50, and GDC 35. Specifically, ECCS design, reliability, and operating requirements exist to ensure that the system can successfully mitigate postulated loss-of-coolant accidents (LOCAs). Consideration of an instantaneous break with a flow rate equivalent to a double-ended guillotine break (DEGB) of the largest primary piping system in the plant generally provides the limiting condition in the required 10 CFR Part 50, Appendix K analysis. However, the DEGB is widely recognized as an extremely unlikely event. Therefore, the staff is establishing a risk-informed revision of the design-basis break size requirements for operating commercial nuclear power plants.

A central consideration in selecting a risk-informed design basis break size is an understanding of the LOCA frequency as a function of break size. The most recent NRC-sponsored study of pipe break failure frequencies is contained in NUREG/CR-5750. Unfortunately, these estimates are not sufficient for design basis break size selection because they do not address all current passive-system degradation concerns and they do not discriminate among breaks having effective diameters greater than 6 inch. There have been two approaches traditionally used to assess LOCA frequencies and their relationship to pipe size: (i) estimates based on statistical analysis of operating experience and (ii) probabilistic fracture mechanics (PFM) analysis of specific postulated failure mechanisms. Neither approach is particularly suited to evaluate LOCA event frequencies due to the rareness of these events and the modeling complexity. In this study, LOCA frequency estimates have been calculated using an expert elicitation process to consolidate operating experience and insights from PFM studies with knowledge of plant design, operation, and material performance. This process is well-recognized for quantifying phenomenological knowledge when data or modeling approaches are insufficient.

The principal objective of this study was to develop separate boiling water reactor (BWR) and pressurized water reactor (PWR) piping and non-piping passive system LOCA frequency estimates as a function of effective break size at three distinct time periods: current-day (25 years fleet average), end-of-plant-license (40 years fleet average), and end-of-plant-license-renewal (60 years fleet average). These estimates are based on the responses from an expert panel and one aim of this study was to obtain estimates that represent a type of group consensus. Additionally, another objective was to reflect both the uncertainty in each panelist's estimates as well as the diversity among the individual estimates.

The elicitation focused on developing generic, or average, estimates for the commercial fleet and the uncertainty bounds on these generic estimates rather than bounding values associated with one or two plants. This approach is consistent with prior studies that did not consider plant-specific differences in developing LOCA frequencies for use in probabilistic risk assessment (PRA) modeling. Consequently, the elicitation panelists were instructed to consider broad differences among plants related to important variables (i.e., plant system, material, geometry, degradation mechanism, loading, mitigation/maintenance) in determining both the generic LOCA frequencies and especially the estimated uncertainty bounds. It is the broad differences in these important variables that contribute most to passive system failure and there is generally sufficient commonality among plants to make such a generic assessment valuable.

The elicitation was solely focused on determining LOCA frequencies that initiate by unisolable primary system side failures that can be exacerbated by material degradation with age. Therefore, active system failures (e.g., stuck open valve, pump seals, interfacing system LOCAs) and consequential primary pressure boundary failures due to either secondary side failures or failures of other plant structures (e.g., crane drops) were not considered. Active system frequency contributions should be combined with the passive system LOCA frequencies to estimate the total system risk. The effects of safety culture on primary side failure (i.e., LOCA) frequencies were also considered.

This study developed LOCA frequency estimates consistent with historical small break (SB), medium break (MB), and large break (LB) flow rate definitions. Additionally, three larger LOCA categories were defined within the classical LB LOCA regime to examine trends with increasing break size, up to and including, a DEGB of the largest piping system in the plant. Contrary to earlier studies, the six LOCA categories are defined in terms of cumulative thresholds rather than break intervals because this definition was more conducive to the elicitation structure. However, the differences between the interval and cumulative estimates are typically much smaller than the estimates' uncertainties. It is therefore recommended that the cumulative threshold estimates from any set of tabular results be used (e.g., Table 1) if interval-defined LOCA frequencies are desired.

Because the LOCA frequency estimates were intended to be both generic and consistent with historical internal-event PRAs, the elicitation primarily considered normal plant operational cycles and loading histories. The loads include representative constant stresses (e.g. pressure, thermal, residual) and expected transient stresses (e.g. thermal striping, heat-up/cool-down, pressure transients) that occur over the license-renewal period. The elicitation implicitly considered all modes of operation based on the loading or operational experience associated with each piping system or non-piping component. Rare event loading from seismic, severe water hammer, and other sources was also not considered in this generic evaluation because of their strong dependency on plant-specific factors. However, a separate research study was conducted to assess the potential impact of seismic loading on the break frequency versus break size relationship. As part of that study, both unflawed and flawed seismic piping contributions were considered. The results of that seismic LOCA analysis are summarized in Section 7.2.

Several important assumptions were made to guide the elicitation process. One such assumption is that plant construction and operation comply with all applicable codes and standards required by the regulations and the technical specifications. The specific impact on the LOCA frequencies of purposefully violating these requirements was not considered. However, it was assumed that regulatory oversight policies and procedures will continue to be used to identify and mitigate risk associated with plants having deficient safety practices. While deviations from these requirements do represent some percentage of the events included in the passive-system failure data, extrapolation of this data implicitly assumes that similar future deviations will continue to occur with similar frequency.

Another important assumption is that current regulatory oversight practices will continue to evaluate aging management and mitigation strategies in order to reasonably assure that future plant operation and maintenance has equivalent or decreased risk. A related assumption inherent in this elicitation is that all future plant operating characteristics will be essentially consistent with past operating practice. The effects of operating profile changes were not considered because of the large uncertainty surrounding possible operational changes and the potentially wide-ranging ramifications of significant changes on the underlying LOCA frequencies.

The expert elicitation process employed in this study is an adaptation of the formal expert judgment processes used to evaluate reactor risk (NUREG-1150), to develop seismic hazard curves (NUREG/CR-6372), and to assess the performance of radioactive waste repositories (NUREG/CR-5411). The process consisted of a number of steps. To begin, the project staff identified many of the issues to be evaluated through a pilot elicitation. The panel members for this pilot elicitation were all NRC staff. A group of twelve panel members was then selected for the formal elicitation. The staff gathered background material and prepared an initial formulation of the technical issues which was provided to the panel. At its initial meeting, the panel discussed the issues and, using the staff formulation as a starting point, developed a final formulation for the elicitation structure. This structure included the decomposition of the complex technical issues which impact LOCA frequencies into fundamental elements so that these important contributing factors could be more readily assessed. Piping and non-piping base cases were also defined for use in anchoring the quantitative elicitation responses. The base cases represent a set of well-defined conditions which could cause a LOCA. A subset of the panel was created to develop quantitative LOCA frequencies estimates associated with the base case conditions. At this initial meeting, the panel was also trained in subjective elicitation of numerical values through exercises and discussion of potential biases.

After this initial meeting, the staff prepared a draft elicitation questionnaire and iterated with the panel to develop the final questionnaire. The panelists quantifying the base case conditions also developed their initial estimates. A second meeting was then held with the entire panel to review the base case results, review the elicitation questions, and finalize the formulation of remaining technical issues. At their home institutions, the individual panel members performed analyses and computations to develop answers to the elicitation questionnaire.

The elicitation questionnaire required panelists to assess the following technical areas: the base case evaluation effort, utility and regulatory safety culture effects on LOCA frequencies, piping system LOCA frequencies, and non-piping system LOCA frequencies. The utility and safety culture questions required the panelists to compare future safety culture with the existing culture and predict the effect on LOCA frequencies. These effects were considered separately from passive system degradation because the panelist judged safety culture and degradation to be independent. The base case evaluation required the panelists to assess the accuracy and uncertainty in the base case analyses, and to also choose a particular base case approach for anchoring their elicitation responses. The piping and non-piping LOCA frequency questions required each panelist to first identify important LOCA contributing factors (i.e., piping systems, materials, degradation mechanisms, etc.) and select appropriate base case conditions for comparison. The panelists were then required to provide the relative ratios between their important contributing factors and the base case conditions based on their knowledge of passive system component failure. Each relative comparison required mid value, upper bound, and lower bound values. The mid value is defined such that, in the panelist's judgment, there is a 50% chance that the unknown true answer lies above the mid value. The upper and lower bounds are defined such that there is a 5% chance that the true answer lies above the upper bound or below the lower bound, respectively. Each panelist was also required to provide their qualitative rationale supporting their quantitative values.

A facilitation team consisting of individuals knowledgeable about the technical issues (substantive members), a staff member with extensive experience conducting expert elicitations (normative member) and two recorders met separately with each panel member in day-long individual elicitation sessions. At these sessions, each panel member provided answers to the elicitation questionnaire along with their supporting technical rationales. The panelists were asked to self-select, based on their expertise, the questions that they addressed. Consequently, several panelists only provided responses for either BWR or PWR plants. After the elicitation sessions, the panel members returned to their home institutions where they refined their responses based on feedback obtain during their session. Upon receipt of the updated responses, the project staff compiled the panel's responses and developed preliminary estimates of the LOCA frequencies. Along with the rationales, these preliminary estimates were presented to the panel at a wrap-up meeting. Panel members were invited to fill in gaps in their questionnaire responses and, if desired, to modify any of their responses based on group discussion of important technical issues considered during the individual elicitations[1]. Final individual estimates of the LOCA frequencies were then calculated and provided to the panel members for final review and quality assurance.

The qualitative insights provided by the panel are reasonably consistent. Panelists identified several advantages and disadvantages with determining the base case LOCA frequencies through operating-experience assessment or PFM analyses. However, there was a general consensus that operating experience provides the best basis for evaluating current-day, small-break LOCA estimates. Hence, many panelists used the operating-experience base cases to anchor their elicitation responses. The panel members also generally expressed the opinions that the future safety culture will not differ dramatically from the current culture, and that the utility and regulatory safety cultures are highly correlated. Many panelists do believe that safety culture can significantly affect LOCA frequencies at specific plants, but there is also an expectation that regulatory actions using existing enforcement measures will diminish both the possibility and impact of deficient safety culture at particular plants. However because it was thought that these plant-specific issues do not affect the generic averages, no specific adjustment to the LOCA frequency estimates was applied to explicitly account for safety culture effects. This decision was endorsed by the elicitation panelists.

There were several technical insights that were consistently identified. Many participants believe that the number of precursor events (e.g., cracks and leaks) is generally a good barometer of the LOCA susceptibility for the associated degradation mechanism. Welds are almost universally recognized as likely failure locations because they can have relatively high residual stress, are preferentially-attacked by many degradation mechanisms, and are most likely to have preexisting fabrication defects. Most panelists also agreed that a complete break of a smaller pipe, or non-piping component, is generally more likely than an equivalent size opening in a larger pipe, or component, because of the increased severity of fabrication or service cracking. Therefore, the biggest frequency contributors for each LOCA size tend to be systems having the smallest pipes, or component, which can lead to that size LOCA. The exception to this general rule is the BWR recirculation system, which is important at all LOCA sizes due to IGSCC. Many panelists thought that aging may have the greatest effect on intermediate diameter (i.e., 6 to 14-inch nominal diameter) piping systems due to the large number of components within this size range and the fact that this piping generally receives less attention than the larger diameter piping and is harder to replace than the more degradation-prone smaller diameter piping.

The participants generally identified thermal fatigue, stress corrosion cracking (SCC), flow accelerated corrosion (FAC), and mechanical fatigue as the degradation mechanisms that most significantly contribute to LOCA frequencies in BWR plants. Generally, the most important BWR degradation mechanism is intergranular SCC (IGSCC), although the panelist's recognize that mitigation has greatly reduced the susceptibility of BWR plants to this mechanism over the past 20 years. With the exception of FAC, similar degradation mechanisms and concerns were also deemed to be important in PWR plants. Specifically, primary water SCC (PWSCC) is a principal concern. Many panelists believe that PWSCC will be mitigated in PWR plants over the next 15 years, but that effective mitigation has yet to be developed and implemented.

The panelists generally agreed on the important technical issues and LOCA-contributing factors. However, the individual quantitative responses are much more uncertain and there are relatively large differences among the panelists' responses. This is to be expected given the underlying scientific uncertainty. The analysis of these responses was structured to account for the uncertainty in the

---

[1] Each panelist's quantitative elicitation responses can be found through the "Electronic Reading Room" link on the NRC's public website (http://www.nrc.gov/) using the Agencywide Documents Access and Management System (ADAMS). The document is found in ADAMS using the following accession number: ML080560005.

individual estimates and the diversity among the panelists. The quantitative responses were analyzed separately for each panel member to develop individual BWR and PWR total LOCA frequency estimates. A unified analysis format was developed to ensure consistency in processing the panelists' inputs. The panelists' mid-value, upper bound and lower bound responses were assumed to represent the median, 95th, and 5th percentiles, respectively, of their subjective uncertainty distributions for each elicitation response. The analysis structure was based on the assumption that all the responses correspond to percentiles of lognormal distributions. These distributions were then combined using a lognormal framework. The final outputs for each panelist are estimates for the means, medians and 5th and 95th percentiles of the total BWR and PWR LOCA frequencies. The panelists' estimates were then aggregated to obtain group LOCA frequency estimates, along with measures of panel diversity.

The individual and group estimates for the means, medians, 5th and 95th percentiles of the LOCA frequency distributions were calculated using the following principal assumptions and choices:

(i)   The mid-value, upper bound, and lower bound supplied by each panelist for each elicitation question were assumed to correspond to the median, 95th percentile and 5th percentile, respectively, of a split lognormal distribution, with the mean calculated assuming that the upper tail is truncated at the 99.9[th] percentile.

(ii)   Only those panelists whose uncertainty ranges were relatively small were adjusted using an error-factor adjustment scheme to account for possible overconfidence (Section 7.6.2.2).

(iii)   Split lognormal distributions were summed by assuming perfect rank correlation among the individual terms.

(iv)   The individual estimates are the total LOCA frequency parameters (i.e., mean, median, 5th percentile, and 95th percentile) determined for each panelist.

(v)   The group estimates of the total LOCA frequency parameters were determined using the geometric means of the individual estimates.

(vi)   Panel diversity was characterized with two-sided 95% confidence intervals based on an assumed lognormal model for the individual estimates.

In the report, these six assumptions and choices define what are termed the *summary* estimates. The report also calculates and discusses *baseline* estimates. These baseline estimates are calculated using all the above assumptions, except no overconfidence adjustment (as described in (ii) above) is applied. Because it is well-established that experts tend to be overconfident, the overconfidence adjustment is deemed to result in improved LOCA frequency estimates

The resultant individual and group summary estimates are consistent with the elicitation objectives and structure and are reasonably representative of the panelists' quantitative judgments. In particular, they are not dominated by extreme results, either on the high or low end, and the geometric means of the individual estimates approximate the medians of these estimates. The median is often used to represent group opinion in elicitations, especially when the individual estimates differ by several orders of magnitude, as they do in this study.

The LOCA frequency summary estimates for the current-day (25 years) and end-of-plant-license (40 years) periods are provided in Table 1 for both BWR and PWR plant types. The aggregated group estimates for the median, mean, 5th and 95th percentiles are presented. Frequency estimates are not expected to change dramatically over the next 15 years for any size LOCA, or even over the next 35 years for BWR LOCA Categories 1- 5 and PWR LOCA Categories 1 - 2 (see results in Appendix L). While order of magnitude increases in BWR LOCA Category 6 and PWR LOCA Categories 3 - 6 over the next 35 years are expected, these increases are largely due to uncertainty about the future and the concern that new degradation mechanisms could arise in the operating fleet. However, while aging will continue, the panelists' consensus is that mitigation procedures are in place, or will be implemented in a timely manner, to alleviate significant increases in future LOCA frequencies for existing degradation mechanisms. Because of the predicted stability in these estimates over the near-term, the current-day (25 year) results can be used to represent the LOCA frequencies over the next 15 years of fleet operation.

The current-day median, mean, and 95th percentile estimates are graphically presented in Figure 1. The 95% confidence intervals calculated for these parameters are also illustrated in this figure. The LOCA frequencies as a function of threshold break diameter were estimated only for the six specified LOCA categories in the elicitation. The plotted points are connected with straight lines in the figure for visual clarity, but this should not be construed as a recommended interpolation scheme. Interpolation of frequencies between category sizes can be done at the user's discretion depending on the conservatism required by the application. Some common interpolation schemes are linear, multi-point nonlinear and cubic spline. A step-wise or stair-step interpolation between two categories where the frequency for the lower category size is used for all flow rates or corresponding break sizes between the two categories provides the most conservative interpolation scheme. Note that any interpolation scheme does not reflect the uncertainty in the interpolated frequencies.

A measure of the individual uncertainties in Table 1 and Figure 1 is given by the differences between the medians and the corresponding 5th or 95th percentile estimates. Panel diversity is reflected in the confidence bounds in Figure 1. The large widths of these confidence bounds (as much as 3 orders of magnitude) reflects the significant diversity of the individual estimates in this study As the LOCA size increased, the panel members generally expressed greater uncertainty in their predictions, and the variability among individual panelists' estimates increased. This is to be expected because of the increased extrapolation required from available passive-system failure data for larger LOCA sizes.

While it is acknowledged that operating experience-based estimates do not necessarily reflect the current state, it is informative to compare such estimates with the elicitation frequencies for the smallest LOCAs (Category 1). This comparison requires the least extrapolation of passive-system failure data since no larger LOCAs have occurred. The BWR and PWR Category 1 LOCA frequencies (including the steam generator tube rupture (SGTR) frequency for PWRs) were estimated up through December 2006. For BWR plants, the average SB LOCA frequency based solely on the number of reported events was estimated to be 5.5E-04 per calendar year. The mean elicitation BWR SB LOCA estimate is 6.5E-04 per calendar year. The BWR elicitation estimate is less than 20 percent higher than the operating-experience estimate which, given the uncertainty of these estimates, is not statistically significant.

For PWR plants, the average SB LOCA frequency estimated analogously is 3.6E-03 per calendar year. The mean elicitation PWR SB LOCA estimate is 7.3E-03 per calendar year. The PWR elicitation estimate is about 100 percent (or a factor of two) higher than the operating-experience-based estimate. Additional insight into this difference can be gained by partitioning the PWR passive-system failure data into frequencies for SGTRs and for all other passive-system SB LOCAs.

Based on reported failures, the mean SGTR LOCA frequency is 3.2E-03 per calendar year. This result is almost identical to the current-day elicitation estimate of 3.7E-03 per calendar year. The frequency of all other Category 1 PWR passive-system failures is 4.0E-04 per calendar year based on operating experience. The corresponding elicitation estimate is 1.9E-03 per calendar year. While this value is 5 times greater than the operating-experience-based estimate, this difference is explained by the elicitation panelists' estimation of the effect of PWSCC on small diameter component failures.

There are several prior studies that also estimated LOCA frequencies. Some care is needed when comparing the elicitation LOCA frequency estimates with these earlier studies because the LOCA categories are defined differently. Specifically, the current-day LOCA Category 1 and 2 estimates in Table 1 are comparable to total system SB, MB, and LB LOCA frequencies, respectively, reported in

NUREG/CR-5750. Additionally, current-day SGTR frequencies (Table 7.18) and PWR LOCA frequencies for all other passive system failures (i.e., frequencies for breaks greater than 100 gpm (380 lpm) in Table 7.19) are comparable to NUREG/CR-5750 SGTR and PWR SB LOCA frequencies when these failure modes are analyzed separately. The NUREG/CR-5750 LB LOCA frequency estimates are best compared to the elicitation LOCA Category 4 frequency estimates because the pipe break sizes are most similar.

After accounting for these differences, the elicitation LOCA frequency estimates are generally much lower than the WASH-1400 estimates and more consistent with the NUREG/CR-5750 estimates. The SB LOCA PWR elicitation estimates after subtracting the SGTR frequencies are approximately 3 times greater than the NUREG/CR-5750 estimates, due to the aforementioned PWSCC concerns. However, the total BWR and PWR SB LOCA frequency estimates are similar once the SGTR frequencies are added to the NUREG/CR-5750 PWR results. The elicitation MB LOCA estimates are higher than the NUREG/CR-5750 estimates by factors of approximately 4 and 20 for BWR and PWR plant types, respectively. These increases are partly due to concerns about PWSCC of piping and non-piping (e.g., CRDM) components as well as general aging concerns with piping in this size range. The NUREG/CR-5750 LB LOCA frequency estimates are slightly higher (less than a factor of 3) than the current elicitation results for both PWR and BWR plants. The generally good agreement between the NUREG/CR-5750 and current elicitation estimates is somewhat surprising given the markedly different methodologies used to arrive at these results.

Table 1 Total BWR and PWR LOCA Frequencies
(After Overconfidence Adjustment using Error-Factor Scheme)

Plant Type	LOCA Size (gpm)	Eff. Break Size (inch)	Current-day Estimate (per cal. yr) (25 yr fleet average operation)				End-of-Plant-License Estimate (per cal. yr) (40 yr fleet average operation)			
			5th Per.	Median	Mean	95th Per.	5th Per.	Median	Mean	95th Per.
BWR	>100	½	3.3E-05	3.0E-04	6.5E-04	2.3E-03	2.8E-05	2.6E-04	6.2E-04	2.2E-03
	>1,500	1 7/8	3.0E-06	5.0E-05	1.3E-04	4.8E-04	2.5E-06	4.5E-05	1.2E-04	4.8E-04
	>5,000	3 ¼	6.0E-07	9.7E-06	2.9E-05	1.1E-04	5.4E-07	9.8E-06	3.2E-05	1.3E-04
	>25K	7	8.6E-08	2.2E-06	7.3E-06	2.9E-05	7.8E-08	2.3E-06	9.4E-06	3.7E-05
	>100K	18	7.7E-09	2.9E-07	1.5E-06	5.9E-06	6.8E-09	3.1E-07	2.1E-06	7.9E-06
	>500K	41	6.3E-12	2.9E-10	6.3E-09	1.8E-08	7.5E-12	4.0E-10	1.0E-08	2.8E-08
PWR	>100	½	6.9E-04	3.9E-03	7.3E-03	2.3E-02	4.0E-04	2.6E-03	5.2E-03	1.8E-02
	>1,500	1 5/8	7.6E-06	1.4E-04	6.4E-04	2.4E-03	8.3E-06	1.6E-04	7.8E-04	2.9E-03
	>5,000	3	2.1E-07	3.4E-06	1.6E-05	6.1E-05	4.8E-07	7.6E-06	3.6E-05	1.4E-04
	>25K	7	1.4E-08	3.1E-07	1.6E-06	6.1E-06	2.8E-08	6.6E-07	3.6E-06	1.4E-05
	>100K	14	4.1E-10	1.2E-08	2.0E-07	5.8E-07	1.0E-09	2.8E-08	4.8E-07	1.4E-06
	>500K	31	3.5E-11	1.2E-09	2.9E-08	8.1E-08	8.7E-11	2.9E-09	7.5E-08	2.1E-07

Sensitivity analyses were also conducted to examine the robustness of the quantitative results to the analysis procedure used to develop the summary estimates. These sensitivity analyses investigated the effect of distribution shape on the means as well as the effects of correlation structure, panelist overconfidence, panel diversity measure, and aggregation method on the estimated parameters. The mean calculation in the analysis procedure used a split lognormal distribution truncated at the 99.9th percentile to obtain reasonably conservative values compared with other possible choices. The correlation structure in the analysis procedure assumed maximal correlation, which is reasonably representative of the elicitation structure. The structure provides conservative 95th percentile estimates. However, based on selected Monte Carlo simulations, an independent correlation structure leads to larger median and 5th percentile estimates. The means are unaffected by the choice of the correlation structure. The analysis procedure also used confidence intervals for the aggregated estimates to measure panel diversity. An alternative approach used quartiles of the individual estimates, leading to comparable, but narrower intervals.

The analysis procedure also adjusted those panelists' responses that had relatively narrow uncertainty bands using an error-factor scheme to account for a known tendency for people, including experts, to be overconfident when making subjective judgments. Sensitivity analyses examined the effects of other overconfidence adjustments of the nominal subjective confidence levels supplied by the panelists. No overconfidence adjustment was also investigated. While blanket overconfidence adjustments can result in large, unsupportable increases in the mean and 95th percentile frequency estimates, no adjustment results in modest decreases in these estimates. Therefore, the error-factor scheme, which adjusts only those panelists who are most overconfident, is deemed to the most appropriate.

Finally, the largest sensitivity is associated with the method used to aggregate the individual panelist estimates to obtain group estimates. The baseline and summary estimates were developed using geometric-mean aggregation. In this study, the geometric-mean aggregation produces frequency estimates that approximate the median of the panelists' estimates and therefore effectively leads to consensus-type results. Therefore, the summary estimates in Table 1 are believed to be a reasonable representation of the expert panel's current state of knowledge regarding LOCA frequencies.

The sensitivity analyses evaluated the effect of using alternative aggregation methods to calculate the group estimates. Specifically, mixture-distribution and arithmetic-mean aggregation were evaluated. For the panelists' responses in this study, these alternative aggregation methods can lead to significantly higher mean and 95th percentile estimates than those obtained using geometric-mean aggregation. Alternative LOCA frequency estimates that are higher than the summary estimates (Table 1) can be derived by using either the summary estimates with 95% confidence bounds (Tables 7.8), the arithmetic-mean aggregated results (Table 7.13), or the mixture-distribution results (Table 7.16). These estimates also incorporate the same overconfidence adjustment as the summary estimates.

Because alternative aggregation methods can lead to significantly different results, a particular set of LOCA frequency estimates is not generically recommended for all risk-informed applications. The purposes and context of the application must be considered when determining the appropriateness of any set of elicitation results. In particular, during the selection of the BWR and PWR transition break sizes for the proposed 10CFR50.46a rule making, the NRC staff considered the totality of the results from the sensitivity studies rather than only the summary estimates from this study The NRC anticipates that a similar approach will be used in selecting appropriate replacement frequencies for NUREG/CR-5750 estimates and for other applications where frequencies for break sizes other than those in NUREG/CR-5750 are required. While the lack of clear application guidance places an additional burden on the users of the study results, those users are in the best position to judge which study results are most appropriate to consider for their particular applications.

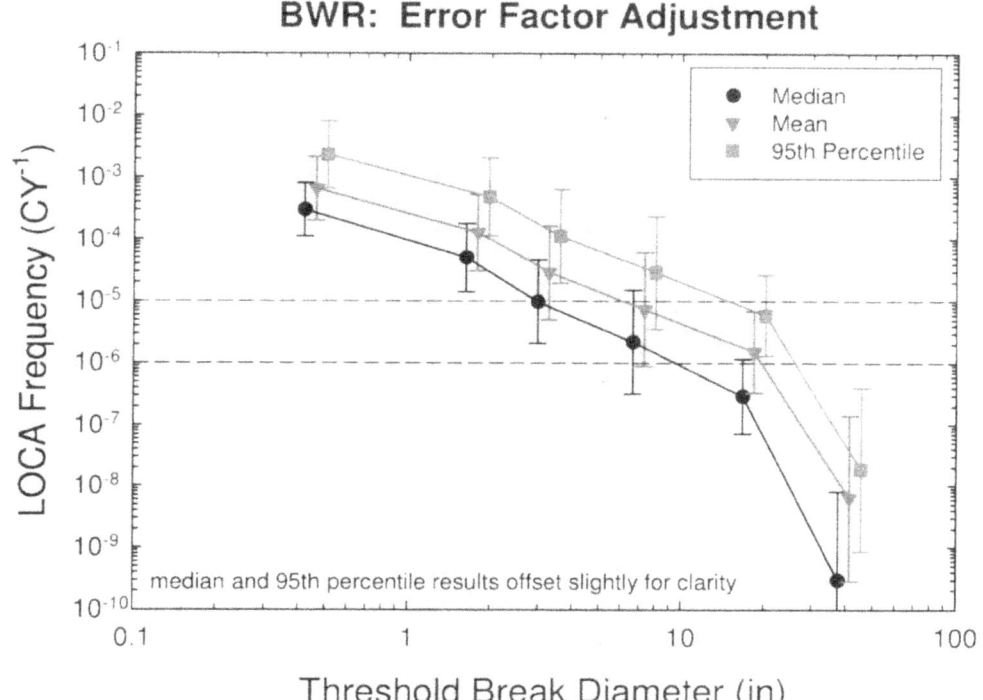

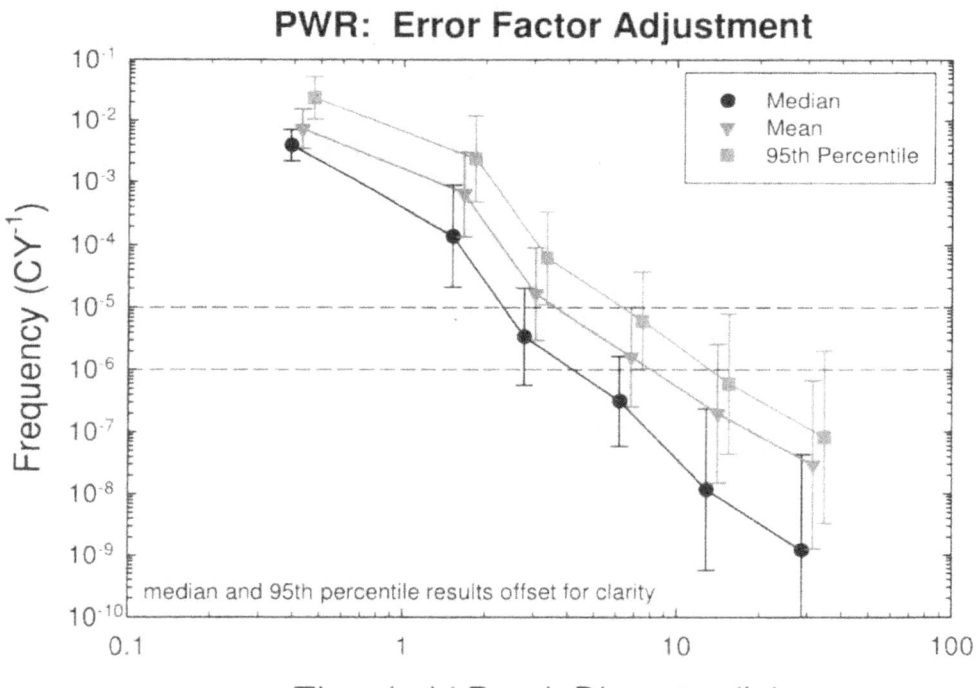

Figure 1 BWR and PWR Error-Factor Adjusted LOCA Frequency Estimates

ACKNOWLEDGMENTS

This work was supported by the United States Nuclear Regulatory Commission (NRC) through the Component Integrity Branch of the Division of Engineering of the Office of Nuclear Regulatory Research under several different contracts. Battelle Memorial Institute support was provided under NRC Contract NRC-04-02-074 (NRC Job Code Y6538). Mr. William Galyean of Idaho National Engineering Environmental Laboratory was supported under NRC Job Code Y6332. Dr. Cory Atwood was supported under NRC Job Code Y6492. Dr. Alan Brothers was supported under NRC Job Code Y6604.

The authors would first like to thank the members of the expert elicitation panel whose insights formed the basis for the results and conclusions reached in this report. The elicitation panel members also provided valuable editorial and technical comments that have been incorporated into this summary document. The panel members included:

- Mr. Bruce Bishop   Westinghouse Electric Co. LLC
- Dr. Vic Chapman   OJV Consultancy Limited
- Mr. Guy DeBoo   Exelon Nuclear
- Mr. William Galyean   Idaho National Engineering Environmental Laboratory
- Dr. Karen Gott   Swedish Nuclear Power Inspectorate
- Dr. David Harris   Engineering Mechanics Technology, Inc.
- Mr. Bengt Lydell   ERIN® Engineering and Research, Inc.
- Dr. Peter Riccardella   Structural Integrity Associates, Inc.
- Mr. Helmut Schulz   Gesellschaft für Reaktorsicherheit (GRS) mbh
- Dr. Sampath Ranganath   Formerly GE Nuclear Energy/Now XGEN Engineering
- Dr. Fredric Simonen   Pacific Northwest National Laboratory
- Dr. Gery Wilkowski   Engineering Mechanics Corporation of Columbus

A special debt of gratitude is expressed to the base case team members (Dr. David Harris, Mr. Bengt Lydell, Mr. William Galyean, and Dr. Peter Riccardella) for the extra effort they provided to conduct the base case and sensitivity analyses discussed in Sections 3 and 4 of this report. In addition, Dr. Riccardella provided base case frequencies for a series of non-piping components and Mssrs. Galyean and Lydell developed the non-piping precursor database. Summaries of the base case team members' analyses are contained in Appendices D – I. Each appendix has been written by the responsible base case team member and these contributions are much appreciated and are vital to the technical basis of this NUREG. Mr. Lydell and Dr. Harris also tirelessly discussed and corrected the authors' characterization of their efforts for accuracy and clarity.

The authors would also like to recognize the facilitation team members who assisted with the development of the technical issues and associated elicitation questions and who also participated in the individual elicitations sessions. Along with the authors, Dr. Ken Jacquay of Casco Services, Inc. was a principal member of the facilitation team

Other members included:

- Ms. Bennett Brady   NRC
- Mr. Frank Cherny   NRC
- Mr. Alan Kuritzky   NRC
- Mr. Arthur Salomon   NRC

The authors are also appreciative of Dr. Corwin Atwood of Statwood Consulting and Dr. Alan Brothers of Pacific Northwest National Laboratory, who provided an external peer review of the elicitation approach and analysis of results. They suggested a number of sensitivity studies to further validate and clarify the conclusions reached in this report. They also provided many helpful comments that have been used to clarify this report. Dr. Atwood, in particular, provided many valuable insights which were used to develop and refine the techniques used to analyze the elicitation results.

Many NRC staff also provided valuable comments and contributions that were used to revise the draft report, including Dr. Arthur Buslik, Mr. Stephen Dinsmore, Mr. Gary Hammer, Mr. Glenn Kelly, and Mr. Arthur Salomon,. In particular, the authors would like to thank Dr. Arthur Buslik for his contribution to Section 5.6.4.4 and Mr. Arthur Salomon for his reviewing and editing assistance. The authors are also grateful to Dr. David Rudland of Engineering Mechanics Corporation of Columbus for assisting with selected Monte Carlo calculations to evaluate the analysis of individual panelist responses.

Finally, the authors would like to thank Dr. Al Csontos of the NRC, who served as the program manager over the final months of this program, and to Ms. Charlotte Matthews and Ms. Patricia Zaluski, both of Battelle-Columbus, for their invaluable assistance in the preparation of this report.

NOMENCLATURE

1.   Symbols

a	Flaw depth
a^*	Crack depth having 50% chance of being detected
b	Value of the upper bound supplied by a panelist
b'	Value of the lower bound supplied by a panelist
$b_p(Y)$	pth percentile of distribution Y
C	Parameter in fatigue crack growth relationship
C_1, C_2	Coefficients used in probability of detection curve definition
D	Diameter
DN	Nominal pipe diameter
E	Young's modulus
$E(Y)$	Expected value of distribution Y
$EF(Y)$	Error factor of distribution Y
f	Fraction of welds inspected for cracks or flaws
f	Inspection coverage/scope
g	Median of the lognormal distribution U
g'	Group estimate of g
J	J-integral fracture parameter
J_D	Deformation J
J_{Ic}	Plane strain J at crack initiation by ASTM813
J_M	Modified J
J-R	J-resistance
k_a	p_ath percentile of the standard lognormal distribution
k_p	pth percentile of the standard lognormal distribution
K	Stress intensity factor
K_I	Mode I stress intensity factor
K_{Ic}	Plane strain stress intensity factor at crack initiation
f_{POD}	Probability of detection function
I	ISI effectiveness factor
L	Length
L_{BC}	Likelihood of a leak due to any degradation mechanism (base case)
L_{PL}	Likelihood of a perceptible leak
L_{TSL}	Likelihood of a technical specification leak
L_{50}	Likelihood of a crack 50% through wall deep
m	Median value
m	Parameter in fatigue crack growth relationship
$m(Y)$	Mean of distribution Y
n	Number of panelists
n_c	Number of cracks or flaws
n_f	Number of failure events
N	Normal operating stress
N	Number of components
N	Number of stress cycles
p	Percentile
P	Pressure
p_a	Value of the percentile in the adjusted distribution corresponding to the original error factor
P_{BC}	Conditional failure probability for the chosen seismic piping base case

P_{FD}	Probability inspected welds will find existing flaw
P_{ND}	Probability of not detecting a crack
P_{PL}	Conditional failure probability of a crack that has just formed a perceptible leak
P_{TSL}	Conditional failure probability of a crack leaking at the technical specification limit
$P_{TSL@SLB}$	Conditional failure probability of a crack leaking at the technical specification limit assuming a Service Level B load
$P_{TSL@SLD}$	Conditional failure probability of a crack leaking at the technical specification limit assuming a Service Level D load
P_{50}	Conditional failure probability of a crack with a maximum depth of 50% of the wall thickness
$p_{L/F}$	Conditional failure probability
p'	Value of the percentile of the panelist's assumed lognormal distribution corresponding to b'
r	Error factor
r_p	Error factor of the panelist's adjusted lognormal distribution
R	Stress ratio
$R_{C/F}$	Number of non-through-wall cracks per leak event
S	Sum of cyclic stress
S	Seismic
S^2	Sample variance
$SD(Y)$	Standard deviation of distribution Y
t	Wall thickness
T	Thermal
T	Tearing modulus
T	Total time
$V(Y)$	Variance of distribution Y
Z_1	Anchoring factor
Z_2	Adjustment ratios
ε	Strain
ε	Probability of not detecting a crack regardless of depth
λ	Pipe failure frequency (through-wall crack)
μ	Mean
ν	Parameter controlling slope of P_{ND} curve
ϕ	Total frequency (cracks and leaks)
σ	Stress
σ	Standard deviation
σ^2	Variance
σ_{DW}	Dead weight stress
σ_{NO}	Normal operating stress
σ_{te}	Thermal expansion stress

2.   Acronyms and Initialisms

ACRS	Advisory Committee on Reactor Safeguards
ADAMS	Agencywide Document Access and Management System
AM	Arithmetic mean
ANL	Argonne National Laboratories
ASME	American Society of Mechanical Engineers
ASTM	American Society for Testing and Materials
BINP	Battelle Integrity of Nuclear Piping

BWR	Boiling water reactor
B&W	Babcock and Wilcox
CBP	Conditional break probability
CD	Compact disk
CDF	Cumulative distribution function
CE	Combustion Engineering
CFR	Code of Federal Regulations
CFP	Conditional failure probability
COD	Crack opening displacement
CRD	Control Rod Drive
CRDM	Control Rod Drive Mechanism
CS	Carbon steel
CV	Correct value
CVCS	Chemical Volume and Control System
CY	Calendar year
DEGB	Double ended guillotine break
DVI	Direct Volume Injection
DW	Dead weight
ECCS	Emergency Core Cooling System
ECSCC	External chloride stress corrosion cracking
EDY	Effective degradation years
EF_a	Error factor after overconfidence adjustment
EF_i	Error factor of distribution Z_i
EF_0	Error factor before overconfidence adjustment
$EF(Y)$	Error factor of Y
EFPY	Effective full power years
Emc^2	Engineering Mechanics Corporation of Columbus
EMT	Engineering Mechanics Technology
EPRI	Electric Power Research Institute
EQ	Elicitation question
FAC	Flow accelerated corrosion
FAD	Failure assessment diagram
FDR	Fabrication defect and repair
FS	Flow sensitive
FW	Feed water
GALL	Generic aging lessons learned
GC	General corrosion
GDC	General Design Criterion
GE	General Electric
GL	Generic letter
GM	Geometric mean
GRS	Gesellschaft für Reactorsicherheit
HAZ	Heat affected zone
HPCS	High Pressure Core Spray
HPI/MU	High Pressure Injection/Make-up
HWC	Hydrogen water chemistry
IAEA	International Atomic Energy Agency
IC	Independent correlation
ICI	In-core Instrumentation
ID	Inside diameter
IGSCC	Intergranular stress corrosion cracking

IHSI	Induction heat stress improvements
INEEL	Idaho National Engineering and Environmental Laboratory
IPE	Individual plant evaluation
IPIRG	International Piping Integrity Research Group
IQR	Interquartile range
IRS	Incident Reporting System
ISI	In-service inspection
IS LOCA	Interfacing system loss of coolant accident
LAS	Low alloy steel
LB	Large break
LB	Lower bound
LBB	Leak before break
LC	Localized corrosion
LEF	Lower error factor
LER	Licensee Event Report
LERF	Large early release frequency
LIV	Loop Isolation Valve
LLNL	Lawrence Livermore National Laboratory
LOCA	Loss of coolant accident
LOOP	Loss of offsite power
LPCS	Low Pressure Core Spray
LPHSW	Last pass heat sink welding
LQ	Lower quartile
LTOP	Low temperature over pressurization
MA	Material aging
MB	Medium break
MERIT	Maximizing Enhancements in Risk Informed Technology program
MF	Mechanical fatigue
MITI	Ministry of International Trade and Industry (Japan)
MRP	Materials Reliability Program
MSIP	Mechanical Stress Improvement Process
MSIV	Main Steam Isolation Valve
MV	Mid value
NB	Nickel-based weld
NDE	Non-destructive examination
NG	Nuclear grade
NPP	Nuclear power plant
NPS	Nominal pipe size
NRC	Nuclear Regulatory Commission
NSSS	Nuclear steam supply system
NUPEC	Nuclear Power Engineering Test Center (Japan)
NWC	Normal water chemistry
OBE	Operational basis earthquake
OD	Outside diameter
OECD	Organization for Economic Co-operation and Development
OPDE	OECD Piping Data Exchange
ORNL	Oak Ridge National Laboratories
PFM	Probabilistic fracture mechanics
PIFRAC	Pipe fracture mechanics material property database
PIV	Pressurizer isolation valve
PNNL	Pacific Northwest National Laboratories

POD	Probability of detection
PORV	Power operated relief valve
PRA	Probabilistic risk assessment
PRC	Perfect rank correlation
PSA	Probabilistic safety assessment
PSI	Pre-service inspection
PSL	Pressurizer Spray Line
PTS	Pressurized thermal shock
PVP	Pressure Vessel and Piping
PWHT	Post weld heat treatment
PWR	Pressurized water reactor
PWSCC	Primary water stress corrosion cracking
QA	Quality assurance
QC	Quality control
RCIC	Reactor Core Isolation Cooling
RCP	Reactor Cooling Piping
RCPB	Reactor Coolant Primary Boundary
RCS	Reactor cooling system
RES	Office of Nuclear Regulatory Research
RH	Reactor head
RHR	Residual Heat Removal
RI-ISI	Risk informed in-service inspection
RPV	Reactor Pressure Vessel
RR	Rolls Royce
RS	Residual stress
RV	Random variable
RWCU	Reactor Water Cleanup
SAM	Seismic anchor motion
SB	Small break
SCC	Stress corrosion cracking
SCSS	Sequence Coding and Search System
SG	Steam generator
SGTR	Steam generator tube rupture
SI	Stress improvement
SIS	Safety Injection System
SKI	Swedish Nuclear Inspectorate
SLB	Service Level B
SLC	Standby Liquid Control
SLD	Service Level D
SRM	Staff requirements memorandum
SRV	Safety relief valve
SQUIRT	Seepage Quantification of Upsets in Reactor Tubes
SS	Stainless steel
SSE	Safe shutdown earthquake
TBS	Transition break size
TF	Thermal fatigue
TGM	Trimmed geometric mean
TGSCC	Transgranular stress corrosion cracking
TMI	Three Mile Island
TS	Thermal stratification
TSL	Technical specification leakage

TWC	Through-wall crack
UA	Unanticipated mechanism
UB	Upper bound
UEF	Upper error factor
UQ	Upper quartile
US	United States
USNRC	United States Nuclear Regulatory Commission
VTC	Video Teleconference
WH	Water hammer
WO	Weld overlay
WOR	Weld overlay repair
WOG	Westinghouse Owners Group

1. BACKGROUND

1.1 Motivation

The emergency core cooling system (ECCS) requirements are contained in 10 CFR 50.46, Appendix K to Part 50, and GDC 35. Specifically, ECCS design, reliability, and operating requirements exist to ensure that the system can successfully mitigate postulated loss-of-coolant accidents (LOCAs). Loss-of-coolant accidents are defined in 10 CFR 50.46(c) as hypothetical or postulated *"accidents that would result from the loss of reactor coolant, at a rate in excess of the capability of the reactor coolant makeup system, from breaks in pipes in the reactor coolant pressure boundary up to and including a break equivalent in size to the double-ended rupture of the largest pipe in the reactor coolant system"*. In addition, 10 CFR Part 50, Appendix K, paragraph (I)(C)(1), states that *"In analyses of hypothetical loss-of-coolant accidents, a spectrum of possible pipe breaks shall be considered. This spectrum shall include instantaneous double-ended breaks ranging in cross-sectional area up to and including that of the largest pipe in the primary coolant system."* The LOCA definition in 10 CFR Part 50, Appendix A, expands this definition to consider "breaks in the reactor coolant pressure boundary, up to and including a break equivalent in size to the double-ended rupture of the largest pipe of the reactor coolant system." The consideration of instantaneous breaks with a flow rate equivalent to a double-ended guillotine break (DEGB) of the largest primary system piping generally provides the limiting condition in the required 10 CFR Part 50, Appendix K analysis.

A DEGB of the largest primary system piping is widely recognized as an extremely unlikely event. Furthermore, the consideration of this event in nuclear plant design and operation requires significant resources that may not be commensurate with the associated risk. Focusing resources on more risk-significant events in a manner consistent with Regulatory Guide 1.174 [1.1] could potentially improve plant safety and allow those regulatory requirements that have negligible impact on plant safety to be relaxed. In an effort spur a risk-informed reevaluation of the regulatory requirements, the Commission provided NRC staff with direction to proceed with a study of risk-informing the technical requirements of 10 CFR Part 50 [1.2]. The staff provided its plan and schedule for this work in SECY-99-264, "Proposed Staff Plan for Risk-Informing Technical Requirements in 10 CFR Part 50," dated November 8, 1999. In this plan, the risk-informed reevaluation of 10 CFR 50.46 requirements was prioritized. The Commission approved this plan on February 3, 2000 [1.3]. Since that time, the staff has provided five status reports to the Commission: SECY-00-0086, SECY-00-0198, memorandum to the Commission dated February 5, 2001, SECY-01-0133, and SECY-02-0057 [1.4]. A staff requirements memorandum (SRM) was provided on March 31, 2003 [1.5] addressing the latest status report, SECY-02-0057. This SRM provides explicit staff direction on risk-informing 10 CFR 50.46, Appendix K, and GDC 35. Most relevantly, the SRM directed the staff to consider a risk-informed revision of the design-basis break size requirements and certain non-functional changes to the design basis of operating commercial nuclear power plants.

A central consideration in selecting a risk-informed design basis break size is an understanding of the LOCA frequency as a function of break size. The most recent NRC-sponsored study of pipe break failure frequencies is contained in NUREG/CR-5750 [1.6]. These frequencies are currently the basis for initiating event frequencies in many plants' probabilistic risk assessment (PRA) models. However, there are several concerns with utilizing the NUREG/CR-5750 LOCA frequency estimates for the risk-informed reevaluation of 10 CFR 50.46. First, several degradation mechanisms have emerged at plants within the last few years which were not previously evident within the service period covered by the NUREG/CR-5750 estimates. These include primary-water stress-corrosion cracking (PWSCC) of pressurized water reactor (PWR) alloy 82/182 welds, PWR vessel head degradation at Davis Bessie, hydrogen combustion failures of the type experienced at the Hamaoka and Brunsbüttel plants, and control-rod-drive mechanism (CRDM) cracking.

Second, LOCAs can also occur from non-pipe break passive failures (e.g. CRDM nozzles, valve bodies, vessel head degradation, and steam generator tubes). The NUREG/CR-5750 estimates only focused on piping failures. Therefore, LOCA contributions from non-piping passive system failures must also be included in the initiating event estimates. Third, forward-looking LOCA frequency estimates are required to understand the future ramifications associated with possible changes to the 50.46 regulation up to the end of the license-renewal period for approved plants. The NUREG/CR-5750 estimates are unavoidably based solely on the historical operating performance and did not explore the future relevance of these estimates.

Finally, the NUREG/CR-5750 estimates defined LOCA break sizes in a manner consistent with current PRA classification using small break (SB), medium break (MB), and large break (LB) LOCA categories that are loosely based on plant mitigation requirements for each break size. The large break category represents the cumulative frequency of a rupture with an equivalent diameter greater than 6 inches. These frequencies will not be representative of the failure frequency of a DEGB in the largest primary system piping, which is approximately 30 inches in diameter for most PWR plants. Therefore, frequency estimates of breaks larger than an equivalent 6-inch diameter pipe need to be considered when risk-informing 10 CFR 50.46.

A review of the NUREG/CR-5750 and other historical LOCA frequency estimation techniques follows. The strengths and weaknesses of each technique are highlighted with respect to many of the concerns enumerated above. The proper consideration of the role of mitigation in LOCA frequency estimates is discussed as well as a description of non-historical failure modes that are important to consider. The relative merits of expert elicitation are also outlined and are the basis for its use to estimate LOCA frequencies in this study.

1.2 Previous Approaches for Estimating LOCA Frequencies

There have been two approaches traditionally used to assess LOCA frequencies as a function of pipe size: estimates based on statistical analysis of operating experience and probabilistic fracture mechanics (PFM) analysis of specific postulated failure mechanisms. Both of these approaches have unique strengths and weaknesses for determining LOCA frequencies. In many ways, the two methods are complementary although combined or comparative analyses utilizing both approaches are rare.

1.2.1 Operating-Experience-Based LOCA Estimates
There are several distinct advantages to using operating experience to directly calculate LOCA frequencies. Operating experience can provide a historical evaluation of actual initiating event frequencies. This evaluation is more realistic as the LOCA size decreases because actual ruptures of smaller diameter Class 1 piping and non-piping (i.e., steam generator tubes) systems have occurred. Additionally, operating experience can be used to identify the degradation mechanisms in piping systems and non-piping components which have led to defect repair or material replacement under operating plant conditions. Operating experience can also provide, in the long term, information on the effectiveness of specific mitigation techniques and some indication of the likelihood of precursor events (e.g., cracks and leaks) prior to failure for specific degradation mechanisms. Detailed evaluation of operating experience can also be used to define how operating conditions, service environment, design characteristics, and fabrication techniques can lead to degradation. All this information is valuable for estimating LOCA frequencies.

However, some natural deficiencies make it challenging to directly calculate LOCA frequencies from operating data. With the exception of steam generator tube rupture (SGTR), there have been few passive failures of Class 1 piping and non-piping systems. Additionally, most of these failures that have occurred have resulted in leak rates below the SB LOCA threshold. No MB or LB LOCAs have occurred. Because pipe break LOCAs have been sparse, the LOCA frequencies are typically estimated using rupture precursor events such as partial or through-wall cracking. These events, for larger class 1 piping, have also been relatively sparse. Therefore, it is important to capture all precursor failure events for estimating passive-system failure rates. However, it has been difficult to construct a comprehensive database because precursor failure information comes from a variety of information sources. Additionally, the precursor events in the operating experience must be conditionally related to LOCA frequencies, and not all LOCAs evolve from detectable precursor events (e.g., hydrogen explosions). The relationship between precursor and non-precursor events and associated LOCA failure frequencies is a function of the aging mechanism. This relationship, for LOCA determination, cannot be explicitly developed from the operating experience.

Another limitation is that past operating experience is also not necessarily representative of future system performance. New material degradation mechanisms can systematically increase generic LOCA frequencies with time compared with frequencies based on operating experience. Aging mechanisms can require significant incubation time before degradation precursor or failure events are observed. However, once the incubation period is over, degradation can occur relatively rapidly and lead to an increased failure frequency compared with prior operating experience. The failure frequency can then remain elevated until effective mitigation procedures are developed to reduce the failure propensity of the degradation mechanism.

Conversely, industry-wide mitigation programs can produce systematic decreases in LOCA frequencies compared to historical levels. However, it is difficult to judge the effectiveness of the generic mitigation programs using operating experience until sufficient time has past so differences in pre- and post-mitigation precursor event frequencies can be evaluated. The effects on the LOCA frequencies due to the emergence of a new degradation mechanism followed by industry-wide mitigation is illustrated by the intergranular stress corrosion cracking (IGSCC) experience in boiling water reactor (BWR) plants [1.7]. Precursor cracking events increased dramatically, starting in the late 1970s as wide-spread IGSCC was discovered in the BWR fleet. It was then several years before effective mitigation strategies were generically adopted. While the effectiveness of these mitigation strategies was demonstrated through other means, several more years and inspection cycles were required to judge the mitigation effectiveness using operating experience. Recent PWSCC [1.8] in PWR plants is another aging mechanism with similar characteristics to IGSCC. It is anticipated that precursor cracking events will increase as PWSCC is observed, followed by a decreased number of cracking events once effective mitigation measures have been applied throughout the PWR fleet. Because future LOCA frequencies may not be constant, it is crucial to consider the effects that aging mechanisms may have on historical piping precursor and failure rates.

1.2.1.1 WASH-1400 - The first systematic study of piping failure in the nuclear industry is contained within WASH-1400, which was completed in 1975 [1.9]. At that time, the total number of years of reactor operating experience was less than 200. Therefore, the pipe LOCA frequencies were based on failure data from other but similar industries. WASH-1400 examined data from naval nuclear reactors, experimental reactors, United Kingdom military experience, commercial power plants, and the oil and gas transmission pipeline industry. The most comprehensive data was obtained from the oil and gas pipeline industry and formed the basis of the WASH-1400 LOCA frequency estimates, after normalizing to account for nuclear plant pipe lengths.

It was certainly realized in the WASH-1400 study that transmission pipeline materials, quality assurance, in-service inspection (ISI), operating conditions, failure modes, and environments in non-nuclear industries were vastly different from, and in most cases inferior to, commercial nuclear requirements. Therefore, the WASH-1400 analysis provided a conservative estimate of the pipe break LOCA frequency.

However, this estimate was considered the best available, given the relatively scant nuclear experience available at the time of this study. The WASH-1400 pipe break estimates formed the basis of NUREG-1150 estimates in 1987 [1.10]. The NUREG-1150 estimates were calculated by performing a Bayesian update of the WASH-1400 estimates with the information that no additional LOCAs failures had occurred in the interim.

1.2.1.2 NUREG/CR-5750 - The next NRC-sponsored evaluation of pipe break LOCA frequencies was provided by Appendix J of NUREG/CR-5750, "Rates of Initiating Events at U.S. Nuclear Power Plants: 1987 - 1995" [1.6]. In this study, the authors evaluated nuclear piping failures and estimated separate frequencies for BWR and PWR reactors. For BWR plants, only U.S. experience was considered for a total of 710 reactor calendar years. The PWR database combined U.S. and "Western-style" PWR data from international experience for a total of 3,362 reactor calendar years. The authors utilized different methods to calculate pipe break frequencies as a function of break size. The SB LOCA estimates were calculated using a Bayesian update of the WASH-1400 SB LOCA estimates with the information that no SB LOCAs occurred between the WASH-1400 and NUREG/CR- 5750, Appendix J studies. This is analogous to the earlier update for the NUREG-1150 estimates.

The MB and LB LOCA frequencies were derived from precursor leak frequencies determined from operating experience. The leak frequency was multiplied by a Beliczey and Schulz conditional pipe break probability (CBP) which is inversely related to pipe diameter [1.11]:

$$CBP = 2.5/DN \qquad (1.1)$$

In Equation 1.1, DN is the nominal pipe diameter in millimeters. The NUREG/CR-5750, Appendix J analysis capped this expression at 0.01 for piping greater than 250 mm to impart some conservatism to the estimates. The advantage of this approach is that there had been several reported leaks of primary pressure boundary piping, but no failures. Therefore, operating experience could be utilized directly to determine the pipe leak frequency and only the CBP given a precursor leak needed to be estimated. One concern is that the Beliczey and Schulz expression was developed for fatigue crack failures only. It is not expected to be applicable for other aging mechanisms. A bigger concern with this precursor approach is that it ignores failure contributions from existing flaws or degradation that does not result in a leak. There are many potential initiating events which do not exhibit a precursor leak. Recent hydrogen combustion failures of residual heat removal piping at Hamaoka [1.12] and auxiliary coolant system piping at Brunsbüttel [1.13] represent one such mechanism. Flow accelerated corrosion (FAC), which induced a rupture of an 18 inch diameter feedwater suction pipe elbow at Surry 2 in 1986 [1.14], is another mechanism which can lead to rupture prior to precursor leaking.

The NUREG/CR-5750, Appendix J approach also applied a mitigation factor of 1/20 to the BWR leak rate data to account for the effectiveness of IGSCC mitigation strategies. The IGSCC mechanism was the prevalent BWR piping failure mechanism within the precursor database. This mitigation factor was justified in light of analysis of the operating experience [1.15] and quantitative estimates of the improvement in reliability for all mitigation strategies [1.16]. There are several different IGSCC mitigation techniques that have been applied including: hydrogen water chemistry addition, weld overlay, increased inspection, and material replacement with a less susceptible austenitic grade of piping (L or NG grades) [1.16]. Some plants have applied single or multiple mitigation strategies and the effectiveness of each particular strategy will obviously vary and may not be well-characterized by a single mitigation factor. Additionally, the Germans have experienced IGSCC in some of the low carbon, less susceptible steels [1.17]. Therefore, this mitigation factor may also not be applicable to describe the future effectiveness of IGSCC mitigation measures up to the end of the license-renewal period.

1.2.1.3 Barsebäck-1 Estimation - In the same time period as the NUREG/CR-5750 evaluation, the Swedish Nuclear Inspectorate (SKI) initiated an effort to develop an international piping failure database [1.18]. This differed from earlier studies in several fundamental ways. Most importantly, it was the first study to concentrate solely on pipe failure and it included class 2 and class 3 piping failures as well as class 1, or primary system piping. Secondly, several data sources were utilized to corroborate each event and determine the metallographic root failure cause whenever possible. Thirdly, it was constructed to allow specific queries by material engineers and PRA practitioners. The resultant database has been employed to develop initiating event frequencies, evaluate emerging failure trends, and judge the effectiveness of mitigation strategies.

The SKI effort culminated in 1997 with a guide for using the database to evaluate piping reliability in terms of important influence and attribute factors [1.15]. The report stresses the need to consider each failure mechanism separately and develop estimates from the database with respect to the influence factors for that mechanism. For instance, FAC is most severe in carbon steel piping at pipe tees and elbows. Pipe failure frequencies associated with FAC are then estimated per reactor year and per the number of carbon steel pipe tees and elbows in the plant using the database. Once the relevant attributes of a specific plant are determined (i.e. the number of carbon steel pipe tees and elbows), plant-specific event frequencies can be estimated. This process is repeated for all potential failure mechanisms (e.g., IGSCC, vibration fatigue, water hammer, etc.). Finally, all the relevant mechanisms are combined to estimate total LOCA frequencies [1.15].

This framework was followed to develop failure rates in the Barsebäck-1 reactor coolant pressure boundary piping [1.19]. Barsebäck-1 is a third generation BWR by ABB-Atom that was closed in 1999. The study collected all the plant-specific attributes of the Barsebäck-1 plant. The attributes include the pipe materials, geometry, pipe length, number of pipe welds, number of pipe connections, and the connection type. Only medium and large LOCA frequencies were determined by considering those failure mechanisms represented within the database.

While the initial SKI-effort database developments and the Barsebäck study were completed in 1998, the database has been maintained and updated through the current period. There is currently a three-year OECD-sponsored Piping Database Exchange (OPDE) project involving the U.S. and twelve other European and Asian countries that is expanding the coverage of the database for events occurring during the 1990's, adding current events, and improving the accuracy and completeness of all included events. Specific attention is focused on including indications of non-leaking flaws discovered during pre-service and in-service inspections. The 1998 version of the database included 1,880 U.S. piping failures and a total of 2,416 failures worldwide. Currently, in June 2004, the database contains nearly 3,200 U.S. piping failures and 4,600 worldwide failures. Therefore, the database completeness has been enhanced by the inclusion of over 70% more US events and almost double the number of international events since 1998.

1.2.1.4 Comparison of Operating-Experience-Based LOCA Estimates - The pipe break LOCA frequencies from the WASH-1400, NUREG/CR-5750, Appendix J, and the Barsebäck-1 study using the SKI pipe database are compared in Table 1.1. As previously noted, these estimates differ dramatically in the methodology used, the completeness and relevance of the underlying database, and the time period of study. The WASH-1400 estimates are understandably higher because the oil and gas transmission industry does not have the same rigorous design practice, quality assurance, and inspection as the nuclear industry. Also, pipe failure from external damage (e.g., mechanical damage from construction equipment, external corrosion, etc.) represents the largest contributing failure mechanism in this industry. The WASH-1400 estimates also reflect the necessary conservatism because of the lack of operating experience at the time of this study.

The application of actual service data in NUREG/CR-5750, Appendix J and the SKI-pipe database in the Barsebäck-1 study reduced the estimated break frequencies by an order of magnitude compared to WASH-1400 (Table 1.1). There is roughly a factor of two difference between these two studies, with the Barsebäck study predicting more MB LOCA events and fewer LB LOCA events for BWRs. This difference is primarily due to the differences in the analysis methodology and the application of plant specific data for Barsebäck-1. The BWR frequencies in both the NUREG/CR-5750 and Barsebäck-1 studies reflect the successful mitigation strategies adopted to combat IGSCC in the early and mid-1980's.

Table 1.1 Comparison of Mean Results from Previous Studies
(per Calendar Years)

Reactor Type	Analysis	SB LOCA (x 10^{-5})	MB LOCA (x 10^{-5})	LB LOCA (x 10^{-5})
BWR	WASH-1400	300	80	30
	NUREG/CR-5750	40	2.6	2.4
	Barsebäck-1	NA	3.4	0.8
PWR	WASH-1400	300	80	30
	NUREG/CR-5750	40	3.0	0.36

1.2.2 PFM LOCA Estimates

The PFM models are attractive because they can parametrically assess the effects of possible mitigation strategies, including inspection, on future piping system performance for particular degradation mechanisms. This is a valuable feature for assessing the possible severity of emerging degradation mechanisms where there is both a dearth of operating experience and a lack of understanding of mitigation effects. However, it is difficult to utilize PFM models for calculating absolute LOCA frequency estimates because many of the input variables and model assumptions are overly simplistic and may not adequately represent true plant conditions. This oversimplification also leads to increased uncertainty in the results when the input uncertainties are taken into account. It is also difficult to benchmark PFM models using the sparse piping failure information. Instead, it is often necessary to benchmark PFM models using the limited existing data for large non-leaking flaws or for very small leaks. The models can then be used to calculate the probabilities of much larger leaks or complete failures as a function of operating time. The PFM failure rate predictions for future operation can only be considered to be realistic if estimates of current precursor or failure rates are consistent with operating experience.

The international community (e.g., SKI in Sweden and GRS in Germany) has sponsored several PFM-based predictions of passive system failure rates [1.20, 1.21]. The NRC and the nuclear industry have also sponsored PFM-based research over the last 20 years in an attempt to develop LOCA frequency estimates from first principles including an estimate of pipe failure probability in Westinghouse PWR reactor coolant loops [1.22] and a reliability analysis of stiff versus flexible piping [1.23]. Additionally, joint international research and investigation of primary piping system integrity has been conducted by the International Piping Integrity Research Group programs (IPIRG1 and IPIRG2) [1.24, 1.25] as well as the Battelle Integrity of Nuclear Piping program (BINP) [1.26].

There was also a large body of research sponsored during the 1980's to evaluate the leak-before-break propensity of BWR and PWR plants [1.27-1.30]. These studies estimated the frequencies of DEGB in BWR main steam, feedwater, and reactor recirculation piping, and in PWR reactor coolant loop piping. Also, the frequencies of seismic-induced direct and indirect DEGB in PWR plants were estimated. Direct DEGB is induced by fatigue crack growth due to the combined effects of thermal, pressure, seismic, and other cyclic loads. Indirect DEGB occurs when the failure of other passive components due to seismic loading causes a consequential failure in the primary piping system. An example would be the failure of PWR steam generator supports during an earthquake, leading to additional primary coolant piping loads and possible piping failure. Table 1.2 is a summary of the median and 90th percentile estimates of the DEGB frequencies for PWRs. The results reflect contributions from all earthquakes, not just those with SSE or greater magnitudes.

Table 1.2 DEGB Frequencies for PWRs

| Plant Type | DEGB Frequencies [1/Reactor Year] | | | |
| | Direct DEGB | | Indirect DEGB | |
	Median	90th Percentile	Median	90th Percentile
CE PWRs	6 E-14 to 5 E-13	4 E-12 to 7 E-11	5 E-17 to 6 E-6	3 E-14 to 5 E-5
Westinghouse PWRs	2 E-13 to 3 E-11	8 E-10 to 1 E-9	5 E-8 to 5 E-6	1 E-6 to 5 E-5
B&W PWRs	< 1 E-10		6 E-11 to 2 E-7	8 E-9 to 1 E-5

Several of the more well-known US computer codes arising from these various efforts include PRAISE [1.31], SRRA [1.32], PSQUIRT [1.33], and PROLBB [1.34]. Each code has distinct capabilities, and there are marked differences in structure, approach, modeling assumptions, and the degree of benchmarking with failure data. The PRAISE models have been benchmarked with the through-wall fatigue cracks in a PWR feedwater line near the steam generator and with the small leak in a BWR recirculation line inlet nozzle safe end [1.35] and with the observed leak probabilities due to IGSCC [1.36]. The SRRA code [1.32] has been benchmarked with observed repair data for flow-assisted corrosion and directly compared with the PRAISE results for fatigue [1.31]. Crack initiation, sub-critical crack growth, leak rate calculation, and fracture/failure analysis are the four major components required for life prediction using PFM, yet the various codes handle these modules using vastly different approaches, or not at all.

For instance, the SRRA code uses a limit-load failure criterion while the PROLBB code uses an elastic-plastic fracture mechanics based criterion. This single difference can result in extensive output variability in the pipe rupture probability. Figure 1.1 depicts the conditional pipe failure probability for a hypothetical through-wall crack as a function of leak rate. To develop the graphs, the crack size is increased systematically and the associated leak rate and conditional failure probability are calculated. The graph on the left assumes that the pipe fails upon reaching limit load, i.e., the crack is in high toughness wrought stainless steel. The graph on the right assumes that the crack is in lower toughness stainless steel weld metal such that failure occurs below the pipe limit stress. This single difference results in up to a 5 order of magnitude difference in the conditional failure probabilities. For a conservatively assumed rupture loading event frequency of 10^{-3} per year, the failure frequency is about 10^{-10} per calendar year for the base metal and about 10^{-7} per calendar year for the weld metal, assuming loading at 50% of service level A and a 1 gpm (3.8 lpm) leakage detection threshold. This threshold is approximately the level required to locate unidentified leakage within nuclear plants. Either analysis could be realistic for a given weld joint depending on the actual pipe material, fabrication practice, and crack location.

The PFM analysis must also model each potential aging mechanism of concern (FAC, IGSCC, PWSCC, corrosion fatigue, vibration fatigue, etc.) to develop comprehensive LOCA estimates. Each mechanism potentially requires a different degradation/cracking model and corresponding set of material properties to accurately calculate the LOCA frequency. Probabilistic fracture mechanics codes for all of the above mechanisms also utilize different approximations for input parameters such as the applied load magnitude and spectrum, pipe boundary conditions, residual stress contribution, and initial flaw distribution. While the codes are complex, many simplifying assumptions are necessary and the model of actual plant conditions is, at best, approximate. Given these modeling realities, and the inherent sensitivity of fatigue and crack growth calculations to the initial conditions and modeling assumptions, it is not surprising that the results from the various models can vary dramatically.

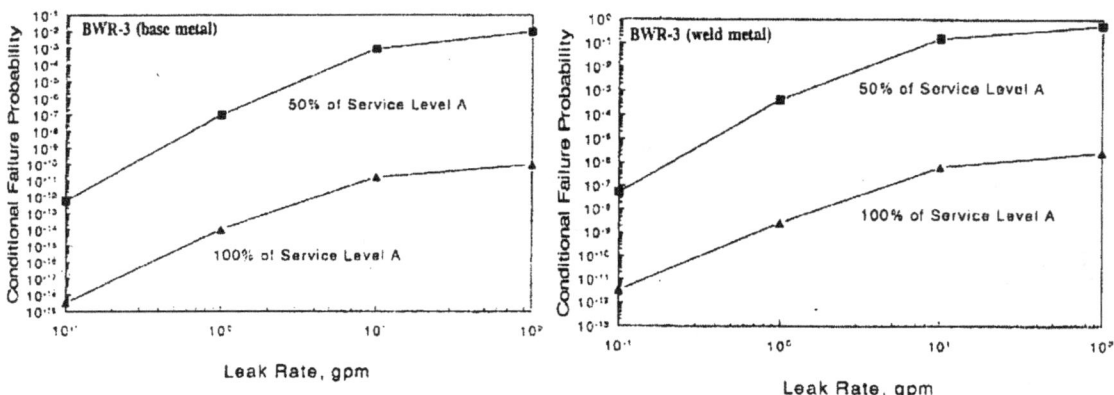

Figure 1.1 Sensitivity of Conditional Failure Probability to Failure Mode Assumption

1.3 Modeling Mitigation Effects

Once degradation mechanisms are discovered and assessed, it is expected that mitigation measures will be eventually developed and applied in order to ultimately decrease the passive system degradation rate to historical levels, or lower. Mitigation includes such measures as inspection, material replacement, repair, and changes in operating conditions. Therefore, the expected impact of a new degradation mechanism on LOCA frequencies is a short term increase followed by a decrease due to mitigation after some time period. This effect, and the benefits of successful mitigation, are demonstrated by the response to the IGSCC cracking phenomena in the early-1980s (Figure 1.2). There was a period of two to three years when the cracking mechanism and its prevalence was discovered in susceptible systems. There was another two to three year period before mitigation was developed and fully implemented. During the approximately four year period of diagnosis and mitigation of the IGSCC problem, the instantaneous discovery rate was approximately one order of magnitude higher than historical levels. However, since mitigation has been employed, the occurrence of IGSCC is less than pre-1980 values [1.19].

It can be difficult to account for the effects of mitigation in either operating-experience-based or PFM LOCA frequency estimates. The pre-mitigation data is part of the operating experience and must be properly screened when estimating frequencies for plants which have implemented mitigation measures. While the general effect of mitigation on IGSCC is apparent, the effect of individual mitigation measures at specific plants is less obvious. This is relevant because several different IGSCC mitigation options are practiced, often in combination, and it is difficult to quantify the effects of individual measures from the data alone. The frequencies estimated by the NUREG/CR-5750 and Barsebäck-1 studies did take into account the general effect of IGSCC mitigation when calculating frequencies (Table 1.1). However, both these studies had the benefit of several years of mitigation experience as a basis for their analyses when the studies were performed.

In PFM-based analysis, an explicit treatment of the effects of individual mitigation measures on the degradation rate is possible. The PFM approach can be especially valuable for comparing the relative differences between pre- and post-mitigation failure probabilities. An ASME-sponsored effort [1.37] examined the effects of mitigation techniques such as stress improvement and weld overlay for BWR recirculation piping subject to IGSCC. However, mitigation measures are often simplistically modeled

1-8

and absolute predictions of post-mitigation degradation rates, and failure frequencies, are subject to the same limitations discussed earlier for PFM models. A common mitigation technique usually considered in PFM analysis is the effect of inspection on the failure frequencies. The impact of inspection is a function of the periodicity, the accuracy of the inspection technique, and operator skill. The accuracy and operator skill are typically modeled by an assumed relationship between the probability of detection (POD) and defect size. Unfortunately, for many inspection techniques, this relationship is highly uncertain and the curves themselves are often developed under idealized laboratory conditions. Therefore, they may not be representative of actual field experience. Furthermore, the POD relationship chosen for the models can significantly influence the predicted failure rates. The possible impact associated with inspection accuracies have been documented by an ASME Research Task Force [1.37].

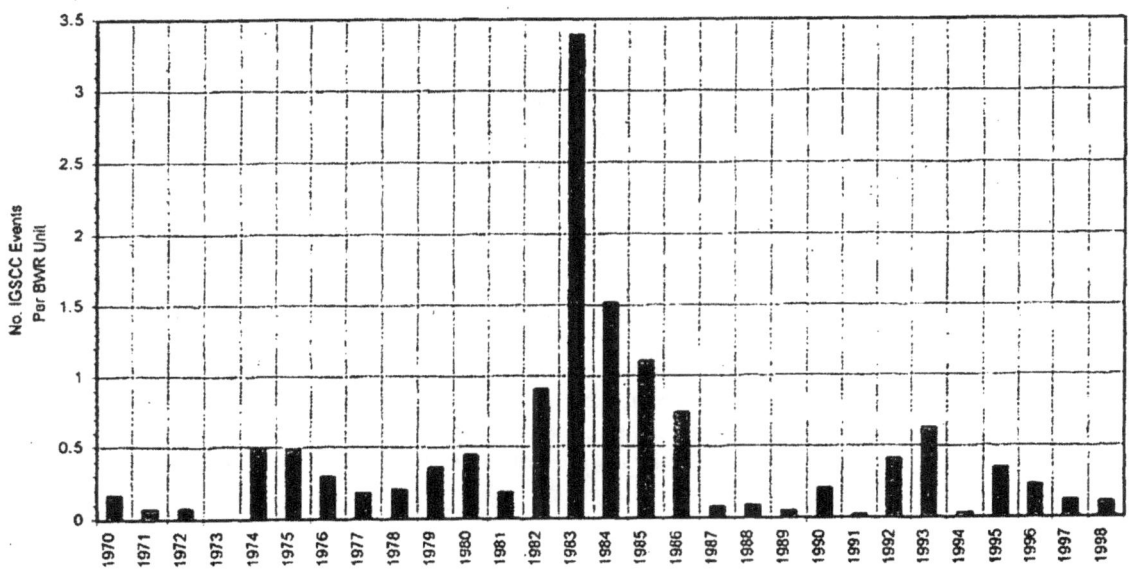

Figure 1.2 IGSCC Events by Year

1.4 Other LOCA Initiating Events

The LOCA frequencies are also influenced by active component failure. Active components are defined as those components subject to periodic maintenance and those that contain moving parts which are subject to either wear or failure (e.g., pump seals, SRV, PORV, etc.). Active component failure was not considered in this study. Therefore, the active-system LOCA frequencies should be combined with the passive system LOCA frequencies from this study in order to estimate total system risk. Because active component failures occur regularly, their frequency can be estimated from the operating experience database as in [1.6].

Steam generator tube rupture is a passive component failure which was also considered in this study. Because tube ruptures have occurred with enough regularity to be represented in the passive-system-failure database, historical rupture frequencies can be estimated as in Reference 1.6. However, the applicability of operating experience to future steam generator tube failure rates is unclear. For instance, it is important to understand how steam generator replacement, secondary and primary side environmental changes, and other factors may affect the future failure rates. For these reasons, these failure modes were also considered in the elicitation.

There is an entire class of other non-piping, passive system failure modes which has recently been recognized and that has not been considered in previous LOCA frequency studies. Recent experience has demonstrated that degradation, and therefore LOCAs, can occur in non-piping components such as CRDM housings, BWR stub tubes (Hamaoka), and the RPV reactor head (Davis Besse). These degradation mechanisms and failure locations have also not been explicitly represented in historical piping-based operating-experience databases. However, the impact of these emerging LOCA initiators is an important consideration in determining total LOCA frequency. Therefore, these non-piping failure frequencies must be combined with historical piping failure frequencies to determine comprehensive passive-system LOCA frequencies.

Finally, in both piping and non-piping components, it is important to identify and assess the LOCA severity of degradation mechanisms with long incubation times which have yet to surface in the operating experience and may not be incorporated in PFM models. The identification and assessment of these future mechanisms is only possible by understanding the long-term interrelationship between the passive component materials, operating environment, and loading conditions.

1.5 Expert Elicitation

Expert elicitation has attributes which can build on the strengths and compensate for the weaknesses associated with purely operating-experience-based or PFM-based approaches to estimate LOCA frequencies. These attributes make this technique a natural choice for estimating LOCA frequencies. Expert elicitation is a formal process for providing quantitative estimates of the frequencies of physical phenomena when the required data is sparse and when the subject is too complex to adequately model. Furthermore, the scientific uncertainty about the phenomena is so large that, in the absence of adequate data, validated models or computer codes cannot be developed. If the issue is also important, i.e., it has significant regulatory implications and may also be controversial, then devoting the substantial resources required for an expert elicitation may be justified.

Expert elicitation is a structured process which enhances its accuracy, consistency, credibility, and thus acceptability compared to informal, less-structured processes. The emphasis on a structured decomposition of the issues improves accuracy and credibility, thus making the results more acceptable to the stakeholders. Expert elicitation reduces the likelihood of bias and enhances the consistency and comparability of the results. The emphasis on documentation leads to improved scrutiny and acceptance of the results. The main drawbacks in using a formal expert elicitation process are the increased time and resources required. Because of the structure, there is also reduced flexibility to make changes as the process proceeds.

Expert elicitation is a well-established technique [1.38] which has been used on a number of occasions to evaluate technical issues related to nuclear safety. Examples include: NUREG-1150 [1.10], the determination of flaw density and size distributions in reactor pressure vessels [1.39], the evaluation of the high level radioactive waste repository [1.40, 1.41], and in the construction of a probabilistic seismic hazard curve [1.42].

Expert elicitation was selected to develop passive system LOCA frequency estimates because data sparseness and subject complexity are characteristic of pipe break LOCA frequency estimation. Data sparseness is evidenced by the fact that no Class 1 nuclear pipe break LOCA events have occurred. Existing pipe break LOCA frequency estimates from NUREG/CR-5750, Appendix J, vary from 4×10^{-4} per calendar-year for SB LOCAs to 4×10^{-6} per calendar-year for PWR LB LOCAs. On the average, this translates into one SB LOCA every 2,500 years per plant and one PWR LB LOCA every 250,000 years per plant.

Complexity is evidenced by the large number of pipe system variables which must be considered to accurately evaluate the full spectrum of pipe breaks using the PFM or statistical methods discussed in previous sections. Variables include piping design and layout; piping fabrication; materials; degradation mechanisms; stress; service environment; application of codes and standards; inspection type, quality and schedule; and plant operating history. These variables serve as input for both operating-experience and PFM models. In particular, the importance of a particular degradation mechanism depends on the effect of the relevant combinations of these variables.

As previously discussed, the number and interaction of these variables and the sparseness of data severely hinders accurate frequency assessment using operating-experience approaches (Section 1.2.1). The PFM approach (Section 1.2.2) suffers because small changes in the input variable assumptions can dramatically affect the predicted piping reliability. Also, the underlying physical modeling of degradation and failure mechanisms (crack initiation, leak rates, complex crack growth, and piping instability) is unavoidably simplistic. These limitations lead to a great deal of uncertainty when utilizing only operating experience or PFM analysis to estimate LOCA frequencies. Using expert elicitation to develop these estimates fosters a more comprehensive approach by using individuals with relevant technical expertise to identify and focus on important variables which affect LOCA frequencies. The panelists then are required to combine insights gained from operating experience, PFM modeling, and other considerations to develop their LOCA frequency estimates.

1.6 References

1.1   USNRC, Regulatory Guide 1.174: An Approach for Using Probabilistic Risk Assessment In Risk-Informed Decisions On Plant-Specific Changes to the Licensing Basis, July 1998.

1.2   Staff Requirements - SECY-98-300 – Options for Risk-Informed Revisions to 10 CFR Part 50 – "Domestic Licensing of Production and Utilization Facilities," dated June 8, 1999.

1.3   Staff Requirements - SECY-99-264 – Proposed Staff Plan for Risk Informing Technical Requirements in 10 CFR Part 50, dated February 3, 2000.

1.4   U.S. Nuclear Regulatory Commission (USNRC), SECY-02-0057, "Update to SECY-01-0133, 'Fourth Status Report on Study of Risk-Informed Changes to the Technical Requirements of 10 CFR Part 50 (Option 3) and Recommendations on Risk-Informed Changes to 10 CFR 50.46 (ECCS Acceptance Criteria)', dated March 29, 2002.

1.5   Staff Requirements – SECY-02-0057 – Update to SECY-01-0133, "Fourth Status Report on Study of Risk-Informed Changes to the Technical Requirements of 10 CFR Part 50 (Option 3) and Recommendations on Risk-Informed Changes to 10 CFR 50.46 (ECCS Acceptance Criteria)", dated March 31, 2003.

1.6   Poloski, J.P, Marksberry, D.G., Atwood, C.L., and Galyean, W.J., "Rates of Initiating Events at U.S. Nuclear Power Plants: 1987-1995," NUREG/CR-5750, U.S. Nuclear Regulatory Commission, February 1999.

1.7   "Investigation and Evaluation of Stress-Corrosion Cracking in Piping of Light Water Reactor Plants," NUREG-0531, U.S. Nuclear Regulatory Commission, February 1979.

1.8   Rao, G.V., Seeger, D.E., Jr., Hoffman, J.A., DeFlitch, C., Rees, R.A., and Junker, W.R., "Metallurgical Investigation of Cracking in the Reactor Vessel Alpha Loop Hot Leg Nozzle to Pipe Weld at the V.C. Summer Nuclear Generating Station," WCAP-15616, Westinghouse Electric Company LLC, January 2001.

1.9   "Reactor Safety Study: An Assessment of Accident Risks in U.S. Commercial Nuclear Power Plants," WASH-1400, U.S. Nuclear Regulatory Commission, October 1975.

1.10   "Severe Accident Risks: An Assessment for Five U.S. Nuclear Power Plants," NUREG-1150, U.S. Nuclear Regulatory Commission, December 1990.

1.11   Beliczey, S., and Schulz, H., "Comments on Probabilities of Leaks and Breaks of Safety-Related Piping in PWR Plants," *International Journal of Pressure Vessel and Piping*, Vol. 43, pp. 219 – 227, (1990).

1.12   "Manual Shutdown of Unit-1 of the Hamaoka Nuclear Power Station," Nuclear Power Safety Press Release Information, ANRE/MITI, Chubu Electric Power Company, November 8, 2001.

1.13   "Unique Brunsbüttel Core Spray was Vulnerable to Gas Explosion," *Nucleonic Week*, Vol. 43, No. 10, March 7, 2002.

1.14   Licensee Event Report (LER)-28186020, "Surry 2 Feedwater Failure," March 31, 1987.

1.15   Nyman, R., "Hegedus, D., Tomic, B., Lydell, B., "Reliability of Piping System Components: Framework for Estimating Failure Parameters from Service Data," SKI Report 97:26, Swedish Nuclear Power Inspectorate, December 1997.

1.16   Danko, J.C., "Boiling Water Reactor Pipe Cracking: The Problem and Solution," *Processing of Materials in Nuclear Energy*, American Society for Metals, Metals Park, OH, 1983.

1.17   Wilkowski, G.M., Rudland, D., Wolterman, R., Krishnaswamy, P., Rahman, S., and Scott, P., "Technical Evaluation of Probabilistic LBB Codes and Approaches," Draft Technical Report, November 30, 2001.

1.18   Nyman, R., Erixon, S., Tomic, B., and Lydell, B., "Reliability of Piping System Components, Volume 1: Piping Reliability – A Resource Document for PSA Applications," SKI Report 95:58, Swedish Nuclear Power Inspectorate, December 1995.

1.19   Lydell, B., "Failure Rates in Barsebäck-1 Reactor Coolant Pressure Boundary Piping: An Application of a Piping Failure Database," SKI Report 98:30, Swedish Nuclear Power Inspectorate, May 1999.

1.20   "Reliability of Piping System Components. Vol. 2: PSA LOCA Data Base Review of Methods for LOCA Evaluation Since the WASH-1400," SKI Report 95:59, Swedish Nuclear Power Inspectorate, 1996.

1.21   Gesellschaft Für Anlagen und Reaktorsicherheit (GRS) mbh; German Risk Study Phase B, GRS-72, Verlag, TÜV; Cologne, 1989.

1.22   Woo, H.H, Mensing R.W. and Benda, B.J.,"Probability of Pipe Failure in the Reactor Coolant Loops of Westinghouse PWR Plants, Volume 2: Pipe Failure by Crack Growth, Load Combination Program," NUREG/CR-3662, Vol. 2, U.S. Nuclear Regulatory Commission, August 1984.

1.23   Lu, S.C. and Chou, C. K., "Reliability Analysis of Stiff Versus Flexible Piping, Final Project Report, NUREG/CR-4263, U.S. Nuclear Regulatory Commission, May 1985.

1.24   Wilkowski, G. M., and others, "International Piping Integrity Research Group (IPIRG) Program," NUREG/CR-6233, Vol. 4, June 1997.

1.25   Hopper, A., and others, "The Second International Piping Integrity Research Group (IPIRG-2) Program," NUREG/CR-6952, March 1997.

1.26   Scott, P., and others, "The Battelle Integrity of Nuclear Piping (BINP) Program Final Report –
       Vol. I: Summary and Implications of Results," NUREG/CR-6837, June 2005.

1.27   T. Loo and R.W. Mensing, "Probability of Pipe Failure in the Reactor Coolant Loops of
       Combustion Engineering PWR Plants," NUREG/CR-3663, U.S. Nuclear Regulatory
       Commission, September 1984.

1.28   Holman, G.S. and Chou, C.K. "Probability of Pipe Failure in the Reactor Coolant Loops of
       Westinghouse PWR Plants," NUREG/CR-3660, U.S. Nuclear Regulatory Commission, July
       1985.

1.29   Holman, G.S. and Chou, C.K. "Probability of Pipe Failure in the Reactor Coolant Loops of
       Babcock & Wilcox PWR Plants," NUREG/CR-4290, U.S. Nuclear Regulatory Commission, May
       1986.

1.30   T. Lo, S.E. Bumpus, D.J. Chinn, R.W. Mensing, and G.S. Holman, "Probability of Failure in
       BWR Reactor Coolant Piping," NUREG/CR-4792, U.S. Nuclear Regulatory Commission, March
       1989.

1.31   Harris, D.O., and Dedhia, "A Probabilistic Fracture Mechanics Code for Piping Reliability
       Analysis (pcPRAISE code)," NUREG/CR-5864, U.S. Nuclear Regulatory Commission, 1992.

1.32   Bishop, B.A., "Westinghouse Structural Reliability and Risk Assessment (SRRA) Model for
       Piping Risk Informed In-Service Inspection," WCAP-14572 Revision 1, Supplement 1,
       Westinghouse Electric Company LLC, October 1997.

1.33   Paul, D.D., Ahmad, J., Scott, P.M., Flanigan, L.F., and Wilkowski, G.M., "Evaluation and
       Refinement of Leak-Rate Estimation Models," NUREG/CR-5128, Rev. 1, U.S. Nuclear
       Regulatory Commission, June 1994.

1.34   Rahman, S., Ghadiali, N., Paul, D., and Wilkowski, G., "Probabilistic Pipe Fracture Evaluations
       for Leak-Rate Detection Applications," NUREG/CR-6004, U.S. Nuclear Regulatory
       Commission, April 1995.

1.35   Simonen, F.A. and Woo, H.H., "Analyses of the Impact of Inservice Inspection Using a Piping
       Reliability Model," NUREG/CR-3869, U.S. Nuclear Regulatory Commission, July 1984.

1.36   Holman, G.S., "Application of Reliability Techniques to Prioritize BWR Recirculation Loop
       Welds for In-Service Inspection," NUREG/CR-5486, U.S. Nuclear Regulatory Commission,
       December 1989.

1.37   "Risk-Based Inspection – Development of Guidelines, Light Water Reactor (LWR) Nuclear
       Power Plant Components," NUREG/GR-0005, Vol. 2, Part 1, U.S. Nuclear Regulatory
       Commission and American Society of Mechanical Engineers (CRTD Vol. 20-2), July 1993.

1.38   Meyer, M.A., and Booker, J.M., "Eliciting and Analyzing Expert Judgment: A Practical Guide,"
       NUREG/CR-5424, U.S. Nuclear Regulatory Commission, January 1990.

1.39   Simonen, F.A., Doctor, S.R., Schuster, G.J., and Heasler, P.G., "A Generalized Procedure for
       Generating Flaw-Related Inputs for FAVOR Code," NUREG/CR-6817, March 2004.

1.40   Bonano, E.J., Hora, S.C., Keeney, R.L., and von Winterfeldt, D., "Elicitation and Use of Expert
       Judgment in Performance Assessment for High-Level Radioactive Waste Repositories,"
       NUREG/CR-5411, U.S. Nuclear Regulatory Commission, May 1990.

1.41   Kotra, J.P., Lee, M.P., Eisenberg, N.A., and DeWispelare, A.R., "Branch Technical Position on
       the Use of Expert Elicitation in the High-Level Radioactive Waste Program, "NUREG/CR-1563,
       U.S. Nuclear Regulatory Commission, 1996.

1.42   Budnitz, R.J., Apostolokis, G., Boore, D.M., Cluff, L.S., Coppersmith, K.J., Cornell, C.A., and Morris, P.A., "Recommendations for Probabilistic Seismic Hazard Analysis: Guidance on the Use of Experts," NUREG/CR-6372, U.S. Nuclear Regulatory Commission, 1997.

2. OBJECTIVE AND SCOPE

The principal objective of the expert elicitation process was to develop separate, generic BWR and PWR piping and non-piping passive system LOCA frequency estimates as a function of effective break size and operating time through the end of the plant license-renewal period. These estimates are based on the responses from an expert panel. Another objective of the elicitation process was to reflect both the uncertainty in each panelist's estimates as well as the diversity among the individual estimates. These objectives are consistent with staff direction [2.1]. This section discusses several important choices and assumptions that were adopted as part of the elicitation process to achieve these objectives.

The staff was directed to provide realistically conservative LOCA frequency estimates rather than bounding values associated with one or two plants [2.1]. Therefore, the elicitation focused on developing generic, or average, estimates for the commercial fleet and uncertainty bounds on these generic estimates. The BWR and PWR LOCA frequency estimates were not partitioned further to describe differences related to design class, vendor, or specific plant operating characteristics. This approach is consistent with prior LOCA frequency studies that did not consider plant-specific differences in developing LOCA frequencies for use in PRA modeling [2.2, 2.3]. However, the elicitation panelists were instructed to consider broad differences among plants related to important variables (i.e., plant system, material, geometry, degradation mechanism, loading, mitigation/maintenance) in determining both the generic LOCA frequency estimates and especially the uncertainty bounds. It is the broad differences in these important variables that are most important to passive system failure and there is generally sufficient commonality among plants to make such a generic assessment meaningful.

Panelists were specifically instructed not to consider differences that only exist at a few plants. For instance, if a particular plant vendor design was judged by a panelist to be more LOCA-sensitive than other designs and encompasses, say, 20 PWRs, then the panelist was expected to consider this in his estimates as well as in the uncertainty bounds. However, if the same plant design only applies to one or two plants, then this should not have been considered in the estimates. It is understood that unique plant features and safety culture may also influence LOCA frequencies and plant-specific estimates may fall outside the generic fleet estimates. However, accident frequency increases stemming from deficient safety practices are expected to be identified and mitigated through regulatory oversight policies and procedures.

The elicitation was solely focused on determining LOCA frequencies that initiate by unisolable primary system side failures that can be exacerbated by material degradation with age. Therefore, active system failures (e.g., stuck open valve, pump seals, interfacing system LOCAs) and consequential primary pressure boundary failures due to either secondary side failures or failures of other plant structures (e.g., crane drops) were not considered. While such LOCA frequency contributions may be an important consideration when evaluating total plant risk, the assessment of these risk contributions was beyond both the original scope and the expertise of the selected panelists.

This study developed LOCA frequency estimates consistent with historical SB, MB, and LB flow-rate definitions. Additionally, three larger LOCA categories within the classical LB LOCA regime were defined for the elicitation. The purpose of these additional LOCA categories was to examine trends with increasing break size, up to and including, a DEGB of the largest piping in the plant. The consideration of the consequences of such a break is a requirement of current 10 CFR 50.46, and associated Appendix K and GDC 35 requirements. The SB, MB, and LB LOCA categories have historically been defined on the basis of flow rate. Simple correlations were developed to relate the rupture size to the expected flow rate required for the ECCS make-up system. Although the correlations developed in this study are different from those used in the past, they provide a mechanism to compare these LOCA estimates with prior estimates.

The elicitation considered three distinct time periods: current-day, end-of-plant-license, and end-of-plant-license-renewal. For purposes of the elicitation, these time periods were quantified as 25 years (approximate current fleet average), 40 years, and 60 years, respectively, after plant operation commences. The current-day estimates were intended to represent current plant conditions and are therefore equivalent to instantaneous LOCA frequency estimates. However, the 40-year and 60-year estimates were not explicitly defined in the elicitation to represent either instantaneous frequencies at 40 and 60 years of operation or averaged frequencies between 25 – 40 years and 40 – 60 years of plant operation. Because the interpretation of the future time period estimates was not explicitly defined, the panelists were free to assess the relevance of each type of estimate and then provide either averaged or instantaneous frequencies as they deemed most appropriate. Each panelist was also asked to discuss possible differences between averaged and instantaneous frequencies for the future time periods. The lack of an explicit definition, while important for understanding the context of the results, is not of practical significance because future changes in generic LOCA frequencies are generally expected to be gradual.

As noted above, the LOCA frequency estimates were intended to be both generic and consistent with historical estimates that have been used to evaluate core damage frequency and large early release frequency (LERF) metrics. For these reasons, the elicitation primarily considered normal plant operational cycles and loading histories consistent with current internal-event PRA. Therefore, separate frequencies for each unique mode of plant operation were not estimated. Rather, the estimated frequencies implicitly consider all modes of operation based on the loading or operational history associated with each piping system or non-piping component. Additionally, consideration of normal plant operational cycles and loading histories was limited to representative constant stresses (e.g., pressure, thermal, and residual) and expected transient stresses (e.g., thermal striping, heat-up/cool-down, and pressure transients) that occur over the extended licensing period. Therefore, only loading events with a frequency greater than approximately 0.01 per calendar year were explicitly addressed.

Rare event loading from seismic, severe water hammer, and other such sources was also not considered in this generic evaluation because of their strong dependency on plant-specific factors. Furthermore, the assessment of these risk contributions was beyond the expertise of many of the selected panelists. As with other consequential LOCAs, the contribution to the LOCA frequency from rare event loading may be an important consideration when evaluating total plant risk. Consequently, a separate research effort was conducted to assess the potential impact of seismic loading on the LOCA break frequency versus break size relationship [2.4]. As part of that study, both unflawed and flawed piping failure contributions were considered. The results of that seismic LOCA analysis are also summarized in Section 7.2.

The generic BWR and PWR estimates were determined by first estimating the separate LOCA frequency contributions associated with BWR piping, BWR non-piping, PWR piping, and PWR non-piping failures for each panelist who provided the required responses. These individual piping and non-piping frequencies were then combined to estimate parameters of the total passive system LOCA frequency distributions for BWR and PWR plants at each distinct LOCA category and time period. The median, mean, 5th and 95th percentiles were the parameters chosen to represent the LOCA frequency distributions and reflect each panelist's uncertainty. Finally, these individual estimates were aggregated to provide group estimates. These aggregated group estimates are intended to provide a consensus-type group estimate. Additionally, confidence intervals were determined to reflect the diversity among the panelists' estimates. The group estimates along with the diversity measures for each LOCA parameter are the primary quantitative results produced by the elicitation process.

Several important assumptions were made to guide the elicitation process. One such assumption is that plant construction and operation comply with all applicable codes and standards. Therefore, all LOCA-sensitive passive-system components were assumed to be designed and fabricated using approved materials in accordance with ASME or similarly applicable requirements. Also, component inspections were assumed to be conducted using applicable ASME Section XI periodicity and quality requirements. Other additional inspections required by the NRC to evaluate specific degradation mechanisms (e.g., IGSCC, PWSCC, or CRDM inspections) were also assumed to occur as required. Plant operation was generally assumed to occur within the expected parameters required by the regulations and the technical specifications.

The specific impact on the LOCA frequencies of using counterfeit materials and fabrication techniques, avoiding inspections, and blatantly deviating from approved operating criteria was not considered. However, it was assumed that regulatory oversight policies and procedures will continue to be used to identify and mitigate risk associated with plants having deficient safety practices such as these. Deviations from good safety practices do represent some percentage of the events included in the passive-system failure data and extrapolation of this data implicitly assumes that similar future deviations will continue to occur with similar frequency.

Another important assumption is that current regulatory oversight practices will continue to evaluate aging management and mitigation strategies in order to reasonably assure that future plant operation and maintenance has equivalent or decreased risk. A related assumption inherent in this elicitation is that all future plant operating characteristics will be essentially consistent with past operating practice. The effects of operating profile changes were not considered because of the large uncertainty surrounding possible operational changes and the potentially wide-ranging ramifications of significant changes on the underlying LOCA frequencies. For instance, significant power upgrade allowances may change plant performance and relevant operating characteristics (e.g., temperature, environment, flow rate) to a degree which significantly impacts future LOCA frequencies. Because operating experience provides fundamental information used to determine the LOCA frequency estimates, these types of changes might undermine the applicability of the operating experience data. Consequently, the relationship between plant operating profiles and LOCA frequencies should be considered when evaluating possible risk-informed changes based on the elicitation study results.

2.1 References

2.1 Staff Requirements – SECY-02-0057 – Update to SECY-01-0133, "Fourth Status Report on Study of Risk-Informed Changes to the Technical Requirements of 10 CFR Part 50 (Option 3) and Recommendations on Risk-Informed Changes to 10 CFR 50.46 (ECCS Acceptance Criteria)", dated March 31, 2003 ADAMS Accession #ML030910476.

2.2 "Reactor Safety Study: An Assessment of Accident Risks in U.S. Commercial Nuclear Power Plants," WASH-1400, U.S. Nuclear Regulatory Commission, October 1975.

2.3 Poloski, J.P, Marksberry, D.G., Atwood, C.L., and Galyean, W.J., "Rates of Initiating Events at U.S. Nuclear Power Plants: 1987-1995," NUREG/CR-5750, U.S. Nuclear Regulatory Commission, February 1999.

2.4 Chokshi, N.C., Shaukat, S.K., Hiser A.L., DeGrassi, G., Wilkowski, G., Olson, R., and Johnson, J.J., "Seismic Considerations For the Transition Break Size," NUREG-1903, U.S. Nuclear Regulatory Commission, February 2008.

3. ELICITATION APPROACH

The expert elicitation process used with the panel is an adaptation of the formal expert judgment processes used in NUREG-1150 [3.1] to estimate core damage frequencies [3.2] and to assess the performance of radioactive waste repositories [3.3]. The elicitation process used for this project consisted of a number of steps (Figure 3.1). To begin, the project staff identified the issues to be evaluated through a pilot elicitation (Block 2 in Figure 3.1) and selected a panel of twelve members (Block 1). The staff then gathered background material and prepared an initial formulation of the technical issues (Block 3) that was provided to the panel. At its initial meeting, the panel discussed the issues and, using the staff formulation as a starting point, developed a final formulation for the elicitation (Block 3). This formulation included an elicitation structure (Block 5) for decomposing the technical issues and the development of base cases (Block 6) which are used in the subsequent anchoring of all the elicitation responses. A base case team was established (Block 7) as a subset of the entire panel to estimate the LOCA frequencies associated with the base case conditions. At this initial meeting, elicitation training (Block 4) was also conducted using exercises and a discussion of biases to educate the panel about the subjective elicitation of numerical values. After the first meeting, the base case team developed preliminary estimates for the base case frequencies. The staff also prepared a draft elicitation questionnaire (Block 8) which was reviewed by the panelists (Block 9) and revised based on comments received (Block 8). A second meeting was held to review the base case estimates, review the elicitation questions, and finalize remaining technical issue formulation issues. The elicitation questions and base case estimates were finalized based on feedback received at that meeting (Blocks 8 and 10).

At their home institutions, the individual panel members performed analyses and computations to develop their answers to the elicitation questionnaire. A facilitation team consisting of substantive experts, a normative expert and two recorders met separately with each panel member in day-long elicitation sessions (Block 11). At these sessions, each panel member provided answers to the elicitation questionnaire along with their supporting technical rationales (Block 12). The panel members then returned to their home institutions where they refined their responses based on feedback from the elicitation session. Upon receipt of the updated responses, the project staff compiled the panel's responses and developed preliminary estimates of the LOCA frequencies (Block 13). Along with the rationales, the preliminary estimates were presented to the panel at a third meeting (Block 14). Panel members were invited to fill in gaps in their questionnaire responses and, if desired, to modify any of their responses after this meeting (Block 15). Based on these updates, final estimates of the LOCA frequencies were calculated and provided to the panel members for review (Block 16). The project staff developed a draft report on the elicitation process and results (Block 17). A fourth meeting via video teleconferencing was held to discuss the draft report which was revised based on feedback received during this meeting (Blocks 19 and 20). Separately, an external peer review of the analysis of the elicitation responses from the panelists (Sections 5 and 6) was conducted (Block 18). In addition, public comment is being solicited (Block 21) prior to the development of the final LOCA frequencies to be used in the redefinition of the ECCS requirements in 10 CFR 50.46 (Block 22). More detail on each of these general steps is provided below.

3.1 Pilot Elicitation

The study was initiated with a pilot elicitation conducted using only NRC staff (Block 2). The primary objectives of this pilot study were to identify technical issues for consideration during the subsequent formal elicitation and to test a possible elicitation framework. Additionally, interim LOCA frequency estimates were developed to support the study conducted by NRC staff on the feasibility of risk-informing 10 CFR 50.46, Appendix K, and GDC 35. Specifically, estimates were sought to explore the potential of eliminating the design requirements to mitigate a simultaneous LOCA and loss-of-offsite-power (LOOP) event. The technical issues identified by the staff and discussed during this pilot study provided the foundation for subsequent discussions during the elicitation. The pilot elicitation was also valuable for identifying strengths and weaknesses in the process. The NRC staff used the lessons learned from this pilot study to help formulate a suggested formal elicitation structure for subsequent consideration by the elicitation panelists. The results of this feasibility study and the staff's pilot elicitation have been reported [3.4], but a synopsis of salient points is provided below.

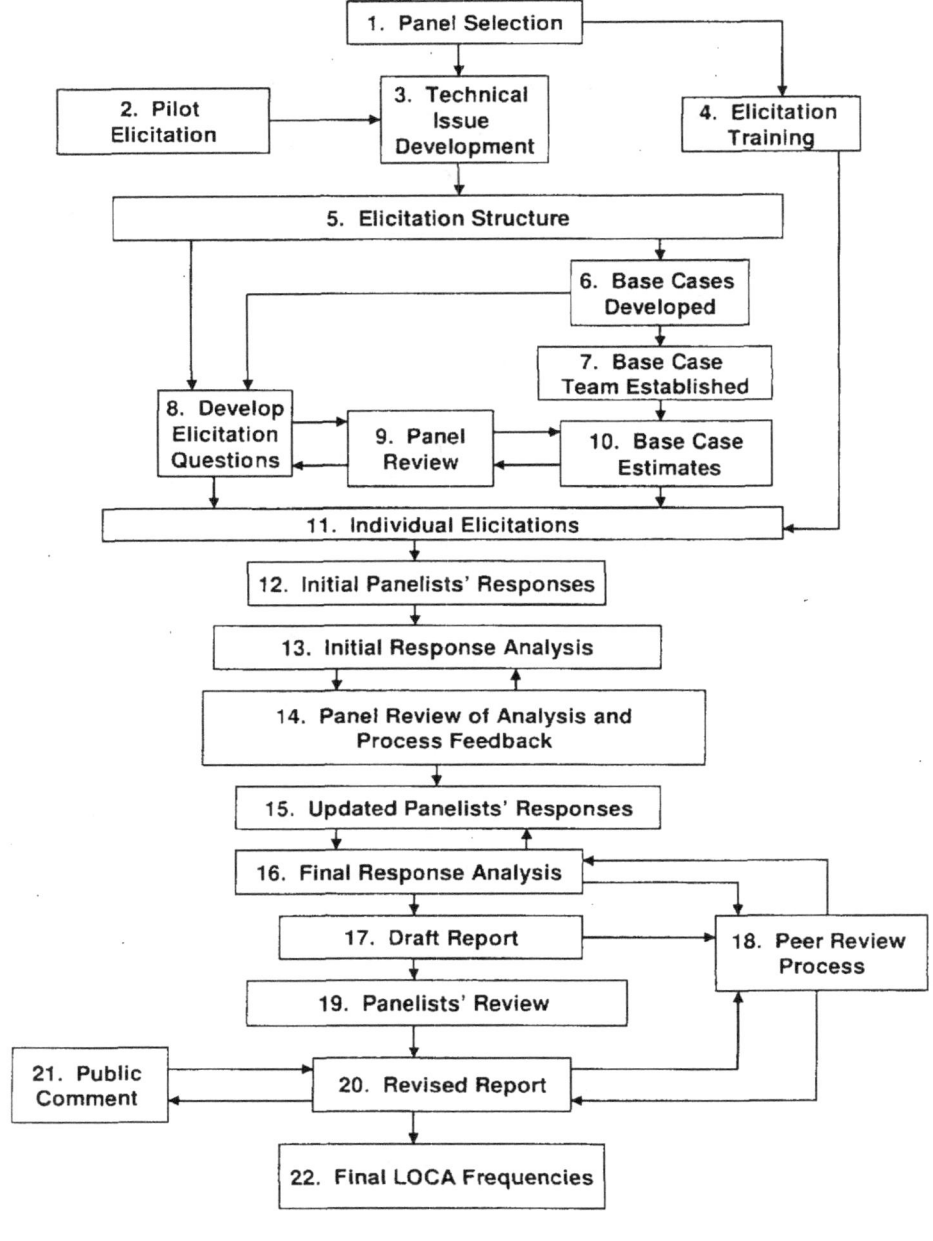

Figure 3.1 Flowchart of the Overall Elicitation Process

The pilot elicitation was structured similarly to the formal elicitation. Panelists were chosen to provide expertise in relevant technical areas. A kick-off meeting was held to discuss objectives and provide background information. An issue development meeting was then conducted to define the problem, identify important technical issues for consideration, develop elicitation questions, conduct elicitation training, and identify baseline LOCA frequencies for subsequent adjustment during the elicitation. The elicitation questionnaire was then independently answered by each panel member, the analyzed results were presented to the panelists, and the panelists provided feedback of the elicitation process.

As part of the issue development, the panelists developed a structure of classifying piping systems and non-piping components separately and identified corresponding materials, degradation mechanisms, and mitigation measures. The panelists then eliminated piping systems, non-piping components, and degradation mechanisms from consideration based on consensus opinion of those issues that were not important. The panelists utilized historical LOCA definitions and size classification from NUREG/CR-5750 [3.5] for consistency and agreed that the NUREG/CR-5750 LOCA frequencies would be the baseline for the exercise. The group evaluated the time period from the current period (2002) up to the end of the plant license-renewal period. The panelists were asked to determine the highest frequencies over this time period, but these maximums were typically associated with the end of the plant license-renewal period. Separate BWR and PWR estimates were developed. It should also be noted that the pilot study considered both active and passive system LOCA frequencies simultaneously.

The elicitation questions were structured so that each piping system or non-piping component was evaluated separately. There were two types of questions that provided redundant information. One type of question asked each participant to provide the relative percentage change between the NUREG/CR-5750 frequency estimates and the frequencies associated with each piping system or non-piping component. The other type of question asked for the relative ratio between MB and SB LOCAs and between LB and MB LOCAs for each piping system or non-piping component. There were also six global questions posed that were not related to any particular piping system. These included a consideration of the importance of the following issues on future LOCA frequencies: current IGSCC mitigation; hydrogen-combustion failures; future mitigation and degradation mechanisms; leak-detection-system accuracy; contribution of non-precursor degradation; and future ISI techniques. All questions were posed in a questionnaire and the results were analyzed and summarized.

There were several features of the pilot elicitation structure that were identified as advantageous for use during the formal elicitation. First, it was evident that classifying issues with respect to piping systems and non-piping components was appropriate. Next, it was valuable to identify those variable classes (e.g., material and degradation mechanisms) that affect the LOCA frequencies. Finally, it was useful to assign all relevant variables within each of these variable classes to each piping system and non-piping component. For example, all relevant materials and degradation mechanisms associated with the surge line were identified.

There were also several deficiencies uncovered in the pilot elicitation which required refinement prior to the formal elicitation. First, the understanding of elicitation questionnaire and the quality of the responses suffered in the pilot elicitation because there was no direct interface with each panelist during the completion of the elicitation questionnaire. Also, it was obvious that not all important variable classes had been appropriately identified during issue development and more comprehension was necessary. Next, the global questions should have been considered as a function of piping system, non-piping component, material, and degradation mechanism; and not independently from these variables. The use of NUREG/CR-5750 estimates to provide baseline values for adjustment was also realized as a limitation because the underlying conditions associated with these estimates were not sufficiently decomposed. Many panelists also thought that projecting estimates over a single 35-year period was too difficult and that breaking up the time period into shorter intervals would have made the assessment easier. Finally, it was evident that active and passive system LOCAs should be considered separately due to their different failure mode characteristics. All these deficiencies were addressed during the development of the formal elicitation process.

3.2 Panel Selection

Panel selection (Block 1) is a critical step in the process. The success of the elicitation is a direct result of the broad expertise of the panel members, and the ability of each member to provide corollary information in their specific areas of expertise. This information exchange enhances the general understanding of the remaining panel members. Potential panel members were sought within industry, academia, national laboratories, contracting agencies, other government agencies, and international agencies. Initially, a pool of 55 nominally qualified people was established by querying knowledgeable sources within the industry and NRC. Twenty-five people were solicited for the panel from this pool. They were sent information about the objective, scope, and approach of the elicitation exercise as background and were asked to submit resumes and also to evaluate their relevant technical areas of expertise for the exercise. Based on this feedback, the final panel of 12 was chosen to achieve both technical and organizational variety, and ensure a diversity of opinion, expertise, and backgrounds.

The elicitation panel members are listed in Table 3.1. The organizational diversity is apparent. Two of the panel members represent the European regulatory community; three of the panel members represent commercial vendors and owner's groups; four members are primarily NRC consultants; and three members have conducted extensive relevant research for both the commercial nuclear industry owner's groups and individual plants. Panel members were also chosen to represent a range of relevant technical specialties: PFM, piping design, piping fabrication, operating experience, materials, degradation mechanisms, operating mitigation practices, stress analysis, nondestructive evaluation, etc. All panel members have at least twenty-five years of experience in these relevant technical areas pertaining to commercial nuclear power applications. Each member's relevant qualifications are summarized in Appendix A.

Table 3.1 LOCA Frequency Elicitation Panel

Panel Member	Organization
Mr. Bruce Bishop	Westinghouse Electric Co LLC
Dr. Vic Chapman	OJV Consultancy Limited
Mr. Guy DeBoo	Exelon Nuclear
Mr. William Galyean	Idaho National Engineering Environmental Laboratory
Dr. Karen Gott	Swedish Nuclear Power Inspectorate
Dr. David Harris	Engineering Mechanics Technology, Inc.
Mr. Bengt Lydell	ERIN® Engineering and Research, Inc.
Dr. Peter Riccardella	Structural Integrity Associates, Inc
Mr. Helmut Schulz	Gesellschaft für Reaktorsicherheit (GRS) mbh
Dr. Sampath Ranganath	Formerly GE Nuclear Energy/Now XGEN Engineering
Dr. Fredric Simonen	Pacific Northwest National Laboratory
Dr. Gery Wilkowski	Engineering Mechanics Corporation of Columbus

A facilitation team was assembled to guide the panel through the elicitation process. The team consisted of one normative member, six substantive members, and two recorders. All but two of the facilitation team members were NRC staff. The substantive members were chosen to provide the same broad relevant technical knowledge and background required of the panel. The facilitation team role was to formulate the elicitation objectives and scope; coordinate and provide background technical information; develop the elicitation questions; guide and record the individual elicitation sessions; analyze and summarize the panel's findings; and develop the final LOCA frequency distributions from the panel's responses.

3.3 Elicitation Training

A basic premise in using an elicitation process is that the panel responses as a whole have no significant systematic bias. While individual responses can be highly uncertain and differ drastically, they do not systematically over- or underestimate the quantities of interest. Many elements of the elicitation procedure are designed to achieve this goal. These include the following:

- Constructing the panel with experts from all relevant technical areas and institutional/organizational affiliations (Section 3.2).
- Conducting elicitation training to identify possible sources of bias (Section 3.3.1) and conduct an exercise involving "almanac-type" questions with known answers (Section 3.3.2).
- Providing operating experience data and base case scenarios for anchoring and validating responses to the panel (Section 3.5).
- Formulating the elicitation questions to avoid response bias (Section 3.8).
- Conducting individual elicitation sessions to eliminate the possibility of group dynamics influencing panelist responses (Section 3.9).

From Appendix C, an analysis of the training exercise results supports the basic premise of no significant systematic bias. Elicitation training is an important tool to eliminate or minimize bias. It can also increase the accuracy and consistency of the responses provided by the panelists.

The elicitation training had three specific purposes:

(i) to discuss sources of bias in the elicitation procedure;

(ii) to familiarize the panelists with the type of responses which they will be asked to make; and

(iii) to provide the panelists with practice in making elicitation responses using a training exercise.

3.3.1 Motivational and Cognitive Biases

The panelists were introduced to sources of bias with the purpose of reducing biases in their individual subjective judgments. There are two sources of bias: motivational and cognitive. Motivational biases are due to emotional and psychological factors while cognitive biases are due to limitations on how information is processed by the human brain [3.6].

Motivational biases can result when a panelist's thoughts and responses are altered by the elicitation process. There are four types of motivational biases.

1. Social pressure can lead to groupthink when panelists may suppress their doubts or differing opinions in order to attain consensus. It can also be manifested in a panelist's response to verbal and non-verbal feedback from an interviewer. In addition, a panelist's responses might be influenced by his perception of what might be acceptable to his employer or society at large.

2. Misinterpretation can occur if the elicitation question structure is inconsistent with a panelist's thought process. For example, a panelist used to thinking in deterministic terms may be unsure how to respond if asked to think in probabilistic terms. Misinterpretation can also occur if a panelist is guided by the interviewer's viewpoint rather than his own.

3. Misrepresentation can be due to incorrect assumptions about the data or the models used to analyze the issue.

4. Wishful thinking can be the result of an institutional bias, e.g., a manager who underestimates the risk of a hazardous activity. This type of bias is relatively uncommon.

Cognitive biases can result when a panelist's thinking is illogical or does not conform to normative rules such as the axioms of probability. There are four types of cognitive biases.

1. Inconsistency can occur when there are multiple issues, assumptions, definitions or algorithms involved and a panelist does not keep them all in mind simultaneously. For example, inconsistency would occur if the panelist's sum of the probabilities of a set of mutually exclusive and exhaustive events does not add to one. Inconsistency is the most common cognitive bias.

2. An anchoring bias can occur when a panelist makes a relative comparison to a base case and does not sufficiently adjust his response with respect to the base case estimates.

3. An availability bias can occur when a panelist's opinion is overly influenced by the recent occurrence of a dramatic event, e.g., the accidents at TMI or Chernobyl.

4. Underestimation of uncertainty can occur when a panelist is asked for uncertainty bounds on his estimates. It is well-established that people are often overly-confident when dealing with highly uncertain issues [3.6], see Section 3.3.2.

3.3.2 Training Exercise

Elicitation training was conducted as part of the first group meeting with the panel. Meeting minutes for the various panel meetings are provided in Appendix B. The training exercise (Block 4) consisted of asking the panelists a number of quantitative questions with known answers, but in a subject area with which they are relatively unfamiliar. The purpose of asking these "almanac-type" questions is twofold:

(i) to accustom the panelists to the types of responses that they will be required to provide in their elicitations, and

(ii) to demonstrate to the panelists that, although individually they may be highly uncertain about their responses, the group response is closer to the correct answer than the individual responses.

For each question, the panelists were asked to supply three numbers: a mid-value (MV), an upper bound (UB) and a lower bound (LB). The MV is defined such that, in the panelist's judgment, there is a 50% chance that the correct answer lays above the MV and a 50% chance that it lays below the MV. The UB and LB are defined such that there is a 5% chance that the correct answer lays above or below them, respectively. In other words, the MV corresponds to the median of the panelist's subjective distribution of the correct answer, and the UB and LB correspond to the 95th and 5th percentiles, respectively. The interval (LB, UB) is a subjective 90% coverage interval for the correct answer. The elicitation exercise consisted of four questions.

1. According to the 2000 census, how many American men age 65 or over were there in the U.S.?

2. How many Americans men age 65 or over suffered from the following chronic conditions in 1995: arthritis, cataracts, diabetes, hearing loss, heart disease, and prostate disease? Express your answer in terms of a rate per 1,000 men.

3. What is the ratio of the rate for men 45-64 years old to the rate for men 65 and older for each of the conditions listed in Question 2?

4. What is the ratio of the rate for men under 45 years old to the rate for men 45-64 years old for each of the conditions listed in Question 2?

Note that Questions 3 and 4 ask about ratios. This corresponds to the type of questions asked in the elicitation, where almost all the questions concern the ratios of LOCA frequencies with respect to the base case conditions.

The results of the elicitation training exercise were consistent with several of the basic premises underlying the elicitation structure and methodology (see Appendix C). Recall that the (LB, UB) intervals are supposed to contain the correct answer 90% of the time. One useful evaluation is to compare the actual coverage interval to its nominally prescribed value of 90%. This provides a measure of how well the panelists are calibrated. As presented in Appendix C, between 15 and 17 responses were provided for each of the four questions. Although there were only 12 members on the panel, members of the facilitation team were also invited to participate in the exercise. The results for the four questions above are as follows.

1. Question 1: Fourteen of 17 intervals (82%) covered the correct answer.

2. Question 2: Fifteen responders provided intervals for each of the six conditions. Of the 90 intervals, 55 (61%) covered the correct answer.

3. Question 3: Sixteen responders provided intervals for each of the six conditions. Of the 96 intervals, 69 (72%) covered the correct answer.

4. Question 4: Sixteen responders provided intervals for each of the six conditions. Of the 96 intervals, 68 (71%) covered the correct answer.

The result for the first question was close to the nominal 90%, most likely because this question dealt with demographic data for which the responders had a relatively good understanding. The results for the other three questions were consistent with the well-established observation that the actual coverage probability for a subjective interval is generally less than its nominal coverage probability [3.6]. In other words, people are generally overconfident and underestimate the uncertainty in their subjective estimates. (Sensitivity analyses examining the potential magnitude of this over confidence are discussed in Section 5.6.1).

The coverage intervals for the last two questions were more accurate than for the second question. This may stem from that fact that Questions 3 and 4 asked for relative ratios rather than absolute numbers. This trend is consistent with the assumption driving the basic structure of this elicitation, namely, that relative values are easier to assess than absolute values (see Section 3.8.1). See Appendix C for the correct answers to the training exercise, as well as a full discussion of the panel and facilitation team responses.

3.4 Technical Issue Formulation

The formal elicitation was begun in February 2003 with a three-day meeting of the panel and facilitation team. The five principal objectives of this meeting were to define the scope and objectives of the elicitation (Section 2); provide background information about previous LOCA frequency estimates (Section 1); construct an approach for determining LOCA frequencies (Section 3); identify significant technical issues affecting LOCA frequencies (Section 3.4); and conduct elicitation training (Section 3.3). Subsequent discussion to finalize technical issues also was held in a follow-up two day meeting in June 2003.

3.4.1 Important Definitions
The first step in the decomposition was to define key technical terms and issues (Block 3), including the definition of a LOCA to ensure consistent understanding within the panel. For the purpose of this elicitation, the panel members defined a LOCA as:

> "A breach of the reactor coolant pressure boundary which results in a leak rate beyond the normal makeup capability of the plant."

The panel next defined various LOCA size categories (Table 3.2). The LOCA size categories are largely consistent with historical definitions developed for small break (SB), medium break (MB), and large break (LB) LOCAs during the WASH-1400 evaluation [3.7]. These definitions were retained in subsequent exercises to characterize plant risk [3.1] and determine initiating event frequencies [3.5]. One distinction is that, historically, break size frequencies were defined over a range of flow rates for SB (100 to 1,500 gpm [380 to 5,700 lpm]) and MB (1,500 to 5,000 gpm [5,700 to 19,000 lpm]) LOCAs. In this exercise, the panel chose to work with threshold values for each LOCA category. The cumulative threshold definitions were chosen instead of interval LOCA definitions because the former are more consistent with the various PFM and operating experience analyses that the panelists performed.[1] Additionally, three additional categories which fall within the historical LB LOCA regime were defined. These additional categories were developed to evaluate frequencies associated with the larger break sizes up to the DEGB of the largest primary system piping. These additional categories also reflect regions where different plant responses may be required to mitigate LB LOCA events of increasing size. LOCA Category 6 was chosen to correspond to the flow rate which would result from the complete rupture of the largest PWR primary piping system in the plant, i.e., a hot leg. For a BWR plant, the LOCA Category 6

---

[1] Interval-based estimates for the mean LOCA frequencies can be determined by simply subtracting adjacent cumulative LOCA categories. However, this technique cannot be used for determining other percentiles of the distribution. Because the differences between the results for consecutive LOCA categories are much smaller than the uncertainties, cumulative threshold (see tabular results in Section 7) and interval-based LOCA frequency estimates are statistically identical.

flow rates are only applicable for a catastrophic rupture of the reactor pressure vessel. There are no primary BWR piping systems that can generate a LOCA of this size upon failure. LOCA Categories 4 and 5 were determined so that the ratios between successive LB LOCA threshold values were approximately equal between Categories 3 to 6. The flow rate increases by either a factor of 4 or 5 for each successive category between Categories 3 to 6.

Table 3.2 LOCA Category Definitions

LOCA Category	Flow Rate Threshold, gpm (lpm)	LOCA Classification
1	> 100 (380)	SB
2	> 1,500 (5,700)	MB
3	> 5,000 (19,000)	LB
4	> 25,000 (95,000)	LB a
5	> 100,000 (380,000)	LB b
6	> 500,000 (1,900,000)	LB c

In addition to considering different LOCA sizes, the panel members also considered the possibility that future LOCA frequencies may be time dependent. Three different time periods were defined: 0-25, 25-40, and 40-60 years of plant operation. The 0-25 year period is representative of the current plant experience considering that the current fleet average age is approximately 25 years. Frequency estimates at 25 years are synonymous with the current-day LOCA frequency estimates. The elicitation panelists were expected to estimate current-day, or instantaneous LOCA frequencies at the end of this first time period.

The 25-40 year period is representative of future operation up to the end of the original design life and plant license. This period allows LOCA frequency changes over the next 15 years to be evaluated. The 40-60 year period is representative of future operation between the expiration of the original plant license up through the plant license-renewal period. This period evaluates LOCA frequency changes within the next 35 years. The 40-year and 60-year estimates were not explicitly defined in the elicitation to represent either instantaneous frequencies at 40 and 60 years of operation or averaged frequencies between 25 – 40 years and 40 – 60 years of plant operation. Panel members could choose to consider either average values over the time periods, maximum values within the time periods, or values at the ends of the time periods. No distinction was made so that the panelists had the most flexibility to assess the relevance of each type of estimate, and provide either averaged or instantaneous values accordingly. Each panelist was also asked to discuss possible differences between averaged and instantaneous frequencies for the future time periods. The lack of an explicit definition, while important for understanding the context of the results, is not of practical significance because future changes in generic LOCA frequencies are generally expected to be gradual.

The panel members considered issues that affect both piping and non-piping passive system failures. The contributions from active system components (e.g. stuck open valves, gasket and seal LOCAs) were not considered. The contributions from active components should be assessed separately and added to the passive system frequencies developed in order to estimate the total LOCA initiating event frequencies from both passive and active component failures. The panel members used existing ASME rules to identify the boundaries between piping and non-piping components. However, this distinction was somewhat academic since total LOCA estimates (i.e., sum of the piping and non-piping contributions) were determined from individual panelist responses. Therefore, piping and non-piping boundary differences among the panelists do not affect the final estimates.

3.4.2 Safety Culture Issue Formulation

Issues related to the effect of safety culture on the LOCA frequencies were often raised during technical issue formulation (Block 3). While none of the panelists are organizational safety culture experts, they all have relevant experience and opinions on how safety culture variability can affect LOCA frequencies. The panelists believe that safety culture affects are only weakly correlated with other variables that influence LOCAs, including the effects of aging mechanisms. These considerations were also thought to be independent of specific piping systems and non-piping components. Therefore, the panelists decided to consider the effect of safety culture LOCA contributions independently from age-related contributions.

For the purposes of the elicitation, safety culture was defined to encompass a number of economic, political, social, and psychological issues. Issues discussed fell into the broad categories of deregulation; decommissioning; human error; new-plant construction; technology improvements; knowledge transfer; lessons-learned from operating experience; public perception; the regulatory environment; and management philosophy. These factors can have either positive, negative, or neutral attributes with respect to their effect on LOCA initiating event frequencies. The objective of the group discussion was to provide the panelists with a clear understanding of the safety culture issues for assessment during their individual elicitations.

Some specific issues discussed include the likelihood that human error can occur during maintenance and mitigation operations and impact LOCA frequencies. An example is the failure to adhere to proper procedures or misinterpretation of obvious inspection indications of degradation. The effect of an aging workforce and the ability to provide sufficient knowledge transfer to the less experienced replacement workers was also discussed. The effect of lessons learned from past experiences on decreasing the response time needed to mitigate the impact of new degradation mechanisms was also debated. For example, the industry experience with mitigating IGSCC in the early 1980's may provide some useful strategies for mitigating current PWSCC concerns. Additionally, with respect to management philosophy, the merits of the adoption and implementation of a risk-informed management strategy were discussed. Some additional discussion topics are summarized in Appendix B.

The elicitation questions asked panelists to individually consider the effect of these and other significant safety culture issues on LOCA initiating-event frequencies. The elicitation structure was defined so that each panelist only addressed relative changes in the future safety culture relative to the existing safety culture (Section 3.8.3). The panel also decided during the initial issue formulation meeting to consider industry and regulatory safety cultures separately. However, it was also identified that this relationship could be strongly correlated. The elicitation questions asked panelists to consider the correlation between the industry and regulatory safety culture in their assessment. Also, panelists were asked to assess the relationship between safety culture and LOCA-size category.

More detail on the safety culture elicitation questions and structure is provided in Section 3.8.3. The qualitative rationale that identifies and discusses important safety culture issues is provided in Section 6.2. This rationale supports the quantitative estimates supplied by the panelists to the elicitation questions. These results are summarized in Section 7.1 along with a discussion on how these results were considered when determining the LOCA initiating event frequencies.

3.4.3 General Issue Development Structure

The remainder of the elicitation focused solely on the contribution of aging-related passive system degradation to the LOCA frequency estimates. The panel developed a structure (Block 5) for considering passive system failures which contribute to LOCAs (Figure 3.2) due to normal operating conditions and loading, and transients expected over the extended plant operating life. The total passive system frequencies were first divided by the panel into piping and non-piping contributions. The panel next agreed that the design and operating characteristics of each piping system and each major non-piping component (e.g., main coolant pumps, steam generators, pressurizer, and valves for PWRs) could impact the underlying LOCA frequencies. Non-piping components were further subdivided into relevant subcomponents (e.g., valve bonnet, valve bonnet bolts, valve casing, etc.) that possess unique operating and design characteristics. For a given LOCA-sensitive piping system or non-piping subcomponent, the panel identified five variable classes (i.e., geometry, loading history, materials, aging mechanisms, and mitigation and maintenance practices) that contain all the principal variables that affect passive system LOCA frequencies.

The elicitation structure (Figure 3.2) developed by the panel provided flexibility to the LOCA frequency estimation process. Panel members could develop LOCA frequencies based on either piping system or non-piping component level information (top-down approach in Figure 3.2), or by considering the individual effects of specific variable combinations within a given system or component and then combining important contributions (bottom-up approach in Figure 3.2). Separate elicitation questions were developed for each analysis approach (Section 3.8.4). The panel members either used one of these two approaches, a combined approach, or an alternative approach in their elicitation responses. More specific details on the issue formulation follow. Other information is available in Appendix B.

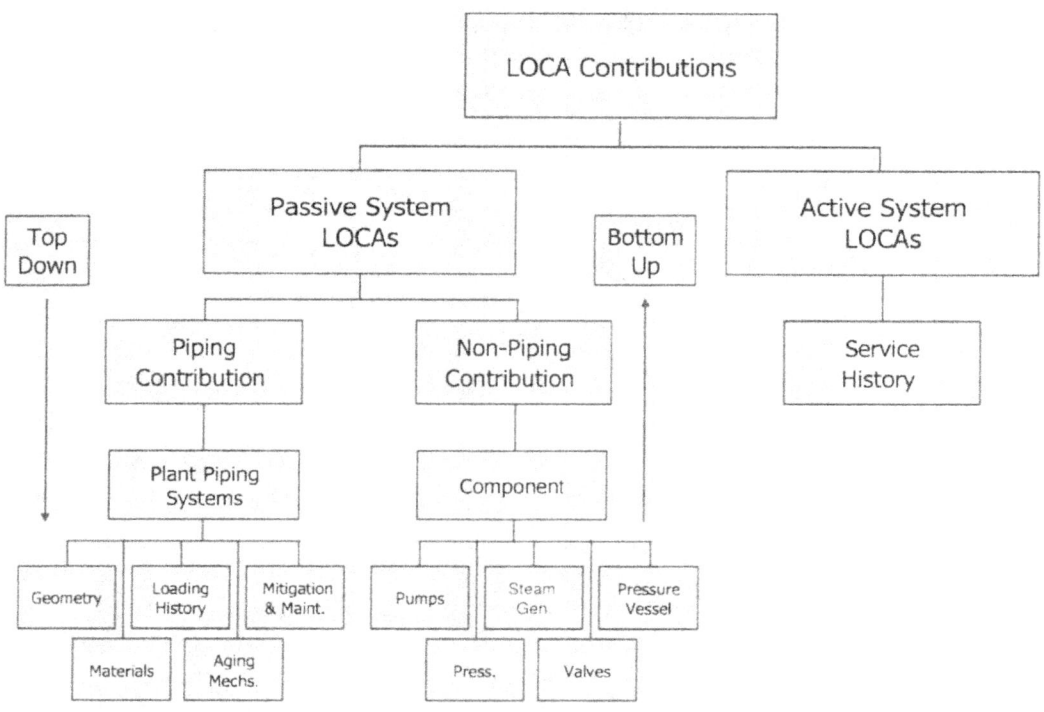

Figure 3.2 Elicitation Structure

3.4.4 Variable Assignments

The panel then developed a list of relevant variables within each of these variable classes that could affect the LOCA frequencies. For instance, under the aging mechanism variable class, mechanisms that both have been experienced or could surface in the future were identified (Table 3.3). Table 3.3 summarizes the mechanisms developed during the group brainstorming sessions. The mechanisms are segregated by the primary mechanism type. Additional sub-categories are used to identify either specific degradation mechanisms under the appropriate main category or features associated with the main category. For example, fatigue degradation was separated into low cycle fatigue which is primarily driven by thermal loading fluctuations due to plant heat-up and cool-down cycles and high cycle mechanical fatigue which could result from general loading functions on the piping, possibly due to an adjacent source of vibration. Both crack initiation and crack growth portions of life are important contributors to fatigue life, although crack initiation occupies a greater percentage of the life in the high-cycle fatigue regime. Stress corrosion cracking (SCC) is listed as a main category and it includes IGSCC which was prevalent in BWRs in the late 1970's, transgranular SCC (TGSCC) which affects casting components, PWSCC which has more recently surfaced in PWRs, and external chloride SCC (ECSCC). Similar discussions ensued at the kick-off meeting for identifying the remaining degradation mechanisms listed in Table 3.3. A synopsis of those discussions can be found in the detailed minutes for the kick-off meeting in Appendix B. In addition, more details on the development of the variables within the other relevant piping system variable classes, i.e., geometry, loading, etc. is provided in Appendix B along with complete tables for all five of these variable classes.

3.4.5 Piping System Variable Combinations

The panel next identified those specific variable combinations that are active for each LOCA-sensitive piping system. For example, the recirculation piping in BWR plants is stainless steel and is susceptible only to certain aging mechanisms, like SCC, thermal fatigue, etc. The geometry, loading spectrum, and maintenance practices applicable to the BWR recirculation system were also defined. A similar exercise was conducted for all the BWR (Table 3.4) and PWR (Table 3.5) LOCA-sensitive piping systems to identify and correlate the relevant variables associated with each piping system.

Table 3.3 Material Aging Degradation Mechanisms

Main Category	Sub-Category 1	Sub-Category 2	Sub-Category 3	Sub-Category 4
Low Cycle Thermal Fatigue	Crack Initiation	Crack Growth		
High Cycle Mechanical Fatigue	Vibration	Pressure	Temperature	
Stress Corrosion Cracking	IGSCC	TGSCC	ECSCC	PWSCC
Localized Corrosion	Pitting	Crevice Corrosion		
General Corrosion	Boric Acid (ID or OD)			
Fretting Wear				
Material Aging	Thermal	Dynamic	Radiation	Creep
Fabrication Defects and Repair				
Hydrogen Embrittlement				
Flow Sensitive	Erosion/ Cavitation	FAC		
Unanticipated (New) Mechanisms				

IGSCC = Intergranular stress corrosion cracking
TGSCC = Transgranular stress corrosion cracking
ECSCC = External chloride stress corrosion cracking
PWSCC = Primary water stress corrosion cracking
FAC = Flow accelerated corrosion

Table 3.4 BWR LOCA-Sensitive Piping Systems

System	Piping Matls.	Piping Size (in)	Safe End Matls.	Welds	Sig. Degrad. Mechs.	Sig. Loads.	Mitigation/ Maint.
RECIRC	304 SS, 316 SS, 347 SS	4, 10, 12, 20, 22, 28	304 SS, 316 SS, A600	SS, NB	UA, FDR, SCC, LC, MA	RS, P, S, T, DW, SUP, SRV, O	ISI w TSL. REM
Feed Water	CS	10, 12 (typ), 12 - 24	304 SS, 316 SS	CS, NB	UA, FDR, MF, TF. FS, LC, GC, MA	T, TFL, WH,P, S. SRV, RS, DW, O	ISI w TSL. REM
Steam Line	CS – SW	18, 24. 28	CS	CS	UA, FDR, FS. GC, LC, MA	WH, P, S, T, RS. DW, SRV, O	ISI w TSL. REM
HPCS, LPCI	CS (bulk), 304 SS, 316 SS	10, 12	304 SS, 316 SS. A600	CS, SS. NB	UA, FDR, SCC, TF, LC. GC, MA	RS, T, P, S, DW, TS, WH, SUP, SRV. O	ISI w TSL, REM
RHR	CS. 304 SS, 316 SS	8 - 24	CS, 304 SS, 316 SS	CS, SS, NB	UA, FDR, SCC, TF, FS, LC, GC. MA	RS, T, P, S, DW, TS, O SUP, SRV	ISI w TSL, REM
RWCU	304 SS, 316 SS, CS	8 – 24	CS, 304 SS, 316 SS	CS, SS, NB	UA, FDR, SCC, TF, FS, LC. GC, MA	RS, TS, T, P. S. DW. SUP, SRV, O	ISI w TSL, REM
CRD piping	304 SS, 316 SS (low temp)	< 4	Stub tubes – A600 and SS*	Crevice A182 to head	UA, FDR, MF, SCC	RS, T, P, S. DW, V, O, SRV	ISI w TSL, REM
SLC	304 SS, 316 SS	< 4	304 SS, 316 SS	SS, NB	UA, FDR, MF, SCC	RS, T, P, S. DW, V, O, SRV	ISI w TSL, REM
INST	304 SS. 316 SS	< 4	304 SS. 316 SS	SS. NB	UA, FDR, MF, SCC, MA	RS, T, P, S. DW, V, O, SRV	ISI w TSL, REM
Drain lines	304 SS. 316 SS, CS	< 4	304 SS. 316 SS. CS	S S, NB	UA, FDR. MF. SCC. LC, GC	RS, T, P, S. DW, V, O, SRV	ISI w TSL, REM
Head spray	304 SS, 316 SS, CS	< 4	304 SS, 316 SS, CS	SS, NB	UA, FDR, SCC, TF, LC, GC	RS, P, S, T, DW, SRV, O	ISI w TSL, REM
SRV lines	CS	6, 8, 10, 28	CS	CS	UA, FDR, MF, FS. GC, LC, MA	RS, P, S. T, DW, SRV, O	ISI w TSL, REM
RCIC	304 SS, 316 SS, CS	6, 8	304 SS, 316 SS	SS NB	UA, FDR, SCC, LC, MA	RS, P, S. T, DW, SRV, O	ISI w TSL, REM

304 SS = 304 series stainless steel
316 SS = 316 series stainless steel
347 SS = 347 series stainless steel
A600 = Alloy 600
CS = carbon steel
CS – SW = seam welded carbon steel
NB = Nickel-based weld (Alloy 82/182)
UA = unanticipated mechanisms
MA = material aging
LC = local corrosion
FDR = fabrication defect and repair
SCC = stress corrosion cracking
MF = mechanical fatigue
TF = thermal fatigue
FS = flow sensitive (inc. FAC and erosion/cavitation)
ISI w TSL = Current ISI procedures with technical
specification leakage detection requirements
considered.
RS = residual stress
P = pressure

S = Seismic
T = Thermal
DW = dead weight
SUP. = support loading
SRV = SRV loading
WH = water (and steam) hammer
O = overload
V = vibration
TFL = thermal fatigue loading from striping
TS = thermal stratification
REM = all remaining mitigation strategies possible
(e.g., not unique to piping system)
HPCS = high pressure core spray
LPCS = low pressure core spray
RHR = residual heat removal
RWCU = reactor water cleanup
CRD = control rod drive
SLC = standby liquid control
SRV = safety relief valve
RCIC = reactor core isolation cooling

Table 3.5 PWR LOCA-Sensitive Piping Systems

System	Piping Matls.	Piping Size (in)	Safe End Matls.	Welds	Sig. Degrad. Mechs.	Sig. Loads.	Mitigation/ Maint.
RCP: Hot Leg	304 SS, 316 SS. C-SS, SSC-CS CS – SW	30 - 44	A600, 304 SS, 316 SS, CS	A82 304 SS, 316 SS. CS	TF, SCC, MA, FDR. UA	P, S, T, RS, DW, O, SUP	ISI w TSL, REM
RCP: Cold Leg/Crossover Leg	304 SS, 316 SS, C- SS, SSC-CS, CS – SW	22 - 34	A600, 304 SS, 316 SS, CS	A82 304 SS, 316 SS. CS	TF, SCC. MA, FDR. UA	P. S, T, RS, DW, O. SUP	ISI w TSL, REM
Surge line	304 SS, 316 SS, C-SS	10 - 14	A600, . 304 SS. 316 SS,	A82 304 SS, 316 SS	TF, SCC, MA, FDR, UA	P, S, T. RS, DW, O. TFL, TS	TSMIT, ISI w TSL. REM
SIS: ACCUM	304 SS, 316 SS, C-SS	10 - 12	A600, 304 SS. 316 SS.	A82 304 SS, 316 SS	TF, SCC, MA, FS. FDR, UA (FAC)	P, S, T, RS, DW, O	ISI w TSL, REM
SIS: DVI	304 SS, 316 SS	2 – 6	A600, 304 SS, 316 SS,	A82 304 SS, 316 SS	TF, SCC, MA, FS, FDR, UA (FAC)	P, S, T. RS. DW, O	ISI w TSL, REM
Drain line	304 SS. 316 SS, CS	< 2"			MF, TF, GC, LC, FDR, UA	P, S, T. RS, DW, O, V, TFL	ISI w TSL, REM
CVCS	304 SS, 316 SS	2 – 8	A600 (B&W and CE)	A82	SCC, TF, MF, FDR. UA	P. S, T, RS, DW, O, V	ISI w TSL, REM
RHR	304 SS, 316 SS	6 – 12			SCC, TF, MA, FDR. UA	P, S, T, RS. DW, O, TFL	ISI w TSL, REM
SRV lines	304 SS, 316 SS	1 – 6			TF, SCC, MF, FDR, UA	P, S, T. RS, DW, O. SRV	ISI w TSL. REM
PSL	304 SS, 316 SS	3 – 6		A82	TF, SCC, MA, FDR, UA	P. S, T. RS. DW, O, WH, TS	ISI w TSL, REM
CRDM	A600	4 – 6			SCC, TF, MF, LC, FDR, UA	P, S, T, RS, DW, O	HREPL, ISI w TSL, REM
RH	304 SS, 316 SS	< 2	A600		MF, SCC, TF, FDR, UA	P, S, T, RS, DW, O, V, TS	ISI w TSL, REM
ICI	304 SS, 316 SS	< 2	A600		MF, SCC, TF, FW, FDR, UA	P, S. T, RS, DW, O, V	ISI w TSL, REM
INST	304 SS, 316 SS	< 2			MF. SCC, TF, FDR, UA	P, S, T, RS, DW, O, V	ISI w TSL, REM

304 SS = 304 series stainless steel
316 SS = 316 series stainless steel
A600 = Alloy 600
A82 = Alloy82
SSC-SC = Stainless steel clad carbon steel
CS = carbon steel
CS – SW = seam welded carbon steel
UA = unanticipated mechanisms
MA = material aging
GC = general corrosion
LC = local corrosion
FDR = fabrication defect and repair
SCC = stress corrosion cracking
MF = mechanical fatigue
TF = thermal fatigue
FS = flow sensitive (inc. FAC and erosion/cavitation)
ISI w TSL = Current ISI procedures with technical specification leakage detection requirements considered.
B&W = Babcock and Wilcox
CE = Combustion Engineering
RS = residual stress loading

P = pressure loading
S = seismic loading
T = thermal loading
DW = dead weight loading
SUP = support loading
SRV = SRV loading
WH = water (and steam) hammer
O = overload
V = vibration
TFL = thermal fatigue loading from striping
TS = thermal stratification
REM = all remaining mitigation strategies possible (eg. not unique to piping system)
TSMIT = thermal stratification mitigation
HREPL = RPV head replacement
RCP = reactor coolant piping
SIS = safety injection system
DVI = direct volume injection
CVCS = chemical volume and control system
RHR = residual heat removal
SRV = safety relief valve
PSL = pressurizer spray line
CRDM = control rod drive mechanism
RH = reactor head
ICI = in-core instrumentation

Table 3.6 Reactor Pressure Vessel (RPV) Failure Modes

Failure Modes	Material	Degradation Mechanisms	Mitigation/ Maintenance	Comment
Vessel Head Bolts	high strength steel	LC	Human error	Removal leading to human error during refueling
CRDM connections	SS			welded, bolted, threaded + seam weld
Nozzles	SSC-CS			
RPV wastage	SSC-CS	LC		Boric acid wastage (upper & lower head, shell)
RPV Corrosion Fatigue	SSC-CS	MF		Initiate at cladding cracks (upper & lower head, shell)

RPV= reactor pressure vessel
SS = stainless steel
SSC-CS = stainless steel clad carbon steel
LC = local corrosion
GC = general corrosion

3.4.6 Non-Piping System Failure Modes

A similar exercise was conducted for the non-piping LOCA-sensitive components and subcomponents. Table 3.6 illustrates the variables considered important for the RPV in BWRs and PWRs. The panel identified failures in the RPV shell (due to wastage or corrosion fatigue), the head bolts, the CRDM connections, and the nozzles as passive system LOCA-sensitive RPV subcomponents. The group generally did not have information on component geometries, loading, and mitigation/maintenance practices at this meeting and salient information was researched later as necessary by each panelist prior to the elicitation. However, several specific failure scenarios were discussed for each of these subcomponents. An RPV shell failure could occur by boric acid wastage from the outer shell or due to corrosion fatigue within the shell emanating from fissures in the stainless steel cladding. The CRDM connections are those that connect to the reactor CRDM nozzles. Failure would imply a failure of this connection which could be bolted, welded, or threaded depending on the RPV design. Vessel head bolt failure is only thought possible due to either human error during refueling as the head is removed and reinstalled or a common cause degradation mechanism such as boric acid corrosion that is occurring simultaneously in several bolts. Nozzle failure includes both the larger main coolant loop nozzles, as well as the smaller CRDM and instrumentation penetration nozzles. Nozzle failure is possible due to SCC or thermal fatigue. Other RPV failures were discussed as well, and similar discussions were held for all the LOCA-sensitive non-piping components and subcomponents. See Appendix B for more details, and for similar tables for other non-piping components. While these initial failure scenarios were discussed for all the non-piping components during the issue development meeting, the panelists were invited to consider other possible failure scenarios that should be considered during the elicitation.

3.4.7 Reference Case Development

Reference cases were developed for each BWR and PWR piping system by assigning specific values to each variable category (e.g., material, geometry, loading, mitigation, and aging mechanism) within Tables 3.4 and 3.5. Similar cases were also developed for the PWR piping systems (Appendix B). The reference cases were defined to create a simplified set of conditions that could be used as a starting point to assess LOCA frequencies within each system. The reference cases are representative of possible conditions that could lead to a LOCA, but they do not necessarily represent the most important LOCA contributor for a given piping system. The actual BWR and PWR reference cases are described in Appendix B. The development of the reference case conditions is analogous to the development of the base case conditions

(Section 3.5). However, the reference cases are not associated with a quantified LOCA frequency. See Section 3.5 for more information on the base-case development. More information on the use of the reference cases within the elicitation is provided in Section 3.8.4.

3.5 Base Case Development

The elicitation structure provided the panel members with a way to prioritize and assess important contributing variables to the generic LOCA frequency distributions. The most challenging aspect for each panelist was to quantify the frequencies associated with the important contributing variables that they individually identified. In order to improve the accuracy of the assessment, piping and non-piping base case frequencies were developed to provide the panelists with quantitative estimates for a set of well defined conditions that they could use for anchoring their responses. A clearly defined set of conditions and assumptions were developed by the elicitation panel for each base case (Block 6). These conditions were then analyzed by a member of the base case team (Block 7) using either PFM or classical statistical and/or Bayesian assessment of operating experience to predict the frequencies associated with each piping and non-piping base case. The assumptions, approach, and results of these analyses were then presented to the elicitation panelists so that they clearly understood the basis for the base case estimates.

This development and use of base cases was employed so that the panelists did not have to provide absolute frequencies during the elicitation. Instead, they chose appropriate base cases and provided a relative ratio to express the difference between the base cases and the important contributing variables that they identified. This technique is based on the premise that relative ratios are easier to assess, and therefore more accurate, than absolute numbers. The rationale for this premise is also discussed in Sections 3.3.2 and 3.8.1. The decision to use and the application of these base cases was made by each individual panelist. Some panelists utilized them extensively, others considered the relative trends expressed by the base case estimates, and others chose not to utilize them at all. However, in general, the panelists using the base case estimates chose to anchor their responses to base case results based on operating experience instead of the PFM results. Therefore, the differences in absolute frequencies between the operating-experience base cases and the PFM base cases (as discussed in Section 4) are not important in assessing variability among individual elicitation responses.

3.5.1 Piping Base Cases
The objective of the piping base case study was to analyze various PWR and BWR piping systems and materials using both operating experience and fracture mechanics analyses with shared information and a common set of conditions and assumptions. The insights from these analyses were provided to the panelists to help them formulate their opinions on the comparative LOCA likelihood between the base case conditions and other systems and degradation mechanisms not explicitly modeled. It was impractical to model all possible LOCA sensitive components, materials, and degradation mechanisms. Therefore, the goal in the base case development was to consider several important variables in BWR and PWR plants including piping size, aging mechanisms, and piping materials. Several aging mechanisms were chosen based on operating experience and a consideration of those mechanisms that provide the greatest LOCA challenge. A range of piping sizes were evaluated to provide information on the spectrum of pipe break sizes considered in the elicitation. The most common piping materials and welds were considered.

Finally, the following sensitivity analyses were conducted to evaluate their impact on the predicted base-case frequencies:
- Effect of different leak detection strategies for ISI
- Effectiveness of ISI assessment,
- Impact of weld overlay on IGSCC mitigation,

3-17

- Effectiveness of IGSCC mitigation techniques,
- Impact of applied loading history,
- Effect of large seismic loading and hydro test loads, and
- Impact of degraded material properties.

See Section 4.3 for a description of these sensitivity analyses and the results.

3.5.1.1 Piping Base Case Definition - Five different base cases were defined for the piping systems: two BWR and three PWR cases (Block 6). The variables established for analysis are summarized in Table 3.7. The loading spectrum for each of these systems was assumed to be the normal steady-state and transient loading histories expected over 60 years of operation. Maintenance and mitigation was assumed to be typical ISI for all systems with leak detection resolution as required by the technical specification limits. The BWR systems were assumed to be operating with normal water chemistry (NWC) and the recirculation system base case assumed that a weld overlay (WO) had been applied after 20 years of operations to mitigate IGSCC[2]. The inspection technique and periodicity for the recirculation base case are as specified in generic letter (GL) 88-01 [3.8]. . It should be noted that both Bengt Lydell and Bill Galyean used the post-IGSCC mitigation cracking rates as the basis for their BWR-1 analysis. This was counter to the original BWR-1 definition in Table 3.7, and results in lower LOCA frequency predictions than the defined case since NWC had largely been eliminated by this time. Separate LOCA frequency estimates associated with each of the piping bases cases were calculated for each defined LOCA category at 25, 40, and 60 years of plant operations

Table 3.7 Piping Base Case Definitions

Plant Type	System	Pipe Size (in)	Pipe Material	Safe End Material	Weld Material	Aging Mechanisms	Maintenance & Mitigation
BWR	Recirculation	12, 28	304 SS	NB	NB	IGSCC	ISI, NWC, WO
	Feedwater	12	CS	NA	NA	FAC, TF	ISI, NWC
PWR	Hot leg	30	304 SS	NB	NB	PWSCC	ISI
	Surge line	10	304 SS	NB	NB	PWSCC, TF	ISI
	HPI make-up	4	CS or SS	NA	NA	TF	ISI

HPI = high pressure injection   SS = stainless steel
CS = carbon steel   NB = nickel-based
TF = thermal fatigue   NA = not applicable

3.5.1.2 Piping Base Case Estimation - Four panelists were selected to separately calculate the frequencies associated with each base case defined in Table 3.7: Victor Chapman, William Galyean, David Harris, and Bengt Lydell (Block 7). These four panelists are referred to as the base case team. Both Chapman and Harris were selected to utilize a PFM-based approach, while Galyean and Lydell were chosen to utilize Bayesian and other statistical analysis techniques to extrapolate frequencies from accumulated operating experience.

---

[2] Most BWRs no longer operate with normal water chemistry as there have been improvements in reactor water conductivity to mitigate IGSCC. Various sensitivity analyses were conducted to evaluate the effectiveness of various IGSCC mitigation techniques (See Section 4.3).

There were several ground rules established by the elicitation panelist for the base case frequency estimations. These are enumerated as follows:

- Base case members should collaborate and use common information as the basis for the estimates.
- All base case calculations should model the established conditions (Table 3.7) as closely as possible so that methodologies and results can be more easily compared.
- Each base case team member could choose any methodology for estimating the base case frequencies.
- The PFM analysis should model service conditions as closely as possible. In particular, the predicted crack leaking rates should be commensurate with operating experience.
- The base case team members should provide best estimate base case frequencies. Explicit uncertainty calculations were not required.
- Frequencies should be calculated for the entire piping system, and not just on a per weld basis.
- Some additional sensitivity analyses should be conducted to evaluate the effect of variables not modeled in the base case calculations.

Separate calculations were conducted by each base case team member to establish a range of base case frequencies. As mentioned, these absolute frequencies are used to anchor panelist responses to the elicitation questions. However, the differences among the results stemming from the four different calculations provided valuable insights as well. The elicitation panelists, by understanding the details of each approach, could identify the qualitative and quantitative effects of important assumptions and modeling differences. The panelists could also assess the underlying uncertainties associated with each calculation methodology. The sensitivity results could also be used to assess the impact of other variables on the base case frequencies.

Periodic meetings were held among the base case team members. The purpose of these meetings was to identify and share necessary background information, resolve technical ambiguities, discuss and resolve problems, and review progress. It was apparent early in the process that the Table 3.7 base case definitions were not sufficient to fully perform the calculations. Therefore, additional technical details were standardized including the piping layout and configuration, weld census, applied stress information, and the model BWR and PWR plant types. Additionally, the failure data associated with the base case piping systems was available to the team members through a searchable database developed by SKI in 1998 [3.9]. The BWR base cases assumed a BWR/4 plant type. The PWR hot leg and surge line base cases were modeled based on a three-loop Westinghouse plant while the HPI make-up line is representative of a Babcock & Wilcox (B&W) plant. More details about the refined base case definitions and a summary of background information are provided in Appendices D – G.

Each base case team member then conducted initial base case calculations and sensitivity analyses. The initial sensitivity analyses included the effect of ISI, the impact of seismic events at 25, 40, and 60 years of plant operation for some of the systems, the effect of proof testing, and the effect of weld overlay on the BWR-1 base case, i.e., recirculation system. More details about the individual calculations and sensitivity analyses are contained in Appendices D – G. Once the initial base case frequencies were obtained, results were shared among the base case team members and reasons for differences were discussed.

Then, each base case team member prepared a presentation to summarize his calculation methodology and results. The goal of these presentations was to make the underlying details of each calculation apparent and provide the remaining panelists with sufficient information to independently assess the calculations in support of their elicitation responses (Block 9). Therefore, the presentations had identical formats so that the various approaches could be easily compared. The presentations explained the general assumptions, general approach, input variable determination and assumptions, detailed calculation procedure, results, and results of sensitivity analyses. Additionally, some major differences among the four methodologies were summarized. A summary of the approaches followed by the four base case team members in developing their piping base case frequencies is provided next.

3.5.1.2.1 Bengt Lydell's Base Case Estimates

- This analysis was one of two base case estimates using available operating experience to develop the base case frequencies. Base Case Report 2 (see Appendix D) develops BWR and PWR LOCA frequency distributions using a "bottom-up approach". Statistical analysis of the relevant operating experience is used to quantify the precursor failure rate and rupture frequency (i.e., LOCA frequency) of individual welds. Here the precursor failure rate is defined as the rate at which particular degradation is found and repaired/replaced in service based on historical data. This is an attribute-influence approach in that the statistical analysis of operating experience is restricted to consideration of the unique material combinations and degradation susceptibilities of primary pressure boundary piping systems.

As there are few relevant events to estimate the rupture frequency in a purely statistical manner, engineering judgment is required in the estimation of the conditional rupture probabilities given that a precursor failure exists. Conditional rupture probability relationships were based on an analysis of operating experience trends. Therefore, these CFPs were not assumed a priori. While Bengt Lydell conditional probabilities vary somewhat as a function of degradation mechanism and LOCA size, the average failure frequency decrease is approximately ½ order of magnitude between successive LOCA categories. See Appendix D for more detail.

The rupture frequency is then simply the precursor failure rate multiplied by the conditional rupture probability. The failure rate and rupture frequency (i.e. LOCA frequency) for an entire system is calculated by concatenating the individual weld failure rates and rupture frequencies. As a final step, Markov model theory is used to evaluate the influence of alternate strategies for ISI and leak detection on the frequency of leaks and ruptures, and to calculate age-dependent LOCA frequencies. See Appendix D for a more detailed description of this analysis (Block 10).

3.5.1.2.2 Bill Galyean's Base Case Estimates

- This analysis also relied primarily on operating experience to develop the base case frequencies, but in contrast to the previous method used a "top-down" approach. The approach (see Appendix E) starts with the straightforward calculation of a best estimate LOCA frequency using the number of failures (i.e., LOCAs) divided by the total number of reactor operating years. The U.S. nuclear power operating experience consists of zero Category 1 (i.e., greater than 100 gpm [380 lpm]) LOCAs over the entire operating experience, which totals 2,647 LWR (calendar) years through April 2003. This operating history data was used to update a non-informative prior distribution in a Bayesian calculation, to produce a posterior probability distribution on the total LOCA frequency. The overall LOCA frequency calculated in this manner is approximately 1.9E-4/LWR-year (0.5/2600 LWR-years). This total frequency is then partitioned using the relative ratios of crack and leak events from the available service data, to pipe and non-pipe passive component contributor categories. Then, within the piping and non-piping categories, the frequency is further partitioned using the available crack and leak data into the base case piping systems and degradation mechanisms.

The partitioned frequencies representative of the base case conditions are used as the base case frequencies for LOCA Category 1. The frequencies for the other LOCA categories were calculated using the common assumption that the frequency of a LOCA decreases as pipe size increases. A ½ order of magnitude factor was used to scale the frequencies for successive LOCA categories. This assumption is consistent with historical LOCA frequency estimates and is also supported by research by Beliczey and Schulz [Beliczey, S., and Schulz, H., "Comments on Probabilities of Leaks and Breaks of Safety-Related

Piping in PWR Plants," *International Journal of Pressure Vessel and Piping*, Vol. 43, pp. 219 – 227, (1990)]. See Appendix E for a more detailed description of this analysis (Block 10).

3.5.1.2.3 Dave Harris's Base Case Estimates

This analysis was one of two base case estimates using primarily PFM to establish base case frequencies. This approach utilized the PRAISE code for all calculations [3.10]. Separate analyses were conducted for the five different piping base cases and a number of sensitivity calculations were also performed. Individual welds were analyzed and then converted into piping system related frequencies. Benchmarking of operating experience for through-wall flaws was conducted where possible. Included below is a synopsis of conditions modeled and the analyses conducted for each base case.

The PWR hot leg (PWR-1) base case was modeled to separately consider fatigue crack growth of pre-existing defects and PWSCC initiation and growth. Stresses from an earlier PRAISE study were used. The case of PWSCC growth from pre-existing defects was selected as the base case. Sensitivity studies were made of the effects of hydrostatic testing, seismic events, and degraded material properties. The PWR surge line (PWR-2) base case considered initiation and growth of fatigue cracks. Stresses provided in NUREG/CR-6674 [3.11] were used although all seismic loads were removed. To obtain results for leaks greater than 100 gpm [380 lpm], an alternative procedure was required that involved extrapolating the lengths of through-wall cracks obtained from a PRAISE run to determine the failure probability given the existence of a through-wall crack. The HPI/make-up nozzle (PWR-3) was modeled with a failed thermal sleeve. Stresses for an intact thermal sleeve from NUREG/CR-6674 [3.11] were employed, but a fatigue crack was considered to initiate immediately with a depth of 0.12 inches (3.0 mm). This depth corresponds to the initiation depth of applicable fatigue crack initiation relationships used within PRAISE.

Two piping sizes were considered for the BWR recirculation system (BWR-1). The smaller lines (12 inch) are more prone to fail due to IGSCC than the larger lines (28 inch), but the larger lines are the only contributor to the larger LOCA categories. Initiation and growth of IGSCC cracks was modeled. Default residual stresses were employed, except for the 12 inch line which was repaired with a weld overlay after 20 years of service. The weld overlay greatly improves the residual stress situation, and employs a material that is less prone to cracking. The application of the overlay was found to provide a large reduction in the failure probability. Applied stresses in the 12 inch line were varied in order to provide agreement with field observations of leaks, with a mean normal operating stress of 12 ksi (83 MPa) providing the best agreement. This stress also provided good agreement with field observations of part-through cracks. While the 12 ksi (83 MPa) mean normal operating stress results in better agreement with field observations of cracking, a 20 ksi (138 MPa) stress value was used for the BWR-1 base case estimates. The 20 ksi (138 MPa) value was chosen for the base case estimates to be consistent with the stresses utilized in the other base case analyses. This choice is somewhat arbitrary however, because the base case frequencies using either 12 ksi (83 MPa) or 20 ksi (138 MPa) normal stresses are similar. This consistency occurs because although the per joint failure frequency is higher for the 20 ksi (138 MPa) stress level, the number of joints that are subject to this higher stress level is much lower than the number of joints that are subject to the 12 ksi (83 MPa) stress level. See Appendix F for more discussion of this phenomenon.

The feedwater system (BWR-2) was modeled considering fatigue crack initiation and growth. Flow accelerated corrosion (FAC) is an important LOCA mechanism, but a probabilistic model is not available for this mechanism. System stresses were again obtained from NUREG/CR-6674 [3.11]. As for the surge line analysis, in order to obtain results for leaks greater than 100 gpm (380 lpm), an alternative procedure was required that involved extrapolating the lengths of through-wall cracks obtained from a PRAISE run for leaks (any through-wall crack). See Appendix F for a more detailed description of this base case analysis (Block 10).

3.5.1.2.4 Vic Chapman's Base Case Estimates - This analysis also relied primarily on a PFM approach for establishing the base case frequencies, but using the RR PRODIGAL code [3.12]. RR PRODIGAL is a basic fatigue failure probability model developed by Rolls Royce (RR) for the British Naval Nuclear program. The PRODIGAL code first simulates the weld construction in order to determine the defect distribution and density for both buried and surface breaking defects at the onset of service life. A failure probability using standard linear elastic fracture mechanics methods is then determined for both the buried and surface breaking defects. Failure is achieved when the defect either exceeds the R6 failure criteria [3.13], or simply grows through to the full thickness of the weld. The failure probability for all initial defects is then combined to form the total failure probability. The specific procedures used to develop the base case frequencies are as follows:

1 Evaluate the basic fatigue failure probability using the RR-PRODIGAL code using the transient data supplied.
2 Evaluate an elastic crack opening displacement (COD) as a function of defect size.
3 Use expert judgment to extend this COD beyond the elastic limit.
4 Evaluate the mean defect cross-sectional area for a given defect size using its associated COD.
5 Evaluate the mean leak rate for a given defect size using correlations developed for the elicitation..
6 Use expert judgment to assess the defect length distribution at failure.
7 Multiply Steps 5 and 6 to obtain the conditional probability of a leak rate greater than the prescribed leak rates for LOCA Categories 1 through 6.
8 Multiply the conditional probability of Step 7 with the basic fatigue failure probability in Step 1 to arrive at the required final probability of a leak greater than the prescribed leak rate for each of the leak categories.

Note, for steps 2, 3, 4 and 5 above, the probabilities for the mean leak rate are determined as a function of assumed defect size for an assumed leak. The defect length probability is multiplied by this to obtain the conditional probability of leakage as a function of LOCA category assuming that a leak exists. The RR-PRODIGAL code supplies the leak frequency. Individual welds were analyzed and then converted into piping system related frequencies using this method. See Appendix G for a more detailed description of this base case analysis (Block 10).

3.5.1.3 Panel Review of Base Case Analysis - The base case team members made separate presentations to the panel at the second elicitation meeting. The presentations consisted of the assumptions, approach, results, and sensitivity analyses utilized by each team member for each base case. The panelists provided feedback and a discussion of the differences among the four chosen methodologies and the subsequent effect on the differences in the base case frequency estimates (Block 9). Refinements to the base case calculations were suggested by the panelists to possibly minimize differences in the initial base case frequency estimates and ensure that the defined base case conditions were appropriately modeled. Refinements included conducting additional calculations to compare operating-experience and PFM predictions of leaking crack frequencies at 25 years of operating life for each base case. Bengt Lydell was selected to estimate leak rates from operating experience because of his access and knowledge of the most extensive piping precursor database. David Harris was selected to calculate leak frequencies using the PRAISE PFM code. It was suggested that additional benchmarking could be performed based on this comparison. Revised stress histories for several base case systems were also developed. Crack initiation and growth rate information for PWSCC cracking was also reevaluated for the appropriate base cases. All refinement calculations were subsequently completed after this meeting and provided to the panelists for use in the elicitation.

The elicitation panelists also identified additional sensitivity analyses to examine the impact of changes in key variable and assumptions on the base case results. These analyses were important to inform the panelists' subsequent elicitation responses. Sensitivity analyses included an evaluation on the effect of different ISI leak detection strategies for all the base cases. For the BWR-1 recirculation system base case, sensitivity analyses compared IGSCC crack growth rates with and without a weld overlay; assessed operating experience to quantify the effectiveness of IGSCC mitigation techniques; and quantified the effect of variations in the applied loading history. For the PWR-1 hot leg base case, sensitivity analyses examined the effect of degraded material properties on base case frequencies and the effect of large seismic and hydro test loading. A study of the effect of ISI on PFM surge line cracking frequencies associated with thermal fatigue (part of PWR-2 base case) was performed. Finally, the effect of applied loading history variations on thermal fatigue cracking in high pressure injection make-up lines (PWR-3 base-case) was evaluated.

The base case frequency results after all refinements were completed are summarized in Section 4. Section 4.3 summarizes these sensitivity analyses and the results. Additional details on the base case refinements and sensitivity calculations specified by the panelists are contained in meeting minutes (Appendix B). More details on the base case and sensitivity calculations are contained in the descriptions provided by each base case team member (Appendices D – G).

3.5.2 Non-Piping Base Cases
In general, the non-piping elicitation structure, available background information and base case development was completely analogous to the corresponding evaluation of piping system contributions to the LOCA frequencies. When the elicitation started, less data had been collected specifically for non-piping system failures. A non-piping failure precursor database was developed from licensee event reports and provided to the panelists to correct this initial deficiency. This database is analogous to the piping precursor failure database that was provided, although it is not as comprehensive. Details of the non-piping precursor database are provided in Appendix H.

Additionally, several panelists had experience evaluating specific non-piping degradation and failure mechanisms that have occurred in service. These evaluations were used to develop specific non-piping base cases that were available for anchoring subsequent elicitation responses in a completely analogous manner to the piping base cases. These non-piping base case evaluations included: SGTR frequency determination, current pressurized thermal shock (PTS) analysis, BWR vessel rupture due to beltline cracking, BWR feedwater nozzle thermal fatigue cracking, and PWR CRDM failure due to PWSCC. Information on the analysis details for each of these non-piping base cases is provided later in this section. Appendix I provides additional details and results for BWR vessel rupture, feedwater cracking, and PWR CRDM failure. Results for the non-piping base case analyses are provided in Section 4.4 of the report. It should also be noted that the piping base case evaluations were also available for anchoring non-piping elicitation responses. Some panelists choose these for anchoring in appropriate situations.

The elicitation structure for evaluating non-piping LOCA frequency contributions was also analogous to the structure chosen to estimate piping LOCA frequency contributions. Discussion of non-piping LOCA frequency contributions can be found in the minutes to the Elicitation Kick-Off meeting in Appendix B, pages B-30 through B-36. The specific failure modes (or scenarios) for the five non-piping components considered to contribute to the LOCA frequencies, i.e., RPV, pumps, valves (BWRs and PWRs) and pressurizer and steam generators (PWRs), can be found in Tables B.1.13 through B.1.17 of Appendix B.

There were only two notable differences between the piping and non-piping evaluations. The first difference is that several panelists performed separate base case evaluations for the defined piping scenarios while a single panelist performed each non-piping evaluation. As mentioned, the multiple piping evaluations were used to identify differences in the estimated LOCA frequencies associated with the base cases, based on the analysis approach and assumptions. It was not necessary to repeat this exercise for similar non-piping scenarios. The second difference is that some of the non-piping failure modes considered were distinct from important piping failure modes and did not lend themselves to classical modeling approaches. Some examples include common cause bolting failures resulting from maintenance that could lead to vessel head, pump or valve bonnet, or steam generator manway failures. It is for these types of failure modes that elicitation is most valuable.

3.5.2.1 Non-piping Precursor Database - Therefore, the first step in the non-piping base case development was to create a separate precursor database for LOCA-sensitive non-piping components. This database was based primarily on licensee event reports (LER) contained in the sequence coding and search system (SCSS) database. The non-piping database identified all LOCA precursor events (e.g., cracks or leaks) between 1990 and 2002. These events were catalogued by the relevant non-piping component (i.e. valve, pump, RPV, pressurizer, or steam generator) and subcomponent (e.g. body, bonnet, nozzle, etc. for valves) identified by the panel during issue formulation (Section 3.4 and Appendix B). The aging mechanism and calendar year associated with each event was identified in the database as was the type of precursor event. Relevant precursor events include leaks, surface cracks, through-wall-cracks, and bolted connection failures. This database of 216 events was provided to the panel for use during the elicitation analogously to the database of precursor piping events. More information about the non-piping LOCA precursor database development and its contents is contained in Appendix H.

3.5.2.2 Steam Generator Rupture Estimates - The historical SGTR frequency was defined as one non-piping base case. Only ruptures resulting in flow rates greater than 100 gpm (380 lpm) were considered. This distinction is consistent with the Category 1 LOCA definition in this elicitation. Ruptures of this size provide the most definitive data on passive system failures in smaller components and represent the majority of pressure-retaining boundary passive system failures that have resulted in flow rates greater than 100 gpm (380 lpm). The non-piping precursor database was used to estimate this frequency. Additionally, steam generator leakage incidents, which are much more prevalent than ruptures, were examined as a function of time to illustrate possible trends. Results of this analysis are contained in Section 4.

3.5.2.3 Non-Piping Base Cases Analyzed using PFM - A number of other specific failure scenarios were also identified as being particularly important. These were analyzed using PFM to provide additional anchoring frequencies. These included ejection of CRDM nozzle tubes, evaluation of low temperature over pressurization (LTOP) events in BWRs, and examination of non-LOCA transient contributions to PTS events in PWRs. Only the non-LOCA PTS transients were considered for anchoring because the LOCA transient frequencies are based on the assumed LOCA initiating event frequencies which are being reevaluated herein.

One of the panelists, Dr. Pete Riccardella, analyzed the LOCA frequencies associated with BWR vessel rupture and CRDM penetration ejection. The BWR analysis considered pressure vessel failures due to both manufacturing flaws within the welds and service-induced flaws from SCC that grow into the low alloy steel. An NRC-approved PFM code (VIPER) was utilized [3.14]. Irradiation embrittlement was modeled and the vessel leak and fracture frequencies associated with both normal operation and LTOP loading were calculated. The most important risk contribution from the calculation is an LTOP event (1,150 psi [7.9 MPa] at 88°F [31°C]) which has an assumed event frequency of 1E-3. The current analysis builds on earlier research [3.15] which was used as technical justification for reducing inspection requirements for BWR reactor vessels. More details on the analysis procedure are contained in Appendix I.

The study of CRDM ejection frequencies utilized a similar PFM program (MRPER) developed under a materials reliability program to evaluate CRDM cracking [3.16]. The analysis predicts crack initiation using a Weibull analysis based on observed leaking or cracked nozzles statistics through the spring of 2003. Probabilistic crack growth rates are based on laboratory PWSCC measurements and analysis was conducted for four characteristic plant types (i.e., B&W, CE, Westinghouse 2-loop, and Westinghouse 4-loop). The effect of inspection type, interval, and POD was also considered. More details on this analysis are contained in Appendix I.

The final non-piping base case built on parallel work being undertaken to risk-inform the PTS screening criteria in 10 CFR 50.61 [3.17]. This study is evaluating PWR vessel failure probabilities as a function of the level of irradiation embrittlement to develop screening criteria for minimizing the PTS risk. Four PWR plants were investigated: Palisades, Beaver Valley, Calvert Cliffs, and Oconee. These plants span the plant vendors and design variables, and also have unique embrittlement characteristics. For each plant, probabilistic risk assessment, including a consideration of human errors, was used to determine the most likely event scenarios that result in safety system actuation. Structural integrity assessment has revealed that both high and low pressure safety system injection can induce high thermal stresses into the vessel. Therefore, the plant-specific thermal-hydraulic transient associated with each safety system actuation scenario was considered.

Vessel failure probability is predicted by first considering the probability of crack initiation from weld and baseplate flaws which are created during vessel manufacturing. The likelihood of crack arrest prior to the crack broaching the vessel thickness is then considered. Finally, the vessel failure frequency is calculated as a function of the level of material embrittlement. The final PTS screening criteria will be chosen such that the vessel failure probability is appropriately low. More details on the PTS reevaluation are available in [3.17].

The PTS challenges are appropriate to consider in the elicitation because they could lead to vessel failure and a resulting large LOCA. However, some care is required in this assessment. The biggest PTS risk contribution is usually from SB, MB, and LB LOCA initiating events, especially as material embrittlement increases. Further, the PTS risk is proportional to the LOCA frequencies. Because of this dependency, the total PTS risk cannot be used as a base case for determining the frequency of other pressure vessel or non-piping failures. However, there are other loading transients which contribute to PTS risk. These transients include stuck open primary and secondary valves and feed and bleed operations. The stuck-open valve and feed and bleed transient contribution to PTS risk ranges from 1% to 80% depending on the plant type and the embrittlement level. The panelists were only asked to consider the frequency associated with PTS risk from these transients for possible anchoring during the elicitation process.

3.6 Elicitation Background Information

The elicitation panel identified certain other generic information necessary for the elicitation. Additionally, individual panel members requested specific information to supplement this generic information. All information generated by these requests was catalogued and made available to all panel members to aid in developing their elicitation responses. In order to facilitate communications among the panel members, a repository of this information was maintained by Engineering Mechanics Corporation of Columbus (Emc^2) on an ftp site. Background information stored on this site includes references on relevant prior studies, a complete set of BWR and PWR piping isometric drawings, weld census information for the different piping systems, information on service stresses and transients, the piping and non-piping databases, and information pertaining to the piping and non-piping base case development. Additionally, all presentations, meeting minutes, elicitation questions, action items, and response tables were included. A complete listing of the ftp site contents is also maintained at the site. This information will be saved and catalogued after the elicitation has been completed.

3.7 Flow Rate Correlation

As previously described (Section 3.4.1) cumulative LOCA size categories[3] were defined in the elicitation as a function of flow rate of the makeup water supply (Table 3.2). This approach is historically consistent with plant response and mitigation action requirements that are a function of the flow rate as well as the break location. However, the panel expertise was concentrated in the areas of structural analysis, materials, and fracture mechanics, not thermal-hydraulics. Therefore, it was more natural for the panelists to develop their responses as a function of the effective break area. In order to facilitate the elicitation responses, it was therefore necessary to develop a correlation between the makeup flow rate and effective break area. The break areas were then converted to equivalent break sizes by assuming circular openings.

Three separate correlations were developed for PWR liquid, BWR liquid, and BWR steam lines. The correlations are a function of reactor type and the pipe length/diameter (L/D) ratio. A large L/D ratio implies saturation at the break plane. At small L/D ratios, the fluid does not reach these conditions. The leakage rate is also assumed to correspond to the flow rate required for the makeup water supply to mitigate the postulated break. Makeup water supply is assumed to exist at 70°F (21°C) under atmospheric pressure.

For the BWR steam lines, the correlations are assumed to be independent of pipe size and are modeled using equations by Todreas and Kazimi [3.18] with the upstream value for specific heat. The BWR liquid correlations were developed by assuming that the upstream pressure is approximately the steady-state operating pressure. For small L/D ratios (used for hole sizes > 9 inches (230 mm) diameter), the Moody correlation was employed [3.19]. For large L/D ratios (used for hole sizes < 9 inches (230 mm) diameter), the Zaloudek correlation was used [3.20]. The PWR liquid correlations were developed under the assumption that the upstream pressure is saturated at a temperature of 600F (315C). Once again, the Moody correlation was used for small L/D (used for hole sizes >7 inches (180 mm) diameter) while the Zaloudek correlation was employed for large L/D (used for hole sizes < 7 inches (180 mm) diameter).

The correlations which relate flow rates and LOCA size categories to the effective break sizes in each PWR and BWR system are summarized in Table 3.8. The break size corresponds to a partial fracture for pipes with larger diameters than the break size, a complete single-ended rupture in pipes with the same inside diameter, or a DEGB in pipes having inside diameters $1/\sqrt{2}$ times the break size. All panelists used these correlations to relate their elicitation responses determined for the effective break sizes to the appropriate LOCA size category. It is important to stress that breaks can occur in either LOCA sensitivity piping or non-piping systems and components. For a BWR plant, the LOCA Category 6 effective break sizes are only applicable for a catastrophic rupture of the reactor pressure vessel. There are no BWR piping systems that can generate a LOCA of this size upon failure.

---

[3] That is the LOCA frequency associated with each LOCA category is the frequency of a break of the cited or larger size.

Table 3.8 Break Size to Leak Rate Correlation

LOCA Category	Flow Rate (gpm)	BWR: Steam		BWR: Liquid		PWR: Liquid	
		Flow Rate Flux (gpm/in^2)	Eff. Break Size (in)	Flow Rate Flux (gpm/in^2)	Eff. Break Size (in)	Flow Rate Flux (gpm/in^2)	Eff. Break Size (in)
1	100	355	1/2	595	1/2	687	1/2
2	1500	355	2 1/4	595	1 3/4	687	1 1/2
3	5000	355	4 1/4	595	3 1/4	687	3
4	25,000	355	9 1/2	595	7 1/4	687	6 3/4
5	100,000	355	19	375	18 1/2	641	14
6	500,000	355	42 1/4	375	41 1/4	641	31 1/2

It should be noted that these correlations are different from those used in NUREG/CR-5750 [3.5] and in other past LOCA efforts. The earlier correlations date to the NUREG-1150 [3.1] plant risk study. The BWR correlations used in NUREG-1150 were developed from a matrix of calculations conducted using existing thermal-hydraulic codes [3.21] specifically developed for each plant studied in NUREG-1150. A variety of break areas were postulated in certain systems and different combinations of mitigation equipment were assumed to be operational. Calculations were performed for each plant. The reported correlation of break area to flow rate is an amalgam of these results. It is assumed that the PWR correlations are similarly based, although their development is not as well-documented.

These prior correlations were not adopted for this exercise because there was concern about their generic applicability given the plant specific nature of the calculations performed. There is also concern about the accuracy of the thermal-hydraulic codes existing at the time of the NUREG-1150 study. Discharge leak rates are highly uncertain. They are difficult to develop and are a function of upstream conditions, break location, plant configuration, and mitigation reliability, as well as break area. Based on these considerations, the simple closed-form approximate solutions are sufficient for the generic correlations required for this elicitation. Application of the elicitation LOCA frequency results may require plant specific calculations to evaluate break flow rate history and required mitigation response for assumed break locations and sizes.

3.8 Elicitation Question Development

The elicitation questions were developed in concert with the panel members (Block 8). One objective was to develop questions that were precise enough to be unambiguous, yet general enough so that the question set could be minimized. Additionally, careful language was chosen to avoid terminology specific to only a subset of the technical specialties represented by the panelists. Another objective was to make sure that a variety of approaches could be used to develop answers to the questions. Initial questions were formulated after the kick-off meeting. These questions were modified based on panel member feedback prior to the base case review meeting. Modifications continued after the base case review meeting and up through the first several individual elicitations. Elicitation questions were posed in the following areas: piping base case evaluation, safety culture, PWR piping, BWR piping, PWR non-piping, and BWR non-piping.

3.8.1 Elicitation Question Philosophy

None of the quantitative questions required the panelists to assess absolute LOCA frequencies. The premise behind this philosophy is that an assessment of the relative likelihoods for comparable LOCA frequency contributors is a more natural reflection of knowledge and experience than is an assessment of the absolute frequencies of each contributor. A relative assessment allows a direct comparison of the effects on LOCA frequencies resulting from specific combinations of materials, loading characteristics, aging mechanisms, and mitigation procedures for all the LOCA-sensitive components. Assessments of these interactions and comparisons of these relationships best match the panel expertise. An assessment of absolute LOCA frequencies would not only require consideration of these relationships, but would also require estimation of the frequencies associated with each postulated set of conditions. However, all of the conditions rarely occur, and there is little supporting data for the members to assess the absolute frequencies. Consequently, assessing relative frequencies should lead to more accurate results. The elicitation questions have been structured accordingly, to require relative rather than absolute assessments. More detail on the elicitation questions pertaining to the base case, safety culture, piping, and non-piping evaluations follows.

3.8.2 Piping Base Case Elicitation Questions

The piping base case evaluation questions required each panel member to address the accuracy and applicability of the four base case calculations. First, each panelist was required to assess how well each calculation modeled the base case conditions defined by the panel (Table 3.7). Each panel member was then asked to comment on the differences among the four base case team members' results for each base case, and assess how modeling differences and variability in model input information led to these differences. Additionally, each panelist was asked to evaluate the reasonableness of the differences in the base case results and determine if the differences provide an accurate measure of the uncertainty inherent in LOCA estimates. Each panel member was then asked to assess the accuracy of each base case calculation and to choose, if desired, a specific set of results for anchoring his or her future elicitation responses. The actual elicitation base case evaluation questions are contained in Appendix J (Block 8).

3.8.3 Safety Culture Elicitation Questions

The first set of quantitative questions focused on the influence of safety culture. As detailed in the issue formulation section (Section 3.4.2), the panelists decided to evaluate the effects of safety culture separately from a consideration of other variables which affect the LOCA frequencies (e.g., piping system, material, degradation mechanism, etc.). This was justified because safety culture effects were judged by the panel to be independent of these other variables, and are strictly a function of general regulatory, general nuclear industry, and plant specific attitudes and practices.

The elicitation questions asked each panelist to assess future safety culture with respect to the current safety culture. The future time period was decomposed into two intervals: the next 15 years (25 to 40 years), and the following 20 years (40 to 60 years). The current safety culture always serves as the baseline and two ratios were required, one for each time period, to predict relative future safety culture changes with respect to this baseline. The questions asked panelists to first separately consider the effects of utility and regulatory safety culture. Then, the panelists were asked to assess the degree of correlation between the utility and regulatory safety cultures. The questions also required the panelists to quantify the relationship between safety culture and LOCA size category, if a relationship exists. Panelists were asked for each question to qualitatively identify and discuss the strongest issues or factors which influence their quantitative estimates. The safety culture evaluation questions are contained in Appendix J.

3.8.4 Piping and Non-piping Elicitation Questions

Two parallel question sets were developed to determine piping and non-piping LOCA frequency contributions due to passive system failure. The question sets were structured to support either a bottom-up or a top-down philosophy (Figure 3.2). The bottom-up approach requires a fundamental consideration of explicit combinations of variables (i.e., loading, geometry, materials, degradation mechanisms, and mitigation), which lead to LOCAs in relevant piping systems and non-piping components (Figure 3.2). The bottom-up approach is consistent with historical PFM analysis of LOCA challenges as well as the Barsebäck-1 analysis [3.22].

The top-down approach allowed a more global consideration of piping system and non-piping component contributions without explicit consideration of individual degradation mechanisms, materials, etc. This approach is more consistent with a LOCA frequency development using classical operating-experience analysis as in NUREG/CR-5750 [3.5]. These two alternative structures were developed to support approaches commensurate with the specific technical expertise and philosophy of individual panelists. Panelists were free to choose the top-down or bottom-up approach, a combination of the approaches, or they could develop their own methodology. Panelists were encouraged to attempt both the top-down and bottom-up question sets to evaluate the consistency of their responses. A few panelists did answer both question sets and were able to iterate their final responses so that both the top-down and bottom-up quantitative results matched their qualitative expectations. Also, several panel members did pursue alternative approaches that they developed. A summary of the various philosophies and approaches from each panelist is provided in Appendix K.

The piping and non-piping top-down and bottom-up questions were identically structured. This commonality allowed the panelists to utilize similar approaches for addressing passive systems using either approach. In each question set, the assessment structure is consistent with the relative assessment philosophy described previously. The panel member first is asked to identify important LOCA-contributing factors. Then, he is asked to select an appropriate base or reference case for comparison with each LOCA-contributing factor. He is then asked to make a relative assessment between the important contributing factor and the chosen base or reference case conditions for each LOCA category and operating time period (25, 40, or 60 years). The assessment continued for all important LOCA-contributing factors. These contributions of each individual factor are then analytically combined to develop the final LOCA frequency estimates for each panelist (Section 5).

Reference cases (Section 3.4.7 and Appendix B) were developed by the panel analogously to the base cases for the purpose of making relative assessments within piping systems that do not have an associated base case. The reference cases, like the base cases, represent a specific set of defined conditions for each important LOCA variable class (Section 3.4.7). Base case conditions have been previously summarized in Table 3.7. If a panelist chose to use the reference cases, he was required to first make a relative assessment between the base case and reference case conditions. This is necessary because only the base case conditions are associated with fundamental, absolute frequencies. There was no explicit calculation of the frequencies associated with the reference cases. Once this assessment was made, the reference case associated with a piping system could be used for assessing issues within that piping system.

3.8.4.1 Top-Down Questions: Specific Considerations - This structure of issue decomposition, selection of important issues, relative comparison with base or reference cases, and final combination of contributions allows the absolute LOCA frequencies to be developed for each panelist starting from only the operating-experience and/or PFM-based estimates of LOCA frequencies associated with the simplified base case conditions. No other absolute LOCA frequency quantification was required.

The only difference between the top-down and bottom-up question sets discussed previously is the decomposition and consideration of the important contributing factors. The top-down approach asked the panelist to identify the most important piping systems or non-piping components. The relative comparison is made between the chosen base case condition and the important piping system/non-piping component. This analysis is continued for all the piping systems and non-piping components.

The following example illustrates the top-down approach. A panel member first lists the important piping systems (or non-piping components) for a PWR plant. He chooses the instrument lines, drain lines, CVCS, hot leg, cold leg, and RHR system and has appropriate rationale supporting this selection. Then, for the RHR system, he chooses a base case. In this example, he picks the hot leg cracking due to PWSCC because he feels that this mechanism is the most likely to cause a LOCA in the RHR. He then compares the LOCA contributions of the RHR to PWSCC cracking in the hot leg as a function of LOCA size. He then determines the following ratios for LOCA Categories 1 – 6, respectively: 100, 50, 20, 10, 5, and 0. These estimates are based on the panelist's determination that PWSCC cracking in the RHR is 100 times more likely to result in a Category 1 LOCA than PWSCC hot leg cracking. However, Category 5 LOCAs are only expected to be five times more likely. The RHR system cannot support a Category 6 LOCA which is why this ratio is zero. The panelist determined these ratios by assessing precursor differences between the RHR and hot leg and the knowledge that most of the precursors were due to mechanical fatigue which would tend to cause smaller LOCAs instead of complete piping failure. The panelist also considered the loading environment, and number of welds in the RHR system compared with the hot leg in making this assessment. This panelist would conduct a similar analysis for all the important PWR piping systems listed above.

3.8.4.2 Bottom-up Questions: Specific Considerations

3.8.4.2 Bottom-up Questions: Specific Considerations - The bottom-up approach asked the panelist to identify important combinations of piping materials, degradation mechanisms, loading, geometry, and mitigation measures within a given piping system or non-piping component. The relative comparison is then made between the base or reference case condition and the selected variable combination. This assessment continues for all important variable combinations for all relevant piping systems or non-piping components.

The following example is applicable to the bottom-up approach. A panelist is assessing BWR LOCA contributions. This panelist believes that IGSCC cracking under typical hydrogenated water chemistry and inspection procedures provides the greatest LOCA contribution. He chooses the recirculation system BWR base case for comparison. This base case evaluated IGSCC LOCA frequencies in normal water chemistry with a weld overlay applied after 20 years of operation. Based on laboratory studies, the panelist believes that crack growth rate of existing IGSCC cracks is similarly retarded using either a weld overlay or hydrogenated water chemistry. However, the weld overlay also contributes an additional margin of two to the failure stress for typical IGSCC crack sizes. In the recirculation system, the lower applied failure stress transient is 10 times more likely than the transient require to fail a joint with a weld overlay. Therefore, the panelist determines that the LOCA frequencies associated with the hydrogenated water chemistries without an overlay are a factor of 10 higher than the base case LOCA frequencies in the recirculation system for all LOCA sizes. The panelist then repeats this assessment for the IGSCC LOCA susceptibility of other piping systems and then similarly considers LOCA contributions from other important degradation mechanisms.

3.8.4.3 Additional Non-piping Considerations - As mentioned, the non-piping questions were structured analogously to the piping questions. However, an additional level of decomposition was necessary for the non-piping considerations. For each non-piping component (i.e., vessel, steam generator, pump, valves, and pressurizer), LOCA-sensitive subcomponents (e.g., tubes, manway, tube sheet, shell for the steam generator) were identified. This subcomponent decomposition is analogous to the piping decomposition by system and was necessary due to the different types of failures associated with each non-piping subcomponent. As with the piping systems, the subcomponents have specific combinations of materials, geometry, degradation mechanisms, loading, and mitigation which affect the LOCA likelihood. The subcomponent decomposition is summarized in Section 3.4.6.

The only other difference between the piping and non-piping questions is that specific failure scenarios were explicitly developed for various non-piping subcomponent variable combinations. The failure scenarios represent a combination of factors that could lead to non-piping LOCA failures. This distinction was necessary because the non-piping failure scenarios were more disparate than the piping considerations and it was not efficient to decompose those failures into single sets of appropriate variable classes as was done for the piping systems. An example is the improper torque application of the reactor head bolts due to human error that is not discovered during either bolt tension testing or initial system pressurization. Another failure scenario could be CRDM nozzle ejection due to PWSCC cracking which is not found during planned inspections. Many different failure scenarios were developed by the panelists during the issue formulation (Section 3.4.6 and Appendix B). Each panelist was also free to develop and evaluate other relevant scenarios.

Other than the additional decomposition and the distinction between the terminologies of variable combinations (piping) and failure scenarios (non-piping), the piping and non-piping questions are identically structured. As previously discussed, the top down non-piping approach focused on assessing those important failure scenarios for relevant non-piping subcomponents regardless of the non-piping component type. The bottom-up approach required a more detailed assessment of possible failure scenarios in each non-piping component and subcomponent.

3.8.5 Elicitation Response Requirements
All elicitation questions, except the base case evaluation, required both a qualitative and quantitative assessment. For each quantitative assessment, the panelists were requested to provide a MV, a LB, and an UB. As in the training exercise (see Section 3.3.2), the MV is defined such that, in the panel member's opinion, the unknown true value for that particular question has a 50% chance of falling above or below the MV. The LB is defined such that the true value has a 5% chance of falling below the bound. The UB is similarly defined such that the true value has a 5% chance of being above the bound. The MV, LB, and UB were interpreted in subsequent analysis (Section 5) as the median, 5th and 95th percentiles, respectively, of a subjective distribution for the true value of the answer.

There were examples created for each elicitation question to clarify the question requirements for the panelists and to illustrate possible approaches that could be used to answer the questions.

3.9 Individual Elicitations

Each panel member had between one and four months to prepare their elicitation responses after the base case review meeting was held and the elicitation questions were finalized. During this time period, individual elicitation sessions were conducted separately between each panel member and the facilitation team (Block 11). With the exception of the last elicitation session with Mr. Helmut Schulz of GRS in Germany, each of the sessions took place at the NRC headquarters in Rockville, MD. Mr. Schulz's elicitation was conducted via video teleconference (VTC). Each session lasted a full day.

The elicitation sessions addressed each elicitation question in order; starting with the base case evaluation questions, then the safety culture questions, followed up by the piping and non-piping related questions. The objectives for the individual elicitation sessions were to:
  1. Obtain and discuss the quantitative and qualitative responses to the elicitation questions.
  2. Identify any inconsistencies between the quantitative and qualitative responses.
  3. Provide additional clarification to the elicitation questions, as necessary.
  4. Identify necessary follow-on work for each panel member.

5. Solicit feedback about the process.

The primary objective of the sessions was to obtain the quantitative responses to each question and understand the qualitative rationale used as the basis for these responses (Block 12).

The elicitation framework (Section 3.8) was structured to provide the panelists flexibility for developing the quantitative questionnaire responses. The flexibility was desirable so that each panelist could select a specific approach --- consistent with his experience --- that would promote the greatest confidence in the supplied responses. However, the framework was sufficiently standardized so that the individual results could be analyzed using a common framework (Section 5) to develop the combined LOCA frequency estimates (Section 7). Not surprisingly, no two panelists used exactly the same approach. Some chose to modify the base case results developed by the panel or even develop their own base case scenarios. A few panelists even provided absolute values instead of the ratios relative to a chosen base case result as requested. In these cases, absolute values were converted to equivalent relative ratios for comparison with other panelist's results.

The philosophy used by each panel member to answer the elicitation questionnaire is summarized in Appendix K. Because no two panel members used identical approaches to obtain quantitative estimates, it was important that the facilitation team fully understood each participant's approach so that their results could be subsequently analyzed. Most of the panelists provided the facilitation team with written responses to the elicitation questions in advance of the interview. These responses were typically the starting point for the review and discussion held during the interview. The facilitation team's role was to evaluate the responses to ensure that they reflected the panel members' judgments and were consistent with the elicitation question requirements. While preparing their elicitation question responses, the panel members were asked to self-assess and not answer questions in areas where they had little or no expertise. In areas where they had limited technical expertise, the panel members were urged to make estimates, but to reflect their uncertainty in the UBs and LBs.

The discussion and review of the panelist's responses quite often revealed inconsistencies between the quantitative estimates and the supporting qualitative beliefs. These inconsistencies were explored to understand their genesis and possible resolution. Additionally, other weaknesses or incomplete areas in the initial question responses were discussed. Sometimes, response deficiencies resulted from a lack of understanding of particular elicitation questions and their requirements. These remaining ambiguities were clarified during the interview.

Each elicitation was taped to provide a record of the elicitation. Minutes from each elicitation were also recorded. The audio tapes and minutes were also used to clarify information and opinions. Approaches were also developed with each panelist for subsequent follow-on work. A list of action items was also developed for both the panel members and the facilitation team to reflect the additional work necessary to resolve inconsistencies, complete missing information, and strengthen weak or ambiguous information provided in these initial elicitation responses. This action item list was used by each panelist to help refine his analysis. Each panel member then had another one to four months to revise his initial input to address the follow-on work identified during the elicitation.

3.10 Final Elicitation Responses

After the individual elicitations were completed, and each panel member's refined elicitation responses were analyzed (Block 13), a third meeting was held with the entire panel. The purpose of this meeting was to summarize and discuss the important quantitative and qualitative results arising from the individual elicitations. The qualitative insights provided for each of the elicitation questions was a focal point. These were discussed among the panelists so that all could share in this combined knowledge.

Each panelist benefited by understanding insights that he may not have explicitly considered in the development of his elicitation responses. Additionally, each panelist gained from the insights in areas outside of his expertise (Block 14).

The quantitative LOCA frequencies associated with important piping systems and all the non-piping components were presented separately. The qualitative insights were also illustrated in the context of the quantitative results. Important degradation mechanisms, loading, and other rationale associated with the LOCA frequencies for each system were illustrated as underlying rationale. Also, the rationale associated with quantitative outliers was also presented in those situations where it was unique and had possibly not been considered by others. This discussion was used to highlight issues and concerns from individual panelists so that the entire panel could assess their merits. Quantitative comparisons among piping systems and non-piping components were also provided so that the panelists could understand the implications of responses that they developed for the decomposed technical issues. Quite often, panelists did not calculate total LOCA frequencies to understand these interdependencies.

Another objective of this meeting was to thoroughly explain the methodology used to combine all the quantitative piping and non-piping LOCA frequencies for the decomposed piping systems and non-piping subcomponents to develop comprehensive LOCA estimates from each panelist's elicitation responses. The facilitation team reviewed these individual calculations with each panelist so that each could understand how his initial responses had been used in this development. Sometimes, the raw input had to be converted to fit the uniform analysis (Section 5.1) and response (Section 5.2) frameworks developed for this calculation. Also, additional assumptions were sometimes necessary to fill in missing information in the panelists' responses. The facilitation team made assumptions or added missing information, whenever possible, in a manner consistent with the philosophy, approach, assumptions, qualitative insights, and other quantitative results already provided by the panelist. The most common reason, by far, for augmenting the results was to provide complete coverage intervals for every elicitation response. While panelists often did not provide every coverage interval, they typically did assess coverage intervals in other places in the questionnaire. If necessary, the facilitation team created the missing coverage intervals to be consistent with these other responses. Imputed data was never used to compensate for an individual panelist's self-reported gaps in expertise. Any imputed data was also subject to panelist review and acceptance prior to final use. Additionally, panel members were also required to ensure that their responses were correctly transferred to the response framework (Section 5.2) and verify the results calculated from their input responses.

Another purpose of this meeting was to discuss the quantitative frequencies associated with the non-piping base cases. The qualitative conditions for all the non-piping base cases had been developed by the panel in previous meetings. Also, the quantitative calculations had been either directly provided to the panelists or made available on the ftp site for consideration. However, these quantitative calculations had yet to receive the same scrutiny as the piping base case results. The analysis methodology and results used to determine the SGTR frequencies, the CRDM ejection frequencies, BWR vessel failures due to normal operation and LTOP loading, and the PWR vessel PTS analysis were shared. There was also a presentation provided for the non-piping database development. The panel asked questions and provided feedback to gain a thorough understanding of the non-piping base case calculations and results similar to the piping base case evaluation in the previous meeting. Some concern was raised during this evaluation that not all panelists may have properly addressed common cause or conditional failures stemming from a single SGTR or CRDM ejection. Subsequent elicitation questions were therefore developed to explicitly address this concern. More information on this meeting is contained in Appendix B.

A number of final action items were developed from this meeting. First, each panelist was asked to review the calculations that the facilitation team performed on his elicitation responses. Each panelist was to ensure that no errors were made in the entry of his responses, and was also asked to comment on the appropriateness of assumptions or missing information added by the facilitation team during the analysis of his results. Each panelist was asked to provide alternative assumptions or missing information if desired. The panelists were also asked to separately consider conditional and common cause failures for steam generator and CRDM ejections.

Finally, all the panelists were given one more opportunity to revise their elicitation responses based on the discussion of the qualitative insights and rationale, a more thorough understanding of the non-piping base case calculations, and a more complete understanding of the analysis procedures used by the facilitation team in processing their raw input. Most panelists did not choose to change their preliminary estimates. However, a few did modify their preliminary responses based on knowledge gained at this final meeting (Block 15). Any necessary iteration was again conducted with each panelist to ensure that the final responses are consistent and reflect the qualitative rationale. These final responses were then analyzed once again to determine the LOCA frequency estimates provided by each panelist (Block 16). Detailed results from elicitation for the various panelists are provided in Appendix L.

3.11 Review and Reporting

A multilayer review was conducted using both the panelists and external reviewers to validate the accuracy and acceptability of the reporting and the analysis of the panelists' responses. First, a draft report (Block 17) was created and circulated to the panelists for review. The panelists provided review and feedback (Block 19) on every report section in order to increase the clarity and understanding of the elicitation process, analysis, and results presentation. However, they conducted the most substantive technical review of the background (Section 1), approach (Section 3), base case results (Section 4), qualitative results (Section 6), and the summary and conclusions (Section 9). Several panelists provided additional background information (Section 1) to ensure that the operating experience and PFM descriptions were comprehensive and balanced. The piping and non-piping base case team members contributed summary descriptions of their approaches (Section 3.5) and ensured that the base case results and sensitivity analyses were accurately and clearly reported (Section 4). The piping base case team members also provided some rationale to explain differences among the results.

The panelist review of and contributions to the qualitative results section (Section 6) was vital to ensure that the results were summarized in a manner that reflects the important insights, yet also presents dissenting views when appropriate. Group consensus of these insights was not explicitly sought during the elicitation process; however, many of the reported insights were shared by the majority. The panelists also reviewed the summary and conclusions (Section 9) that pertain to the base case and qualitative results to ensure that they were complete and accurate. A revised report (Block 20) was developed which addressed the feedback received from the panelists. The panelists were given the opportunity to conduct a final review of this report and additional final comments have been incorporated as warranted.

In parallel with the panelists' review, an external peer review (Block 18) was conducted of the baseline elicitation response analysis (Section 5.5), the sensitivity analyses (Section 5.6), and the corresponding results (Section 7) contained in the draft report (Block 17). These are the portions of the process that the elicitation panelists are not qualified to review. Two peer reviewers were chosen: Dr. C. Atwood of Statwood Consulting and Dr. A. Brothers of Pacific Northwest National Laboratory. Dr. Atwood is a statistician with experience in estimating rare event frequencies for PRA use and is also an author of NUREG/CR-5750 [3.5]. Dr. Brothers is a decision analyst with knowledge of elicitation processes and procedures.

The draft report was initially provided to the peer reviewers so that they could read Sections 5 and 7. A 1 ½ day kick-off meeting was then held. The purpose of the kick-off meeting was to present and clarify the analysis procedures, sensitivity analysis matrix, and results in detail so that the reviewers thoroughly understood the approach. Some initial errors found in the analysis procedure were also discussed at the meeting. Ideas for conducting additional sensitivity analyses and alternative response analysis and aggregation schemes were also evaluated during the meeting. After the meeting, the facilitators were in contact with the reviewers to answer any additional questions. The peer reviewers provided some preliminary feedback within a week of the kick-off meeting and then provided their initial review a month after the kick-off meeting.

The peer reviewers provided several recommendations and insights which are detailed in References 3.23 and 3.24. Dr. Brothers stated that while this was not a focal point of his review, the elicitation process appears adequate and sound for determining the stated objectives. Dr. Atwood found some errors in the baseline analysis procedure (Section 5.5) and provided an exact formulation for means of summed distributions that were initially estimated. These errors were corrected and summation approximations eliminated wherever possible. The accuracy of remaining approximations has been evaluated using selected Monte Carlo analysis as requested in Reference 3.23. Both peer reviewers also suggested additional sensitivity analyses to verify other analysis assumptions and approximations. All important sensitivity analyses have been conducted and the calculational methodology has been updated as warranted to reflect these findings. The updated procedure is reflected in the analysis section (Section 5) of this revised report (Block 20). The elicitation responses were reevaluated using the updated analysis procedure (Block 16) to develop the final LOCA frequency estimates (Block 22).

One focal point of the peer review recommendations pertains to the aggregation of the individual LOCA frequency estimates (Section 5.4). Aggregation schemes other than the baseline methodology (Section 5.5) could be appropriate depending on the assumptions and interpretation of the individual total LOCA frequency estimates. The most reasonable alternate aggregation scheme creates a mixture distribution from the individual responses. This aggregation was conducted as an additional sensitivity analysis (Section 5.6) for comparison with the baseline results (Section 7.6) and is included in the revised report (Block 20).

The revised report (Block 20) was made available for public comment (Block 21). All public comments were addressed and this final version of the report has been revised as necessary to address the more pertinent of those public comments. The public comments and associated responses to those comments are included in this report as Appendix M. Now that the public comment period is over and the public comments resolved, the elicitation process is completed and the LOCA frequency estimates contained in this report are considered final (Block 22).

3.12 References

3.1   "Severe Accident Risks: As Assessment for Five U.S. Nuclear Power Plants," NUREG-1150, US NRC, December 1990.

3.2   T.A. Wheeler, et al., "Analysis of Core Damage Frequency From Internal Events: Expert Judgment Elicitation," NUREG/CR-4550, Vol. 2, Sandia National Laboratories, 1989.

3.3   E.J. Bonano, et al., "Elicitation and Use of Expert Judgment in Performance Assessment for High-Level Radioactive Waste Repositories," NUREG/CR-5411, Sandia National Laboratories, 1990.

3.4   Memorandum from A.C. Thadani to S.J. Collins, Transmittal of Technical Work to Support Possible Rulemaking on a Risk-Informed Alternative to 10 CFR 50.46/GDC 35, dated July 31, 2002.

3.5 Poloski, J. P., Marksberry, D. G., Atwood, C. L., and Galyean, W. J., "Rates of Initiating Events at U.S. Nuclear Power Plants: 1987-1995," NUREG/CR-5750, US NRC, February 1999.

3.6 M.A. Meyer and J.M. Booker, "Eliciting and Analyzing Expert Judgment," NUREG/CR-5424, Los Alamos National Laboratory, 1990.

3.7 "Reactor Safety Study: An Assessment of Accident Risks in U.S. Commercial Nuclear Power Plants," WASH-1400, US NRC, October 1975.

3.8 Generic Letter 88-01, "NRC Position on IGSCC in BWR Austenitic Stainless Steel Piping," January 25, 1988.

3.9 Lydell, B., "OPDE Database Coding Guideline and Quality Control Manual," May 2002.

3.10 D.O. Harris and D. Dedhia, *WinPRAISE: PRAISE Code in Windows*, Engineering Mechanics Technology, Inc. San Jose, California, Technical Report TR-98-4-1, 1998.

3.11 Khaleel, M. A., and others, "Fatigue Analysis of Components for 60-Year Plant Life," NUREG/CR-6674, June 2000.

3.12 Chapman, O. J. V., and Simonen, F. A., "RR-PRODIGAL – A Model for Estimating the Probabilities of Defects in Reactor Pressure Vessel Welds," NUREG/CR-5505, August 1998.

3.13 Milne, I., and others, "Assessment of the Integrity of Structures Containing Defects," CEGB Report R/H/R6 – Revision 3, 1986.

3.14 VIPER Version 1.2, Structural Integrity Associates., Report # SIR-95-098 Rev. 1, Feb. 1999.

3.15 EPRI Report, "BWR Reactor Pressure Vessel Shell Weld Inspection Recommendations (BWRVIP-05)," TR-105697, September 1995.

3.16 Materials Reliability Program, MRP-105, "Probabilistic Fracture Mechanics Analysis of PWR Reactor Pressure Vessel Top Head Nozzle Cracking," EPRI Report 1007834 (EPRI Licensed Material), May, 2004.

3.17 Memorandum from A.C. Thadani to S.J. Collins, "Technical Basis for Revision of the Pressurized Thermal Shock (PTS) Screening Criteria in the PTS Rule (10 CFR 50.61)," December 31, 2002.

3.18 Todreas, N. E., and Karimi, M. S., Nuclear Systems I Thermal Hydraulic Fundamentals, Taylor and Francis, 1993.

3.19 Moody, F. J., "Maximum Two-Phase Vessel Blowdown from Pipes," *J. Heat Transfer*, 88:285, 1966.

3.20 Zaloudek, F. R., "The Low Pressure Critical Discharge of Steam-Water Mixtures from Pipes," HW-68934, Hanford Works, Richland, WA, 1961.

3.21 "Additional Information Required for NRC Staff, Generic Report on Boiling Water Reactors," General Electric, NEDO-24708A, Class I, Revision 1, December 1980.

3.22 Lydell, B., "Failure Rates in Barsebäck-1 Reactor Coolant Pressure Boundary Piping: An Application of a Piping Failure Database," SKI Report 98:30, Swedish Nuclear Power Inspectorate, May 1999.

3.23 Atwood, C. L., "Review of Draft Report on LOCA Frequency Estimates by Expert Elicitation," NRC ADAMS Accession No. ML051430327, May 2005.

3.24 Brothers, A., "Review of Draft LOCA Frequency Estimates by Expert Elicitation," NRC ADAMS Accession No. ML051430431, May 2005.

4. BASE CASE RESULTS

As indicated earlier (Section 3.5), base case frequencies were developed by a subset of the panel. These frequencies were then provided to the other panelists as possible anchoring frequencies for use during their elicitations (Section 3.5). Alternatively, panelists were free to develop a different approach as the basis for their elicitation responses (Section 3.9). Base case frequencies were developed for five piping systems, two BWR systems and three PWR systems. Base case frequencies were also developed for a number of non-piping failure scenarios including BWR vessel failure, CRDM ejection, SGTR, and non-LOCA PTS transients. There were also two databases available containing precursor failure information for LOCA-sensitive piping systems and non-piping components as discussed in Section 3 that could be used for anchoring during the elicitations. The results of the associated base case analysis and sensitivity studies are provided in this section.

4.1 Piping Base Case Through-Wall Cracking Frequencies

Bengt Lydell (BL) and David Harris (DH) developed through-wall cracking (TWC) frequencies associated with each base case up to 25 years of average plant life. The purpose of this exercise was to provide the most direct comparison between operating experience and PFM results and provide information for benchmarking the PFM results. Bengt Lydell used a comprehensive operating-experience data base (Section 3.5.1.2.1) to obtain direct estimates of the TWC frequencies. David Harris (Section 3.5.1.2.3) determined frequencies using the PRAISE code. Some of the PFM results were adjusted, or benchmarked, through manipulation of variable input parameters (Appendix F) in order to more closely match the operating-experience TWC frequencies.

Figure 4.1 is a comparative plot of the leak frequency results after 25 years of service. In this figure, the hot leg (PWR-1), surge line (PWR-2), HPI/MU line (PWR-3), recirculation system (BWR-1), and feedwater system (BWR-2) base cases are all represented. These through-wall cracking frequencies represent LOCA precursor events due to the material degradation mechanisms specified in the base case definitions (Section 3.5.1.1). The flow rate associated with these through-wall cracks range from effectively zero up to the technical specification leak rate limit of approximately 1 gpm (3.8 lpm). Leaks bigger than 1 gpm (3.8 lpm) are assumed to either be detected by the leak detection systems or result in a LOCA.

There is relatively good agreement between the operating-experience (Bengt Lydell) and PFM (David Harris) results for the PWR-3 and BWR-1 base cases. This is to be expected because the PFM results were effectively benchmarked against the operating-experience estimates for both of these base cases. It is worth noting that IGSCC events in the BWR recirculation system (BWR-1) and thermal fatigue cracking in the HPI make-up nozzles (PWR-3) have relatively high service frequencies and there is correspondingly a larger body of TWC data associated with these cracking mechanisms. However, there is significant disagreement between the operating-experience and PFM results for the other three base case results.

The PWR-1 and PWR-2 base cases represent cracking in the hot leg due to PWSCC and in the surge line due to PWSCC and thermal fatigue, respectively. The PFM results are quite sensitive to initial input variable distributions for these base cases. Relatively minor changes can result in TWC frequency differences of several orders of magnitude (Appendix F). The through-wall cracking PFM frequencies (Figure 4.1) could have been calibrated to provide better agreement between the PWR-1 and PWR-2 through-wall cracking operating-experience results. However, there is the least amount of failure data available, and consequently the lowest frequencies and highest uncertainty associated with these operating-experience estimates. Therefore, the PFM through-wall crack frequencies (Figure 4.1) were chosen to provide more realistic and comparable Category 1 LOCA estimates with the operating-experience results (Sections 3.5.1.2 and 4.2).

The most significant difference between the PFM and operating -experience TWC frequencies occurs for the feedwater system base case (BWR-2). The PFM estimate is approximately 4 to 5 orders of magnitude less than the operating-experience estimate. There are at least two factors that contribute to this difference. The most important is that the BWR-2 operating-experience frequency estimate was based on 20 reported incidents in BWR feedwater systems. However, none of these events resulted in through-wall leakage. Therefore, the operating-experience estimate is undoubtedly conservative. Additionally, the PFM result only considered thermal fatigue frequency contributions and not FAC as required for the BWR-2 base case. There is no FAC degradation model in the PRAISE code. However, FAC is not a mechanism that typically results in TWC prior to failure, so differences due to this inconsistency are not expected to be large.

Operating experience could be analyzed more closely in an attempt to resolve this difference, but this was not a principal concern since both the assumptions and approach associated with both estimates were provided for panel consideration. It should be stressed that one principal objective of the base case evaluations was to illustrate the variability which is possible using different estimation schemes. The panelists were asked to consider this variability in their individual elicitations.

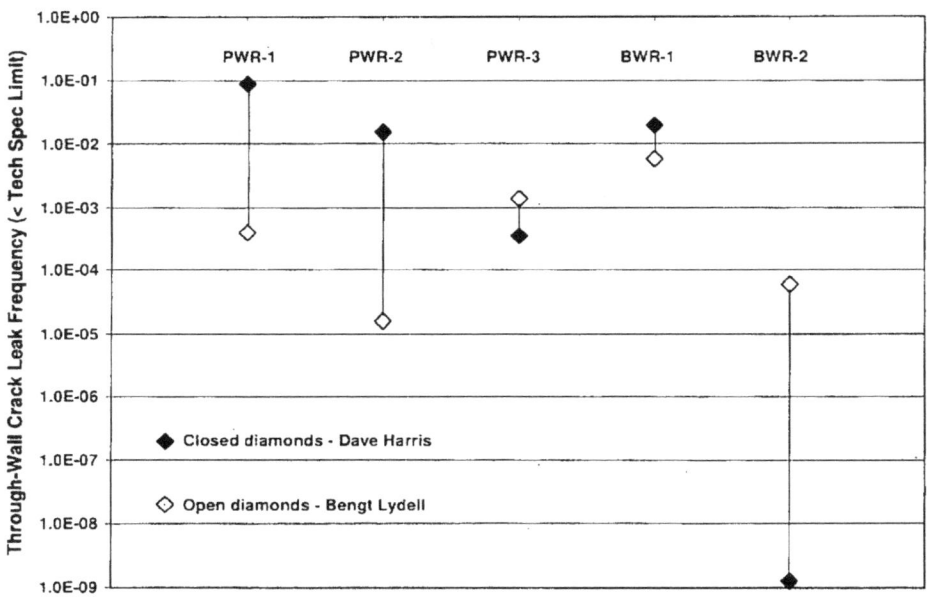

Figure 4.1 Leak Frequency Comparison Between Operating-Experience-Based and PFM-Based Analyses

4.2 Piping Base Case LOCA Frequencies

Piping base case LOCA frequencies were developed by four of the panel members: Bill Galyean (BG), Bengt Lydell (BL), Dave Harris (DH), and Vic Chapman (VC). As previously described in Section 3.5, base case frequencies were intended to be developed for three time intervals: 0-25 years (current-day), 25-40 years (end of plant-license), and 40-60 years (end-of-plant- license-renewal). However, only BL and DH predicted base case frequencies beyond 25 years. Bill Galyean and VC provided estimates only for 25 years, although BG has indicated (Appendix E) that his base case frequencies are assumed to be largely constant with time. Frequencies were developed for all six LOCA categories for the hot leg base case (PWR-1). For the surge line base case (PWR-2), two of the analyses were conducted up to LOCA Category 5 while the other two stopped at LOCA Category 4. It should be noted that a double-ended rupture in the surge line would briefly qualify as a Category 5 LOCA until the pressurizer completely drains. The flow rate for the remainder of this hypothetical transient would be classified as a Category 4 LOCA. Two of the analyses explicitly considered this initial higher flow rate in classifying the biggest PWR-2 breaks as a Category 5 LOCA. For all other base cases, estimates were made up to the largest LOCA size amenable to each piping system.

All four base case team members developed LOCA frequencies for the PWR piping base cases. All team members except VC developed frequencies for the BWR piping base cases. Two of the panelists based their estimates on existing operating experience (BG and BL) while the remaining two (DH and VC) used PFM analyses to develop their estimates. Where possible, DH's PFM calculated through-wall cracking frequencies were benchmarked with values estimated from operating experience (Section 4.1). Vic Chapman's PFM results were not similarly benchmarked. More detailed information on the approaches utilized by the four base case team members is provided in Section 3.5.1.2 and in Appendices D through G.

The piping base case frequencies (per calendar year) are summarized in Table 4.1. Note that the operating experience-based frequency estimates (BL and BG) were based on calendar years of plant operation where various reactor states from full power to shutdown occur. The PFM-based estimates (VC and DH) were based on reactor years of operation where it is assumed that the reactor is at full power over the entire year. The PFM estimates in Table 4.1, and subsequent tables, have been multiplied by a conversion factor of 0.8 to account for the approximate fleet average shut-down time per year [4.1]. This adjustment assumes that the LOCA risk during shut-down time is zero, but it ensures that all frequencies are reported on a consistent per calendar year basis. This adjustment is a minor consideration in light of the uncertainty in these estimates, but is performed for consistency.

It should be noted that these base case estimates (Table 4.1) are those that each base case team member believes most accurately represent the base case definitions (Section 3.5.1). These estimates evolved during the course of the elicitation and were not finalized prior to the onset of the individual elicitations. Therefore, interim estimates were provided to the panel members for anchoring during their individual elicitations. These interim estimates are not reported. The interim results of BL and DH were generally within a factor of 2 of their final estimates (Table 4.1) although a few cases did vary by a factor of 10 or more. In some cases the interim results for BG and VC varied by an order of magnitude from their final estimates. However, the differences between the interim estimates and those in Table 4.1 are insignificant compared to both the uncertainty associated with any single estimate and the variability among estimates for a single base case. Even so. the panel members were provided the option of updating their original elicitation responses once the final base case results (as reported in Table 4.1) were completed.

Several interesting observations can be made with respect to these results. First, some of the base case frequency estimates are extremely low, less than 1E-10 per calendar year. These low frequencies are most often associated with the PFM analyses. The prediction of low frequencies, by itself, does not imply that the analysis is any more flawed than one which predicts higher frequencies. An analogous situation is the following. Suppose someone buys one lottery ticket in each of three successive months, and each ticket has one chance in a million of winning. The probability of winning all three times is 1E-18. This is truly an incredible event, but the calculation of its probability is correct nonetheless. However, lower frequency estimates are typically more sensitive to the accuracy and variability of critical underlying modeling approaches and assumptions because changes are more likely to have a proportionally bigger impact. Therefore, the only direct conclusion that can be drawn from an extremely low LOCA frequency estimate is that the occurrence of such an event is incredible based on the assumptions, input distributions, and modeling approach used for the prediction.

The results in Table 4.1 are also interesting because large differences exist among some of base case team member estimates. Figure 4.2 graphically depicts these differences for the Category 1 piping base case frequencies (per calendar year) at 25 years of plant operations. The variability among the majority of the Category 1 LOCA estimates (Figure 4.2) is approximately one to two orders of magnitude for all five piping base cases. Additionally, the variability between the two operating-experience-based analyses (i.e., the BG and BL results) is always relatively small, less than an order of magnitude for all five base case Category 1 estimates (Figure 4.2). However, the VC's PWR-1 and PWR-3 base case estimates, and DH's estimate for the BWR-1 base case are greater than two orders of magnitude different from the other results.

An understanding of the underlying modeling is necessary to both explain the low frequency estimates and provide some reasons for the large discrepancies among individual base case estimates. For instance, VC's PFM estimates for the PWR-1 base case (hot leg cracking due to PWSCC) predicted Category 1 LOCA frequencies of less than 1E-10 per calendar year. However, VC only considered failures due to fatigue from preexisting flaws, and not PWSCC. The cyclic stresses due to normal startup/shutdown cycles plus an assumed seismic event were very low. Thus, the estimated failure frequency associated with fatigue crack growth from a relatively small number of preexisting flaws is extremely low. Conversely, the PFM evaluations made by DH considered both fatigue and PWSCC crack growth mechanisms. Dave Harris's analysis also allowed flaws to initiate during operation. These additional considerations substantially elevated DH's Category 1 LOCA frequency estimates (Figure 4.2). The implication to draw from these differences is not that LOCA hot leg frequencies are extremely low, but rather, that the LOCA frequency contribution due to fatigue crack growth from preexisting flaws is extremely low.

The large, seven orders of magnitude difference between the VC and DH PFM predictions for PWR-3 (HPI/MU nozzle) can also be attributed to modeling considerations. Dave Harris considered accelerated initiation and elevated stress due to a failed nozzle thermal sleeve. This is typically the condition evident when service cracking has been apparent. The DH results are also benchmarked by the through-wall cracking frequency for PWR-3 (Figure 4.1). However, VC analyzed the main weld of the nozzle assuming that the thermal sleeve was intact. The intact sleeve significantly decreases the thermal transients, thus leading to the extremely low failure frequencies calculated by VC. David Harris conducted a sensitivity analysis by considering an intact thermal sleeve with a similar transient stress history to VC. The LOCA frequencies predicted by DH for these more similar conditions was also extremely low and varied from VC's results by only three orders of magnitude. Other, unexplained differences in the modeling accounts for these remaining differences, but the important conclusion is that these nozzles are not likely to crack if the thermal sleeve remains intact.

The PWR-2 base case considered both thermal fatigue and PWSCC for the surge line. Both DH and VC modeled thermal fatigue, but neither modeled PWSCC for this base case. Thermal fatigue is apparently a much more important contributing factor than PWSCC since this contribution, by itself, is close to the operating-experience estimates. Also, the models have much less uncertainty. The increased impact of thermal fatigue is due to the higher cyclic stresses present in the surge line. Hence, the two PFM estimates are much closer, differing by less than one order of magnitude for the Category 1 LOCAs.

The variability between the two operating-experience-based analyses (i.e., the BG and BL results) is less than an order of magnitude for all five base case Category 1 estimates (Figure 4.2). This consistency is not surprising given the similarity of the approaches. While BG and BL used different operating-experience databases, all the information in BG's database was contained within the more expansive database used by BL. Additionally, both BG and BL used similar conditional pipe rupture relationships (see Appendices D and E) between Category 1 and larger LOCA sizes although, as described in Section 3.5.1.2, each had a distinct basis to justify their relationship.

Therefore, the most significant difference in the approaches was how the initial Category 1 LOCA estimates were determined (Section 3.5.1.2). As previously indicated in Section 3.5.1.2, BG used a top-down approach where the total LOCA Category 1 estimates, based on total events, were partitioned based on the relative cracking frequencies among the LOCA sensitive systems. Bengt Lydell determined cracking frequencies associated with the welds and degradation mechanisms for the base case system, and then determined the system frequencies by scaling with the number of susceptible welds. The fact that these two approaches generally agree implies that the failure rate of the modeled degradation mechanism for a system is similar to the relative failure rates for all degradation mechanisms associated with the same system.

The biggest difference between the LOCA Category 1 operating-experience and DH PFM estimates is for the BWR-1 base case which modeled IGSCC in the recirculation system. Both Bengt Lydell and Bill Galyean used the post-IGSCC mitigation cracking rates as the basis for the BWR-1 analysis. As with the other estimates, their BWR-1 recirculation system LOCA Category 1 base case results are within approximately a half order of magnitude. While the use of post-IGSCC mitigation rate is counter to the original BWR-1 definition, the frequencies are more consistent with current conditions at BWR plants. However, this confounds the comparison with the PFM estimates developed for this base case. The conditions analyzed by DH did consider IGSCC of a normal water chemistry environment and utilized residual stresses to simulate the effect of a weld overlay, although the residual stress distribution was conservative (Appendix F). Not surprisingly, then, his Category 1 LOCA frequency estimates are significantly higher than either BL or BG who effectively modeled the frequencies associated with all the mitigation schemes used for IGSCC.

Table 4.1 Piping Base Case Frequency Results by Participant

LOCA Cat.	25 Years Participant				60 Years Participant			
	BG	BL	DH	VC	BG	BL	DH	VC
PWR-1 Hot Leg (per calendar year)								
1	1.5E-07	7.4E-07	3.2E-08	1.8E-11		9.7E-07	3.2E-08	
2	4.6E-08	7.6E-08	3.8E-11	1.8E-12		1.0E-07	3.8E-11	
3	1.5E-08	2.9E-08	1.1E-12	4.4E-13		3.8E-08	1.1E-12	
4	4.6E-09	1.1E-08	1.1E-12	7.0E-15		1.4E-08	1.1E-12	
5	1.5E-09	3.8E-09	8.8E-16	1.4E-15		4.9E-09	8.8E-16	
6		1.3E-09	8.8E-16			1.6E-09	8.8E-16	
PWR-2 Surge Line (per calendar year)								
1	1.5E-08	1.3E-07	4.8E-07	1.9E-07		1.8E-07	1.4E-05	
2	4.5E-09	1.5E-08	2.2E-09	3.1E-08		2.1E-08	9.6E-08	
3	1.5E-09	5.4E-09	1.5E-10	1.2E-08		7.4E-09	1.4E-08	
4	4.5E-10	1.6E-09	6.3E-14	1.9E-09		2.1E-09	4.0E-11	
5		5.3E-10		3.9E-10		7.3E-10		
6								
PWR-3 High Pressure Injection/Make-Up (HPI/MU) (per calendar year)								
1	2.3E-06	1.6E-05	6.2E-05	6.7E-12		2.0E-05	3.2E-04	
2	7.0E-07	2.3E-06	6.2E-05	7.4E-13		3.3E-06	3.2E-04	
3	2.3E-07	9.2E-07		2.8E-13		9.5E-07		
4								
5								
6								
BWR-1 Recirculation System (per calendar year)								
1	5.3E-05	9.5E-06	9.2E-03			1.9E-05	5.7E-04	
2	1.6E-05	1.2E-06	6.8E-03			2.4E-06	5.0E-04	
3	5.3E-06	4.6E-07	4.8E-03			9.2E-07	5.0E-04	
4	1.6E-06	1.5E-07	3.1E-03			3.0E-07	5.0E-04	
5	5.3E-07	3.0E-08	2.1E-06			6.1E-08	3.0E-06	
6								
BWR-2 Feedwater Lines (per calendar year)								
1	4.0E-06	2.5E-06	<1.0E-7			2.6E-06	1.0E-07	
2	1.2E-06	3.4E-07	<1.3E-11			3.4E-07	1.3E-11	
3	4.0E-07	1.2E-07	<6.1E-13			1.3E-07	6.1E-13	
4	1.2E-07	4.1E-08	<1.1E-17			4.1E-08	1.1E-17	
5		7.3E-09				7.4E-09		
6								

Differences among the base case predictions are also graphically depicted in Figure 4.3 for larger Category 3 piping base case frequencies (per calendar year) at 25 years of plant operations. The variability among the Category 3 LOCA base case estimates (Figure 4.3) when compared with the Category 1 LOCA estimates (Figure 4.2) would be expected to increase as the event frequency decreases. Small analysis variations and differences can result in greater relative disparity among the LOCA frequency estimates as the absolute frequencies decrease. However, for four of the five piping base cases, the range among the LOCA Category 1 and 3 estimates are similar.

The biggest increase in variability with increasing LOCA size occurs for the BWR-2 base case which considered thermal fatigue and FAC in feedwater piping. For this particular base case, the variability between the base case frequencies for the Category 1 LOCA estimates is less than 2 orders of magnitude while the variability among the Category 3 estimates is approximately 6 orders of magnitude. This additional variability results from the substantial LOCA Category 3 frequency decrease predicted by DH. The reason for this large decrease is that DH did not model FAC (Section 3.5.1.2). Flow-accelerated corrosion is a degradation mechanism that is more likely to lead to a complete pipe rupture than thermal fatigue which often surfaces as a leak, or potentially a partial pipe failure (i.e., Category 1 LOCA). Hence, DH predicts that the LOCA Category 3 frequency contribution due to thermal fatigue decreases substantially. Conversely, the FAC contributions, as modeled by BG and BL, become more important contributors for the larger LOCA sizes.

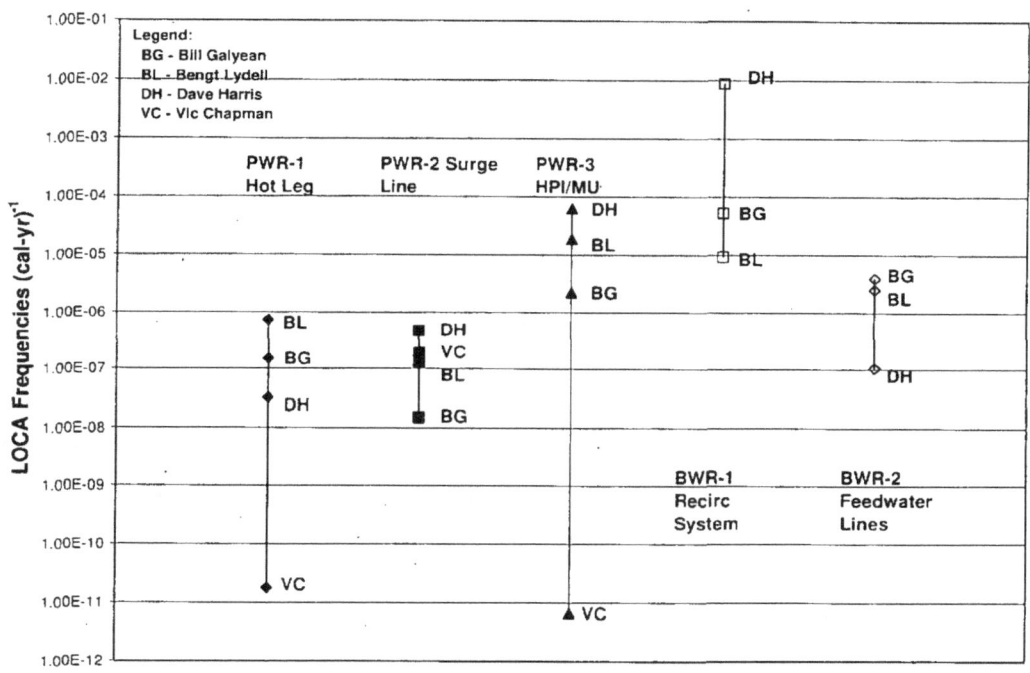

Figure 4.2 Category 1 Piping Base Case Frequencies at 25 Years

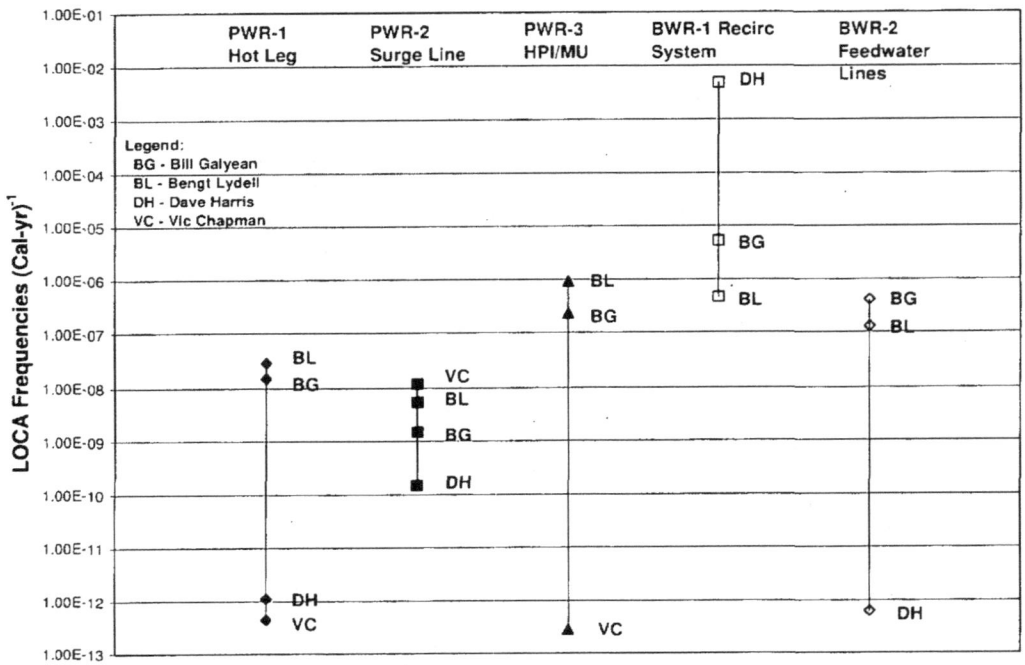

Figure 4.3 Category 3 Piping Base Case Frequencies at 25 Years

The estimated PWR and BWR base case frequencies are illustrated as a function of the LOCA size categories in Figures 4.4 and 4.5. These figures are more useful for evaluating the relative effects of increasing LOCA size than Figures 4.2 and 4.3. The PWR base case LOCA frequency estimates are illustrated in Figure 4.4, while the BWR base case results are shown in Figure 4.5. In both figures, the results for each base case are connected by a vertical line to indicate the variability among the results (as in Figures 4.2 and 4.3). The base case results have been offset slightly and the number above each vertical symbol represents the base case number. For example the 1 above the results in Figure 4.4 indicates that these results are for the PWR-1 base case while the 1 above the results in Figure 4.5 indicates that these results are for the BWR-1 base case. The results for each base case team member are presented using the same symbol type in both figures. The symbol fill is also varied to indicate the base case number.

As indicated in Figure 4.4 for the PWR base case results, both BG and BL predict that the PWR-2 (surge line) frequencies are about an order of magnitude less than the PWR-1 (hot leg) frequencies for all of the LOCA categories. The DH estimates also predict a similar trend. Conversely, VC estimates that the PWR-2 frequencies are about 4 orders of magnitude higher than the PWR-1 frequencies. Note that VC, as discussed previously, did not model PWSCC in either the PWR-1 or PWR-2 base cases. However, the cyclic stresses in the surge line were higher than in the hot leg, which leads to VC's higher predicted failure probability for PWR-2.

It is also apparent in Figure 4.4 (as in Figures 4.2 and 4.3) that the operating-experience-based estimates are most similar. For all LOCA categories, BL's PWR base case predictions are less than an order of magnitude higher than BG's predictions for all three PWR base cases. This is much closer agreement than the PFM results, but is not surprising given the similarity in the approaches discussed previously. There is also (as in Figures 4.2 and 4.3) large discrepancies between the PFM estimates, which are a reflection of modeling differences as discussed previously. However, occasionally the PFM estimates actually become closer with increasing LOCA size (e.g., PWR-1 estimates). Vic Chapman speculates that some of the differences in the estimates may be attributed to the initial flaw distributions and how the transition from a surface flaw to a through-wall crack was modeled. These factors determine the crack size and therefore the leak rate when the flaw penetrates the piping outer surface. Vic Chapman believes that these factors significantly affect smaller LOCA Category estimates but have less effect as the LOCA size increases.

The more interesting observation in Figure 4.4 is that most of the differences among the LOCA Category 1 estimates remain relatively consistent as the LOCA size increases. In fact, three of the base case team members (BG, BL, and VC in Figure 4.4) maintain approximately a ½ to 1 order of magnitude decrease in LOCA frequency with each successive LOCA size increase for all three PWR base cases. This relative decrease in BG's estimates mirrors his assumption of approximately a half order of magnitude decrease in the LOCA frequency for each successive LOCA category. Bill Galyean's assumption is based on an evaluation of the propensity of though-wall fatigue flaws to lead to various LOCA break sizes [4.2].

The other operating-experience estimates, provided by Bengt Lydell, developed conditional failure probability (CFP) relationships based on an analysis of service history trends. Therefore, these CFPs were not assumed a priori. While Bengt Lydell's conditional probabilities vary somewhat as a function of degradation mechanism and LOCA size, the average failure frequency decrease is approximately ½ order of magnitude between successive LOCA categories. See Appendix D for more detail. The relative decrease in VC's results is predicted directly by the PFM analysis.

Only DH's results tend to predict a more dramatic decrease between the Category 1 and bigger LOCA sizes in the PWR-1 and PWR-2 base cases. Big decreases between his Category 1 and 2 predictions occur due to the expectation that the cracking in these base cases is much more likely to result in a smaller break size. However, beyond Category 2, the DH LOCA frequencies for the PWR-1 and 2 base cases decrease, on average, by approximately ½ to 1 order of magnitude per LOCA category up through the largest LOCA Category modeled. It should be noted that DH only performed analyses for LOCA Categories 3 and 6 for the PWR-1 base case and then assumed that the Category 3 and 4 results were equivalent and that the Category 5 and 6 results were equivalent. This treatise explains the step-wise behavior apparent between LOCA Categories 3 through 6 in DH's PWR-1 results.

There are several implications which result from these trends in the relationship between the frequencies and LOCA size.. The first is that the average conditional probability between successively increasing LOCA categories for all these base cases generally falls between 0.1 to 0.5, which is equivalent to a ½ to 1 order of magnitude decrease. Further, the conditional probability associated with each of the four base cases results also typically falls within this range. More importantly, and consistent with previous discussion (Figures 4.2 and 4.3), is the finding that much of the variability among the estimates is apparent for the smallest LOCA sizes. If the LOCA Category 1 estimates were normalized to eliminate variability, subsequent LOCA Categories would not differ as widely.

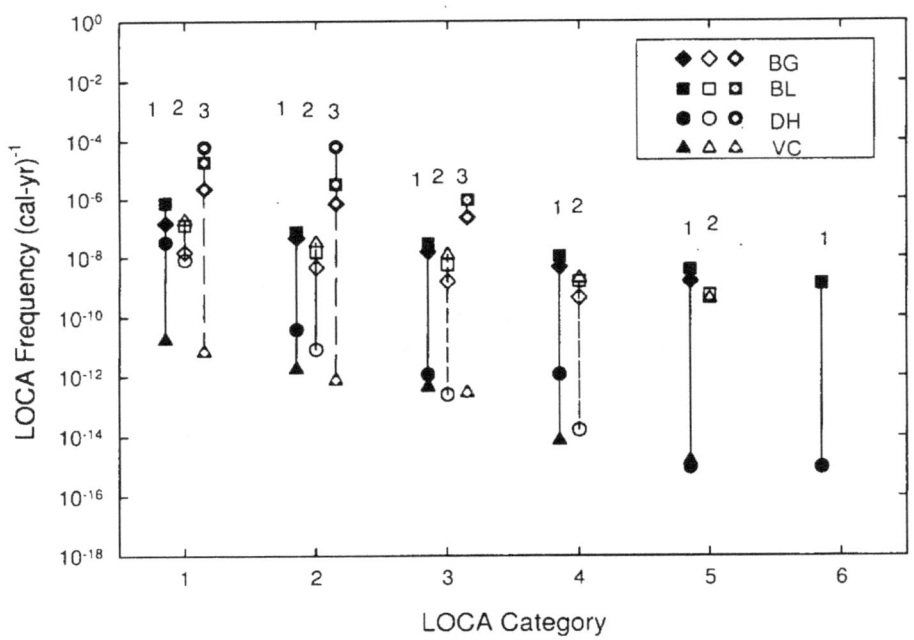

Figure 4.4 PWR Piping Base Case Frequencies at 25 Years of Plant Operation

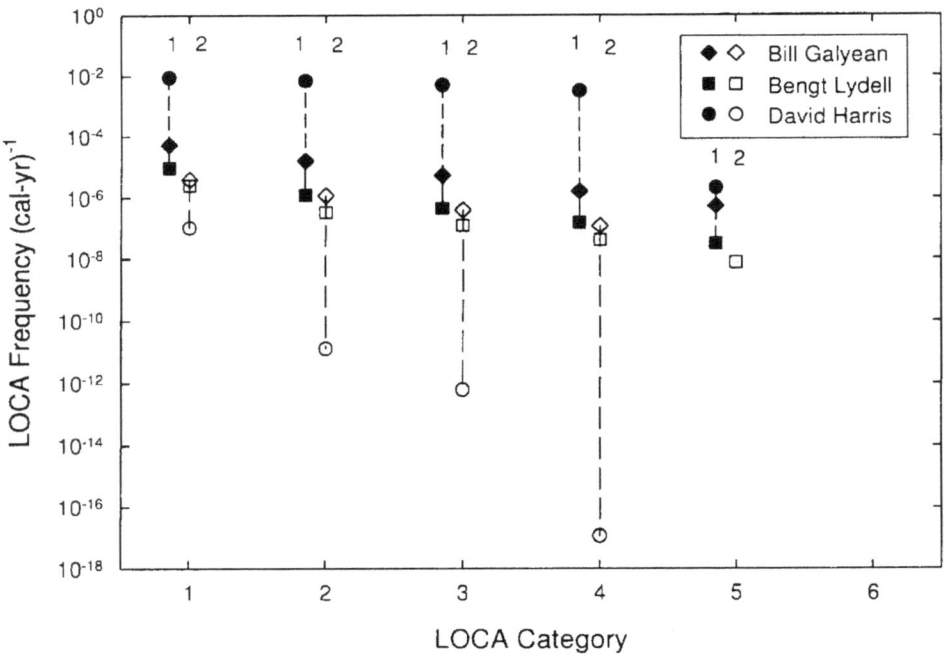

Figure 4.5 BWR Piping Base Case Frequencies at 25 Years of Plant Operation

4-10

The BWR base case results (Figure 4.5) exhibit many similar trends as the PWR results in Figure 4.4). Both BL and BG predict slightly higher frequencies for the recirculation system (BWR-1) than for the feedwater system (BWR-2). Both operating-experience-based predictions are once again consistent but, contrary to the PWR results, the BG predictions tend to be slightly higher. The operating-experience base case frequencies again decrease by approximately one half order of magnitude for each successive LOCA size increase because similar conditional failure probability relationships are used as in the PWR analyses.

However, the DH results for BWR-1 LOCA Category 1 are two orders of magnitude higher than the other estimates. This is a consequence of using the operating-experience-based TWC frequency (Figure 4.1) to calibrate these estimates. There is also very little expected frequency decrease between the TWC frequencies and LOCA Category 4 estimates. However, a precipitous, three orders of magnitude decrease occurs between LOCA Categories 4 and 5. The implication is that DH predicts that through wall cracking and LOCAs are equally likely up to a break size of 7 inches (178 mm) diameter (i.e., LOCA Category 4). Only breaks larger than this become much less likely. As in the PWR comparisons, variability among the three BWR-1 estimates is largely a function of the initial variability present in the Category 1 results. Normalization of using the LOCA Category 1 estimates would substantially decrease the variability in subsequent LOCA categories.

The DH BWR-2, LOCA Category 1 estimate is much closer to the operating-experience estimates than the BWR-1 base case. This result is also lower than the BG and BL estimates. Further, the DH results for larger LOCA categories decreases precipitously compared to the other BWR-2 estimates. This increased variability with increasing LOCA size is contrary to the other base case results. The reason for this large decrease, as discussed previously, is that DH did not model FAC (Section 3.5.1.2). Hence, DH predicts that the LOCA Category 1 frequency contribution due to thermal fatigue is high, but then decreases substantially for each successively larger LOCA size. Conversely, the FAC contributions, as modeled by BG and BL, become more important contributors for the larger LOCA sizes.

The base case frequency estimates, and the corresponding variability, are not significantly affected by the time period (i.e., 25 versus 60 years) of operation. From Table 4.1, BL predicts that the LOCA frequencies for all base cases will increase by less than a factor of 2 between 25 and 60 years of operation. The DH 25 and 60 year results are identical for the PWR-1 and BWR-2 base cases. Additionally, the PWR-3 60 year estimates increase only by ½ an order of magnitude. The biggest changes are predicted for the PWR-2 and BWR-1 estimates. There is approximately a two order of magnitude increase between the 60 and 25 year PWR-2 estimates for each LOCA Category. The BWR-1 LOCA Category 1 through 4 estimates decrease by approximately one order of magnitude between 25 and 60 years. However, there is virtually no change with time between the BWR-1 LOCA Category 5 estimates. The leak frequencies in DH's analysis are determined by averaging the cumulative LOCA probability over the time period being evaluated, either 25 or 60 years. Instantaneous frequencies may therefore differ somewhat from these simplified average estimates for the BWR-1 and PWR-2 analyses. The LOCA frequency increases with time predicted in the PWR-2 estimates are caused by continued crack growth from thermal fatigue. Decreases in the BWR-1 frequencies with time are expected because cracks in the analysis with the highest probability of leading to a LOCA have already been detected and repaired by 25 years of operation.

The significant variability among the PFM and operating-experience base case estimates reflects the current uncertainty in the state of knowledge in making these predictions. While reasons and trends associated with this variability are explored in this section, the elicitation did not attempt to resolve these discrepancies. Rather, one of the elicitation's goals was to reflect the uncertainty in the technical community through the use of these base cases. Therefore, differences between the base case analyses were highlighted and possible causes (as discussed in this section) were supplied to the panelists to inform the selection of their base case results for anchoring their elicitation responses and in providing uncertainty estimates.

The other purpose of the base case analyses (Section 3.5) was to provide anchoring points as a basis for panel members' elicitation responses. However, decisions regarding the use and application of these base case results were made by each individual panelist. The expectation was that each panelist's use of the base case results would be informed by the assumptions, limitations, and approaches used to develop each estimate; the variability among estimates; and the difference between the base cases and the actual plant operating systems and conditions. Some panelists utilized the base case results extensively, others considered the relative trends expressed by the base case estimates, and others chose not to utilize them at all.

In general, those panelists using the base case estimates chose to anchor their responses to base case results based on the operating-experience estimates instead of the PFM results. The operating-experience estimates were most often used for anchoring the LOCA Category 1 responses. The PFM results were used only by some panelists to extrapolate their assessments beyond 25 years or to LOCA sizes greater than Category 1, where no passive-system failure data exists.

As seen in this section, the PFM assumptions are less critical for assessing relative trends for increasing LOCA sizes than in determining absolute frequencies. Therefore, the differences in absolute frequencies between the operating-experience base cases and the PFM base cases are not considered significant in accessing variability among the individual elicitation responses. Additional discussion of the use of the base case estimates is provided in Section 6 while a more complete description of anchoring philosophy chosen by each panelist is found in Appendix K.

4.3 Piping Base Case Sensitivity Studies

A number of sensitivity studies of the base case systems were conducted to evaluate the effects of various parameters on the piping base case frequencies. These studies were also conducted by various base case team members. They were used to evaluate the effect of certain variables and analysis choices on the base case estimates. The sensitivity analyses and results were presented to the panelists during the base case review meeting so that they could inform subsequent elicitation responses. One sensitivity analysis evaluated the effect of various leak detection strategies as part of an ISI program on all of the base cases (Section 4.3.1). All other sensitivity analyses were conducted for a specific base case.

The BWR-1 base case analysis assumed that the model plant followed the Generic Letter 88-01 inspection technique, used weld overlay to reinforce the flawed piping, and utilized normal water chemistry. This base case definition was chosen to best compare the operating-experience and PFM estimates by only considering a single, less-complex mitigation strategy, i.e., weld overlay. It was well recognized by the panelists that this base case is not representative of current BWR operating conditions which generally employ hydrogenated water chemistry in addition to mitigation strategies other than weld overlays (e.g., noble metal chemistry addition, mechanical stress improvement, etc.). Therefore, several additional sensitivity analyses were conducted for this base case to determine the effect of mitigation and applied loading variables. Specifically, the effectiveness of weld overlays (Section 4.3.3) was evaluated; the general effectiveness of IGSCC mitigation through analysis of operating experience (Section 4.3.4) was assessed; and the effect of variations in the applied plant loading history on the LOCA frequencies (Section 4.3.5) was examined.

The PWR-1 base case analysis considered PWSCC cracking in a hot leg bimetallic weld subject to typical ISI inspection procedures. The sensitivity analyses for this base case included an examination of the influence of hydro-testing and safe shutdown earthquake (SSE) seismic loading on the LOCA frequencies (Section 4.3.6), and considered the effect of degraded material properties (Section 4.3.7). The effect of ISI on the surge line cracking frequencies resulting from thermal fatigue was evaluated to illustrate the effect on the PWR-2 base case (Section 4.3.2). Finally, the effect of applied loading history variations was also studied for the high pressure injection make-up line cracking defined by the PWR-3 base case.

4.3.1 Effect of In-Service Inspection with Different Leak Detection Strategies

Bengt Lydell (BL) used a Markov model approach to investigate the sensitivity of the base case LOCA frequencies to different ISI strategies. This approach determines the ISI effectiveness factor, I, which is defined as the ratio of the time-dependent LOCA frequencies with inspections to the applicable frequencies if no inspections were conducted. Three different inspection strategies were investigated: (1) no ISI, (2) Section XI inspection with POD equal to 0.5, and (3) risk-informed ISI with the POD equal to 0.9. See Appendix D for more details.

The effect of ISI was evaluated for all the base case systems. Figure 4.6 illustrates the results for the BWR-1 evaluation for Category 1 LOCAs. Here, baseline frequencies are depicted as a function of operating time assuming that no active ISI is conducted. However, plant walk downs are performed with varying periodicity to search for primary system leaks. Four different walk down periodicities are considered: none, after the hydro test at each refueling outage, weekly, and during each shift. As can be seen in Figure 4.6, the Category 1 LOCA frequency decreases by about one order of magnitude for each successive periodicity increase between no walk down, refueling outage walk downs, and weekly walk downs. However, there is little additional benefit to increasing the walk down frequency to more than a weekly basis. Also, most of the benefits associated with increased inspection periodicity occur over the first 25 years of service. Beyond 25 years of plant operation, the slopes of the LOCA frequency results (Figure 4.6) are not a strong function of inspection periodicity. Specifically, the 60 year frequencies are about ½ an order of magnitude greater than the 25 year estimates for the range of reported walk-down periodicities. This increase is consistent with BL's 25 and 60 year base case results (Table 4.1).

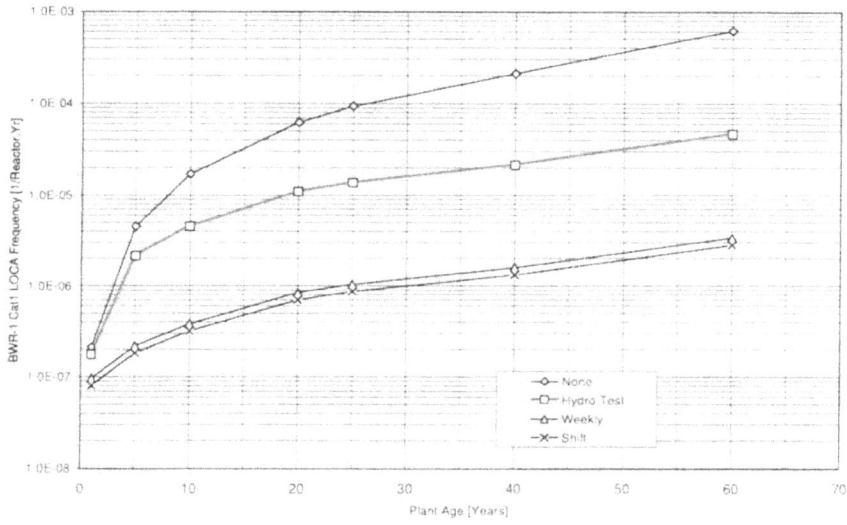

Figure 4.6 BWR-1 LOCA Frequencies as a Function of Plant Operating Time for the Case of No ISI with Different Leak Detection Strategies

4-13

Figure 4.7 illustrates the effect of different walk down periodicities on the BWR-1 base case Category 1 LOCA frequencies when ISI is routinely performed and the POD is 0.9. The same four leak detection strategies as portrayed in Figure 4.6 are included. Comparing Figures 4.6 and 4.7, one can see that the ISI program does not significantly decrease the LOCA frequency estimates. Without performing walk downs, the LOCA frequencies after 25 years of plant operations are 9×10^{-5}/reactor year without ISI (Figure 4.6) and 5×10^{-5}/reactor year with the assumed ISI (Figure 4.7). This less than a factor of two difference is not terribly significant. Differences are even less when the POD drops to 50%. For example, the BWR-1 Category 1 LOCA frequency after 25 years of plant operations is 6×10^{-5}/reactor year when the POD is 50% and no walk downs are performed. The effect of ISI is similar when other walk down periodicities are considered. Additionally, these trends are consistent for the other base case piping systems. This analysis therefore suggests that conducting frequent plant walk downs is a better strategy for decreasing LOCA frequency risk than is performing ISI. More details are provided in Appendix D.

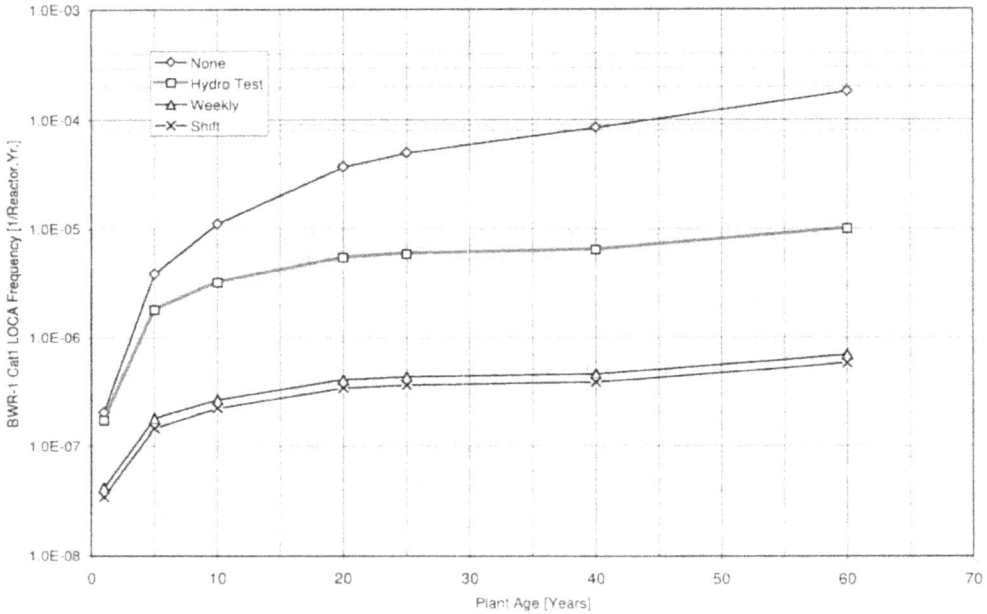

Figure 4.7 BWR-1 LOCA Frequencies as a Function of Plant Operating Time for the Case of ISI with 90% POD with Different Leak Detection Strategies

4.3.2 PFM ISI Assessment

An assessment of the effect of ISI was also carried out using PFM analysis by VC for the surge line elbow weld in the PWR-2 base case. The defect distribution and density was generated by RR-PRODIGAL. The POD relationship as a function of the crack depth to component thickness ratio (a/t) was defined by the following equation:

$$f_{POD} = \Phi\left(c_1 + c_2 \ln\left(\frac{a}{t}\right) \right) \qquad \text{where } c_1 = 1.526 \text{ and } c_2 = 0.533 \qquad (4.1)$$

4-14

This POD curve is shown in Figure 4.8 and the c_1 and c_2 parameters were chosen so that the POD for defects that are 70 percent of the component thickness is approximately 0.90. This POD is believed to conservatively represent historical detection resolution, but for future inspections that conform to more modern standards, this POD could be much better. This POD relationship is more complex than the constant POD assumed in the previous ISI sensitivity analysis (Section 4.3.1). More details are provided in Appendix G.

The results of this ISI sensitivity study on the 60 year cumulative failure probabilities are summarized in Table 4.2 and Figure 4.9 for various ISI intervals. The reduction factor in the table and figure is defined as the ratio of LOCA frequency without ISI to the LOCA frequency with the assumed ISI periodicity. Periodicity varies from no ISI to inspections conducted every 10 years. A special case assuming only a pre-service inspection (PSI) is also considered. These results suggest that even with this rather poor inspection capability, and for a weld with a high failure probability, reductions in the cumulative probability of failure of a decade or more can be achieved with two or three inspections during the life of the plant. The results also indicate that going beyond three inspections leads to little additional benefit assuming that indications are not uncovered. In fact, if a fourth inspection is carried out at forty years and no repairable indications are found, there is almost no additional benefit in conducting an inspection at fifty years. It should be noted that this analysis of the thermal fatigue risk to the surge line assumes that no new cracks are formed during service. More details are provided in Appendix G.

It is interesting to compare the two ISI sensitivity studies. Both results predict LOCA frequency reductions of about a factor of 2 to 3 due to inspections carried out between 25 and 60 years of plant operation on the applicable 60 year LOCA frequency estimates. However, the VC study in this section predicts a much bigger reduction due to inspections carried out prior to 30 years of service. Conversely, the BL study (Section 4.3.1) does not predict similar reductions due to ISI during the first 30 years of service. However, the BL study does predict that periodic plant walk down inspections during the first 30 years of service result in the biggest LOCA frequency reductions.

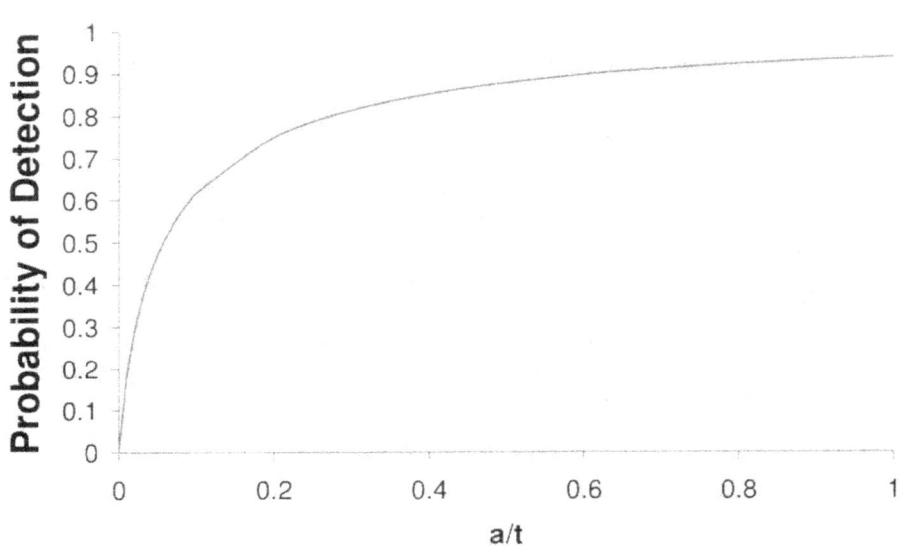

Figure 4.8 Probability of Detection (POD) Curve Used by Vic Chapman in ISI Sensitivity Analysis

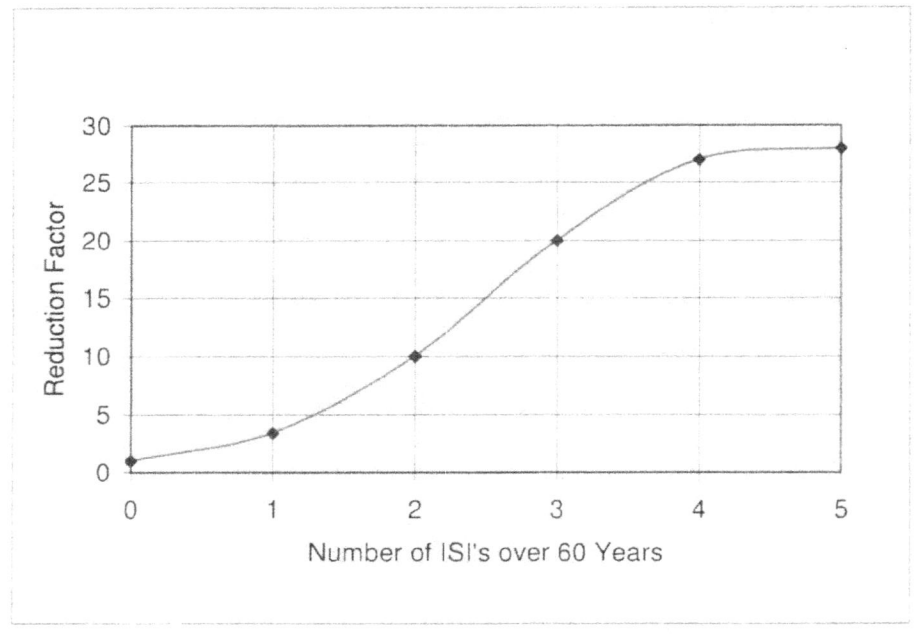

Figure 4.9 Graphical Results from Vic Chapman's ISI Sensitivity Analysis

Table 4.2 Tabular Results from Vic Chapman's ISI Sensitivity Analysis

ISI case	Cumulative Probability of Failure at 60 years	Factor for General Use
No ISI	1.3×10^{-4}	1
0 years (PSI)	4.2×10^{-5}	3
10 years	3.8×10^{-5}	3.4
10, 20 years	1.3×10^{-5}	10
10, 20, 30 years	6.5×10^{-6}	20
10, 20, 30, 40 years	4.8×10^{-6}	27
10, 20, 30, 40, 50 years	4.7×10^{-6}	28

4.3.3 Effect of Weld Overlay on IGSCC Mitigation

The BWR-1 base case assumed that a weld overlay was installed at 20 years of service to mitigate the effects of IGSCC. In order to examine the effect of this mitigation strategy on the estimated LOCA frequencies, a sensitivity analysis was performed assuming that no weld overlay was applied. The piping system chosen for this analysis was the 12-inch diameter section of the recirculation system defined within the BWR-1 base case. There are two similar recirculation loops in this base case with a total of 121 field, shop, and safe end welds. The piping is assumed to be fabricated from Schedule 80 Type 304 stainless with a nominal wall thickness of 0.687 inches (17.4 mm).

Because IGSCC is the degradation mechanism being considered, the time at stress primarily influences the cracking rate while the number of stress cycles is of secondary importance. For the baseline condition, the default residual stress pattern in PRAISE was assumed. These default residual stresses for intermediate size lines vary around the circumference and linearly through the thickness. The residual stress at a given angular location at the ID is normally distributed with a mean of 1.86 ksi (12.8 MPa) and a standard deviation of 2.89 ksi (19.9 MPa). These values were derived from benchmarking the PRAISE results with field observations of leaks for the recirculation system. The mean residual stress at the OD is -1.86 ksi (-12.8 MPa), and the standard deviation is the same as at the ID. For the case where the weld overlay was applied at 20 years, an alternative residual stress field [4.3] was assumed (Figure 4.10). The PRAISE code cannot model the nonlinear gradients illustrated in this figure, so a linear approximation was used. The linear gradient generally underestimates the beneficial effect of the weld overlay within the inner half of the pipe wall thickness. More details are available in Appendix F.

The cumulative probability for a Category 1 LOCA, both with and without weld overlay application, are illustrated in Figure 4.11. The results, of course, are identical up to 20 years when the overlay is applied. The derivative of the curves in Figure 4.11 is the LOCA frequency as a function of time. After 40 years, the LOCA frequency (i.e., derivative of curve) with no overlay is about 7 times greater than the frequency with the overlay applied. The effect of the weld overlay also leads to consistent LOCA frequency reductions for LOCA Category 2 – 4 (not shown) at both 40 and 60 years. More detailed results are provided in Appendix F.

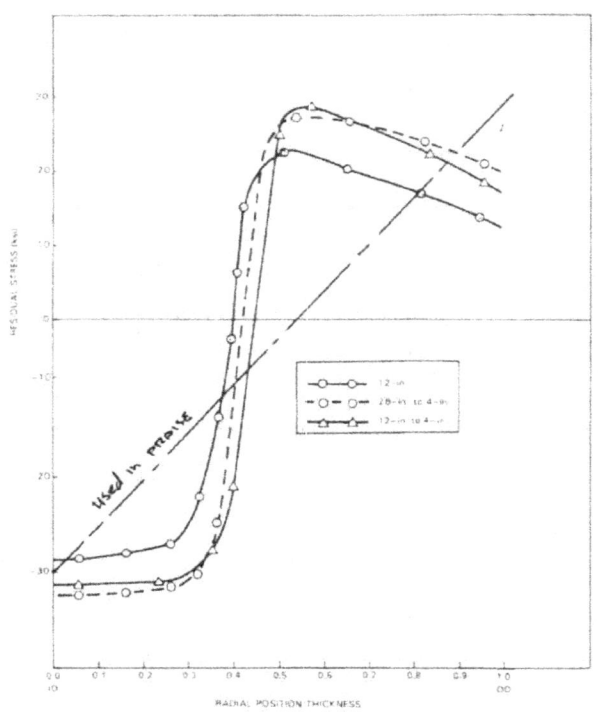

Figure 4.10 Through-Wall Residual Axial Stress Distribution Used in Weld Overlay Analyses

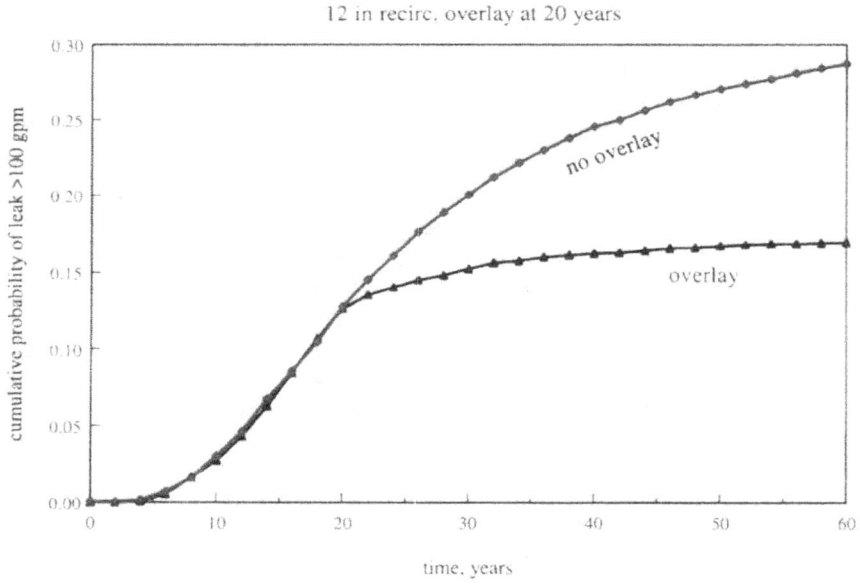

Figure 4.11 Effect of Weld Overlay Repairs on the Cumulative Category 1 LOCA Frequencies for 12-inch Diameter BWR Recirculation System Piping

4.3.4 General IGSCC Mitigation Effectiveness

Weld overlays are just one of several IGSCC mitigation strategies for BWR plants. Other effective strategies include hydrogen and noble metal water chemistry improvements, pipe replacement with more crack resistant materials, and post-weld heat treatment of susceptible welds. These strategies are coupled with higher quality inspections to identify cracking before it gets severe. All of these methods have been used by various plants for mitigating IGSCC. The passive-system failure data contains information on IGSCC cracking events both before and after these various mitigation techniques were generally applied.

While it is more difficult to assess the effectiveness of individual mitigation techniques, it is straightforward to examine the general effect of mitigation on the IGSCC cracking frequency. Bill Galyean simply segregated the operating experience into events prior to 1985 and events after 1985 to represent the pre-mitigation and post-mitigation periods, respectively. He then calculated the average IGSCC cracking frequency for the pre-mitigation and post-mitigation time periods. This LOCA precursor frequency is then converted into LOCA frequencies for each category using the generic conditional LOCA probability relationship that he used to develop his base case frequency estimates. This analysis reveals that the pre-mitigation cracking frequency is approximately 60 times higher than the post-mitigation frequency for all LOCA categories (Figure 4.12). In his analysis, the cracking frequency is assumed to be directly proportional to the LOCA frequency, so that this factor separates the pre-mitigation and post-mitigation LOCA frequencies as well. More details are provided in Appendix E. Bengt Lydell also considered the general effects of mitigation (e.g., hydrogen water chemistry, weld overlays, etc.) on post-1985 through wall IGSCC cracking frequencies through Bayesian analysis. See Appendix D for more details. As indicated in Section 4.2, both Bengt Lydell and Bill Galyean used the post-mitigation cracking rates as the basis for the BWR-1 analysis. This was counter to the original BWR-1 definition (Section 3.5).

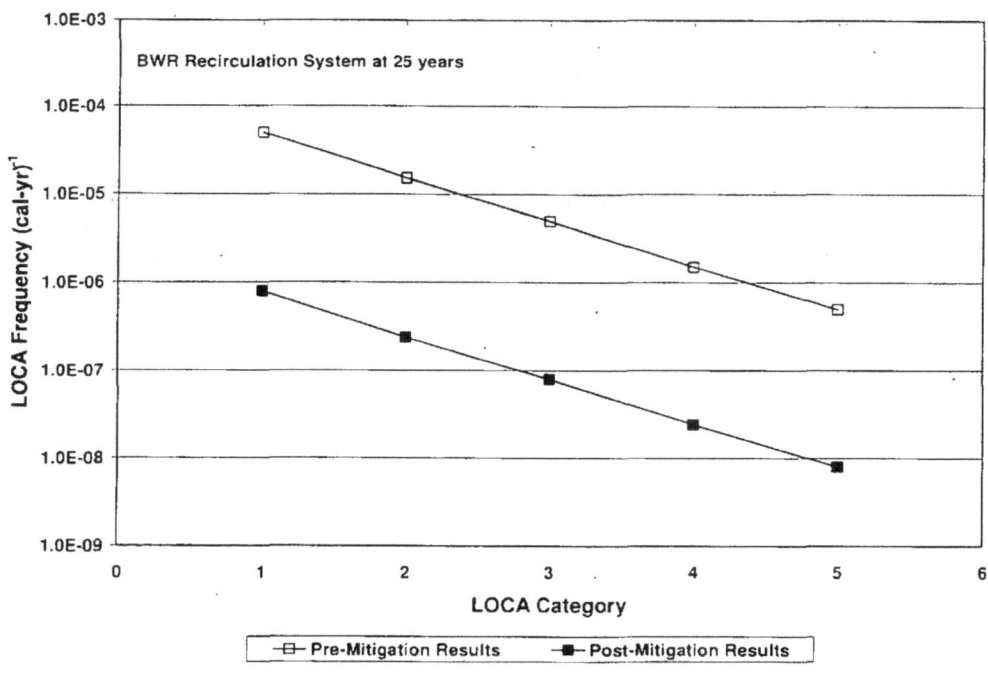

Figure 4.12 Pre and Post-Mitigation Recirculation LOCA Frequencies

4.3.5 Effect of Applied Loading Magnitude

The load history and applied stress magnitude significantly influence the LOCA frequencies calculated using PFM analysis of components. David Harris (DH) conducted a sensitivity analysis for the BWR-1 base case to examine the effect of the normal operating stresses on precursor leak and LOCA frequencies. The analysis was conducted for a 12-inch diameter recirculation system for which a weld overlay repair was installed at 20 years of plant operations (part of the BWR-1 base case). The PRAISE code was used for this analysis and frequencies for a single weld joint are summarized in Figure 4.13. Obviously, the precursor leak and LOCA frequencies for a modeled weld increase as the normal operating stresses increase.

The predicted precursor leak frequency for the 12 ksi (83 MPa) mean normal operating stress is 6×10^{-4} per weld per reactor year. This frequency coincides quite well with the pre-mitigation cracking rates experienced in service (see Appendices D and F). Increasing the normal operating stress from 12 to 15 ksi (83 to 103 MPa) (Figure 4.13) leads to approximately an order of magnitude increase in the frequencies associated with all break sizes. A smaller frequency increase occurs when the applied stress increases from 15 to 20 ksi (103 to 138 MPa).

The selection of the applied stress for the base case analysis is not as important as Figure 4.13 would indicate because as the assumed applied stress increases, the number of weld joints in a system which will be subject to that stress level decreases. The 20 ksi (138 MPa) applied stress was chosen for the BWR-1 base case analysis because this was more consistent with the approach used for the other systems. However, it was assumed that this high applied stress only exists at 2 weld joints in the recirculation system. The lower 12 ksi (83 MPa) stress is likely to exist at all 49 welds in the simulated system. Therefore, both stresses lead to similar recirculation system leak frequencies (Table F.23 of Appendix F) and both stresses correspond to the operating-experience-based leak frequencies (Figure 4.1) used for benchmarking the base case the analysis. In actual service, variability in the residual stress magnitude and distribution also influences the LOCA susceptibility of a particular weld joint. Possible residual stress variability has not been considered in this analysis, but higher tensile stresses at the inside pipe surface than assumed will increase the LOCA susceptibility of that weld.

The effect of the applied loading magnitude was also evaluated for the PWR-3 (HPI/MU nozzle) base case. Thermal fatigue is being modeled in this base case and the cyclic loading magnitude is therefore the principal consideration. In these results, the applied cyclic stress is 25 ksi (172 MPa) in order to properly benchmark the operating-experience cracking frequencies associated with this base case (Figure 4.1). However, while these stresses lead to the highest leak frequencies, the largest LOCA frequency occurs for lower applied cyclic stresses. The reason is that higher cyclic stresses promote leak-before-break in this component by fostering through-wall crack growth instead of surface crack growth that is associated with higher LOCA frequencies. See Appendix F for additional details and results of this analysis.

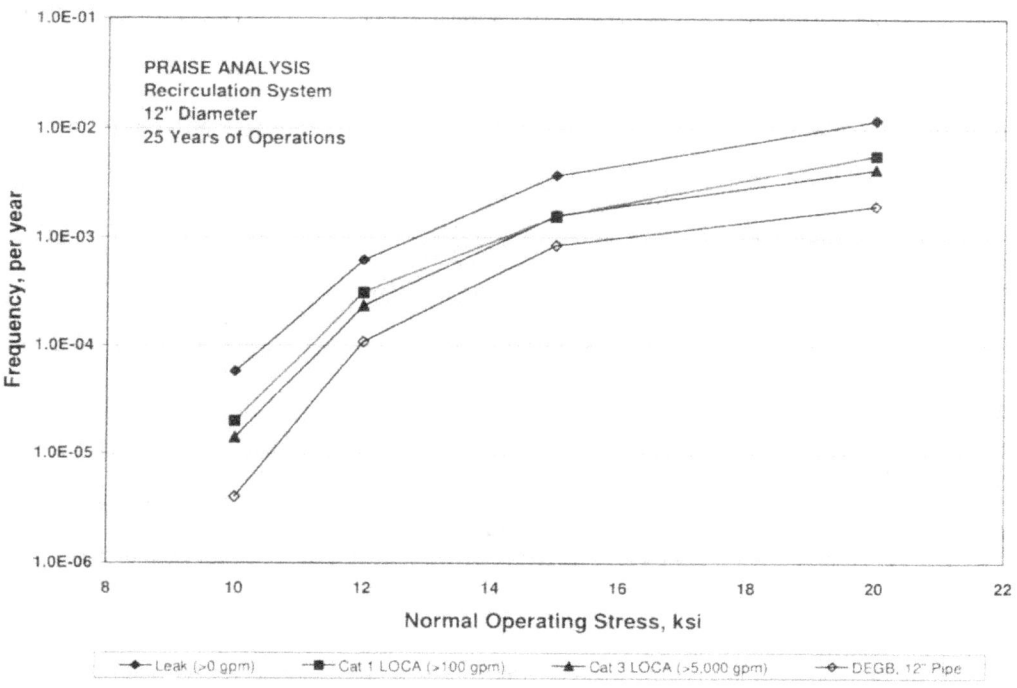

Figure 4.13 PRAISE Analysis Showing Effect of Normal Operating Stress on Various Size Precursor Leak and LOCA Frequencies for a Single Weld Joint in 12-inch Diameter Section of Recirculation System

4.3.6 Effect of Seismic Loading and Hydro Test

The impact of seismic stress was not explicitly considered within the elicitation because the principal objective of this elicitation (Section 2) is to develop LOCA frequencies under normal operating loads and expected transients. However, it is relatively straightforward to examine the effect of a prescribed seismic loading on the LOCA frequencies. The elicitation panel indicated that these sensitivity calculations would be beneficial and could serve as a surrogate for the effect of other rare emergency-faulted loads. The hydro test is conducted prior to initial plant operation. While the base case estimates include this test, it is also straightforward to evaluate the effects that this test has on LOCA frequencies.

The role of an assumed SSE seismic loading and the hydro test was studied for hot leg (PWR-1) failure frequencies. However, while hot leg PWSCC cracking is the PWR-1 base case (Table 3.7), only the impact of seismic and hydro test loading on the LOCA frequencies associated with pre-existing defects growing by fatigue was modeled. These frequencies are only one component of the total LOCA frequencies associated with this base case. The effect of seismic loading or hydro testing on cracks initiating and growing due to PWSCC was not considered. However, the initial hydro test does not significantly alter the failure probabilities associated with service-induced crack initiation and growth. The PRAISE code was used by DH for this analysis and seismic stresses up to 5 SSE were applied at either 25, 40, or 60 years of operation. Only one seismic load application was applied in any single analysis. The seismic magnitude for this sensitivity analysis was selected from the design seismic load spectrum developed for a specific plant. This sensitivity analysis utilized similar inputs and assumptions as this base case analysis (Appendix F). The effect of the hydro test is summarized in Figure 4.14 while the effect of SSE loading is depicted in Figure 4.15. In general, the low absolute frequencies depicted in Figures 4.14 and 4.15 imply that failure is unlikely due to the seismic and/or hydro test loading of preexisting defects under the assumptions and conditions modeled. The main purpose of these results is to examine the relative changes.

The initial hydro test moderately affects the LOCA frequencies (Figure 4.14). The Category 1 LOCA frequencies attributed to thermal fatigue decrease by an order of magnitude while the Category 6 frequencies decrease by approximately ½ an order of magnitude with the application of the hydro test. These trends are generally consistent even if SSE loading is not included (Appendix F). The general decreases occur because hydro testing can identify flaws before the component is place in service that could eventually lead to thermal fatigue failures during service. Bigger Category 1 decreases exist because the thermal fatigue cracks modeled that grow from the pre-existing flaws, which are discovered through hydro-static testing, more likely lead to smaller LOCAs. Hence, the hydro test results in a bigger decrease of the Category 1 LOCA frequencies. Also, of note from Figure 4.14 is the fact that the effect of the hydro test is slightly more pronounced at 60 years than it is at 25 or 40 years.

Seismic SSE loading magnitudes (Figure 4.15) have a minimal effect on the Category 1 and Category 6 LOCA frequencies. There is also no substantial LOCA frequency increase for 5SSE loading. This minimal effect is attributed to the small SSE stresses that are applicable for the hot leg configuration modeled. These stresses are plant specific and other configurations, resulting in higher stresses, could have a greater impact on the LOCA frequencies. The effect of seismic loading appears to decrease slightly with operating time (Figure 4.15). This trend is likely due to identification and repair of cracks in each subsequent inspection.

However, in all these analyses, the LOCA frequencies attributed to fatigue from preexisting defects are extremely small regardless of the application of seismic loading or the hydro test. These frequencies are much less than the earlier PWR-1 base case estimates made by DH (Table 4.1 and Figure 4.4). It is not known if these trends related to seismic stress and hydro test application have similar effects when PWSCC failure is considered and the LOCA frequencies are higher. The greater crack propensity of PSWCC due to the elevated crack initiation and growth rates may cause more significant LOCA frequency increases due to the same applied seismic loading. Additionally, the hydro test may be less effective in screening for PWSCC cracks than thermal fatigue cracks that initiate from pre-existing defects because the PWSCC cracks initiate after service commences.

An additional sensitivity study was performed for seismic events for the PWR-2 base case. There are differences in the treatment of seismic events between PWR-1 and PWR-2 sensitivity analyses, because of the level of detail of the seismic stresses that was available. For the PWR-1 case, specific stresses for the seismic events were available (as discussed in Appendix F), and the seismic stresses were applied once - at 25, 40 or 60 years as previously detailed. For the PWR-2 case, a list of cyclic stress amplitudes was available, which included seismic events. The stress history without seismic loading was determined using rainflow counting with seismic stresses set to a low value. Therefore a different set of stress amplitudes was considered to evaluate the seismic loading effects (see Tables F.8 and F.9 of Appendix F). For the PWR-2 base case, when seismic events are considered, the seismic stresses do not occur at a specific time. Due to the large number of high stress cycles for PWR-2, fatigue crack initiation was considered in this sensitivity analysis. This contrasts the PWR-1 sensitivity analysis which only evaluated the seismic and hydrotest loading impact on LOCA frequencies. The baseline LOCA frequencies are much higher in the PWR-2 analysis than in the PWR-1 and the SSE loading does increase the LOCA frequencies by a factor of 3. This is a larger effect than previously predicted in the PWR-1 sensitivity analysis, but the increase is still not substantial. More details are provided in Appendix F.

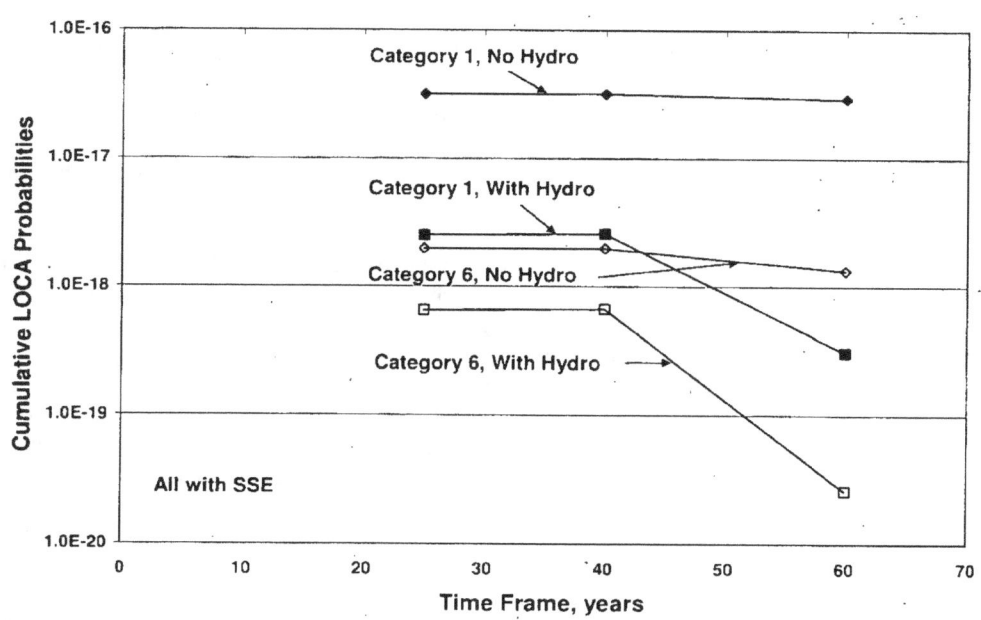

Figure 4.14 Effect of a Hydro Test on the Cumulative LOCA Frequency for the Hot Leg with a SSE Load Included

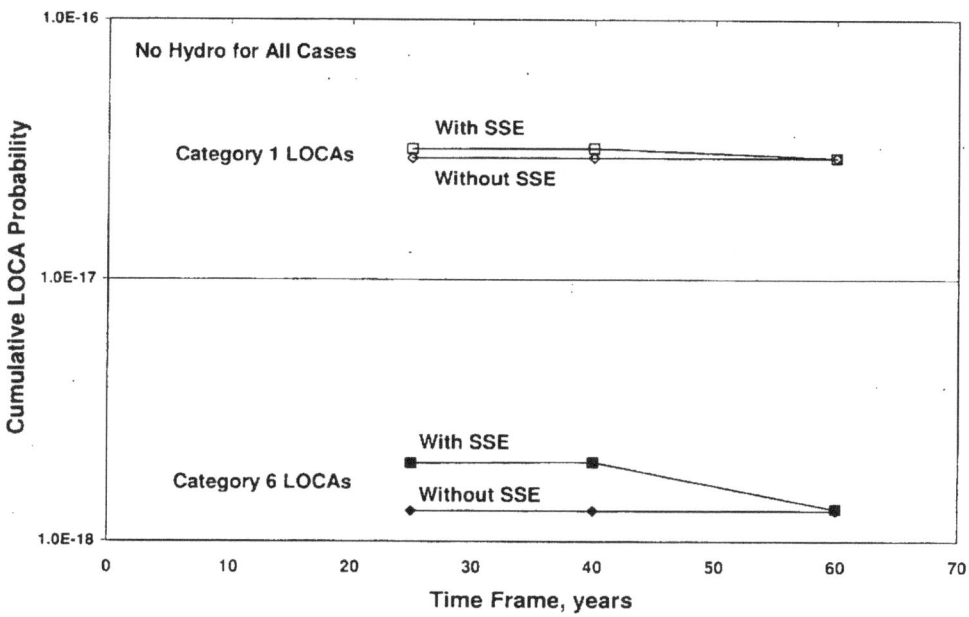

Figure 4.15 Effect of a SSE Load on the Cumulative LOCA Frequency for the Hot Leg Without a Hydro Test

4.3.7 Effect of Degraded Material Properties

Once again, the hot leg was analyzed by DH to examine the effect of degraded material properties on the estimated LOCA frequencies. The effect of thermal aging was simulated by increasing the material flow stress and decreasing the fracture toughness. Several different aging conditions were evaluated using the embrittled properties consistent with a highly aged, cast austenitic stainless steel alloy (CF8M). Once again, only thermal fatigue from pre-existing defects was modeled, using the PRAISE code with inputs similar to the base case analysis. See Appendix F for more details.

Results for Category 1 LOCA frequency estimates are summarized in Figure 4.16. As with the hydro test and SSE hot leg sensitivity studies, the absolute failure frequencies are extremely low because only thermal fatigue due to pre-existing defects was modeled. The only cyclic stresses considered in this fatigue analysis were 3 heat-up and cool-down cycles (with a 6.5 ksi (45 MPa) stress range) each year plus the seismic stresses which were generally quite low, i.e., 1.96 ksi (13.5 MPa) for an SSE event if the seismic event contained 200 stress cycles all of the same magnitude. These stresses result in very little crack growth over the operating history which causes the low absolute frequencies depicted in Figure 4.16. The implication is that failure is unlikely due to this specific mechanism under the assumptions and conditions modeled in the analysis.

The principal point of Figure 4.16 is to demonstrate the relative changes in the frequencies due to changes in material fracture toughness for this one set of analysis conditions. As can be seen in Figure 4.16, the LOCA frequency increases due to fracture toughness reductions are much more significant than the effects of a hydro test or SSE seismic loading (Figures 4.14 and 4.15). In this case, the limiting aging properties for CF8M causes a 5 order of magnitude increase in the LOCA frequencies compared with the unaged, baseline LOCA estimates. More modest toughness decreases still result in an order of magnitude increase in the estimated LOCA frequencies. The results of Figure 4.16 are not time dependent. Time dependency is a function of material property degradation with continued aging. However, the PRAISE code can only consider time-invariant material parameters. Continuing material property degradation with aging would be expected to result in increasing LOCA frequencies with time.

It is not known if similar order of magnitude increases would result if these aging properties were used in the complete PWR-1 base case analysis. This consideration is largely irrelevant because significant aging toughness decreases are not expected for the hot leg weld material. However, this analysis does illustrate the importance of the weld fracture toughness on the LOCA susceptibility when a cracking mechanism is present. These results are consistent with relative differences described in Section 1.2 for a previous study on the effects of material toughness [4.4].

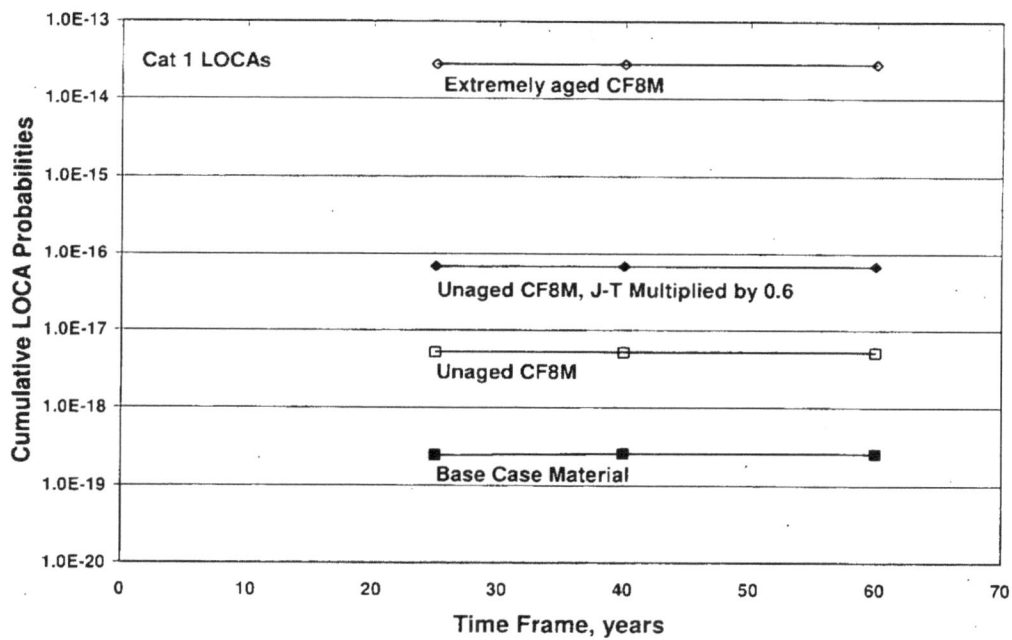

Figure 4.16 Effect of Material Properties (Fracture Toughness) on the Cumulative LOCA Frequency for the PWR-1 Base Case

4.4 Non-Piping Base Case Results

In addition to developing base case frequencies for piping systems, base case frequencies were also developed for a number of non-piping components. A non-piping precursor database based on a review and analysis of LERs was developed. This database is described in more detail in Appendix H. In addition, a series of separate studies were conducted to develop base case frequencies for a number of non-piping components and subcomponents. These studies included: a SGTR frequency study, an overview of the PTS re-evaluation effort, and a BWR vessel rupture and PWR CRDM ejection analyses. More information on the details associated with each of these base cases is provided in Section 3.5.2. The results from each of these studies are provided in this section. Also, additional details and results for the BWR vessel rupture and PWR CRDM ejection analyses can be found in Appendix I.

4.4.1 Steam Generator Tube Rupture Base Case

The LER non-piping database was used to conduct this study as explained more fully in Section 3.5.2.2. From 1990 to 2002 there were 15 reports of steam generator tube leaks. There is a total of 929 reactor calendar years represented in this period, so the mean leak frequency over this period is 16×10^{-3} per calendar year. The leak events plotted as a function of calendar year are summarized in Figure 4.17. Two year increments have been selected for binning purposes. While the number of tube leaks per year is relatively small, these events appear to be relatively constant over this time-frame. The base case SGTR frequency was also assumed to constant over this time period under the presumption that the leaking event frequency is directly proportional to the tube rupture frequency.

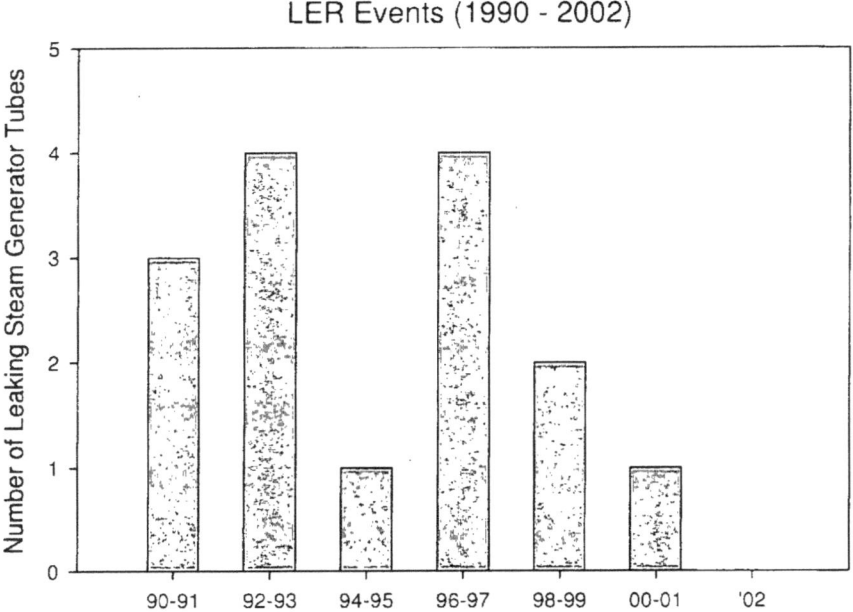

Figure 4.17 Steam Generator Tube Leaks as a Function of Event Date

In assessing the SGTR data, the database was queried over a 15 year time period between 1987 and 2002. This search was expanded over the prior 1990 to 2002 study in order to capture a couple of known SGTRs that occurred in 1987 and 1989. During this 15 year period, there were a total of 4 SGTRs with resultant leak rates greater than 100 gpm (380 lpm) (LOCA Category 1). These 4 ruptures occurred at North Anna in 1987, McGuire in 1989, Palo Verde in 1993, and Indian Point in 2000. This same 15 year time period represents 1,133 calendar years of reactor operation. Therefore, the frequency of steam generator tube Category 1 ruptures (with resultant leak rates greater than 100 gpm [380 lpm]) was 4/1,133 calendar years, or 3.5×10^{-3} per calendar year. NUREG/CR-5750 [4.1] conducted a similar assessment of SGTRs, and estimated a frequency of 7×10^{-3} per calendar year. These two frequencies are consistent because there has been one additional SGTR since the completion of NUREG/CR-5750 at Indian Point. However, the number of reactor operating calendar years is more than twice in the NUREG/CR-5750 estimate

This steam generator rupture Category 1 LOCA frequency of 3.5×10^{-3} per calendar year was used as the base case value for this elicitation exercise. This base case frequency estimate was provided to each panelist. Each panelist then, as part of their elicitation, assessed the adequacy of this Category 1 LOCA SGTR frequency estimate for both the current-day and future operating time periods. This frequency estimate was also provided for anchoring the relative LOCA contributions from other Category 1 LOCA sources, multiple steam generator tube failures resulting in Category 2 and higher LOCAs, and larger LOCA (Category 2 – 6) failures in other piping systems and non-piping components per the discretion of each panelist.

4.4.2 Pressurized Thermal Shock (PTS) Base Case
The potential LOCA contribution due to PTS of the RPV was analyzed using available data from the NRC's PTS Re-Evaluation Project [4.5] as discussed in Section 3.5.2.3. The total PTS through-wall

cracking frequency results for 3 plants (Oconee, Beaver Valley, and Palisades) at 32 effective full-power years (EFPY) was first obtained from current results. The EFPY estimate most closely corresponds to the current fleet average of 25 years used to represent current-day frequencies within the elicitation. The total PTS through-wall cracking frequencies for each plant were then adjusted to consider only the contributions due to stuck open primary side valves, stuck open secondary side valves, and feed and bleed operations. This adjustment removes the frequency contributions due to passive system LOCA failures so that the resulting PTS risk contributions are independent of the LOCA frequency estimates. The total PTS risk is proportional to the SB and MB LOCA frequencies. More information on the total PTS risk and changes resulting from differences in the assumed initiating event SB and MB LOCA frequencies is contained in [4.5].

The adjusted results for each plant are summarized in Table 4.3. The geometric mean (GM) of the 32 EFPY results for these three plants is also determined along with the 5th and 95th percentile of the mean estimates (Table 4.3). The 5th and 95th percentile are estimated by determining the values that lay two standard deviations from the GM estimates using the results from the three plant-specific calculations. These percentiles provide some indication of the plant specific variability of the PTS through-wall cracking estimates. While these base case estimates are developed from the 32 EFPY results, the PTS through-wall cracking frequencies at 60 EFPY increase by less than a factor of two for each plant. Therefore, this calculation is not particularly sensitive to the future operating time period considered in this elicitation. The results from this PTS assessment were provided to the panelists for benchmarking other large piping and non-piping failures with expected low LOCA frequencies.

Table 4.3 Average and Average Plus and Minus Two Standard Deviations of the Adjusted Through-Wall Crack (TWC) Frequency from the PTS Re-Evaluation Project

Plant	Adjusted Category 6 LOCA frequencies from PTS study (per calendar year)	Geometric Mean (per calendar year)	95th Percentile (per calendar year)	5th Percentile (per calendar year)
Oconee	3.6E-11			
Beaver Valley	7.5E-10	5.1E-10	7.0E-08	3.7E-12
Palisades	4.8E-09			

4.4.3 BWR Vessel Rupture and PWR CRDM Base Cases

BWR vessel beltline weld failure was modeled to calculate failure frequencies due to normal operational loading and LTOP events. In addition, the large, predominantly feedwater, nozzles (6 to 28 inch) were analyzed to predict failure frequencies due to thermal fatigue. Both contributions are combined to provide an estimate of the total BWR RPV LOCA frequency as a function of LOCA size for use as a base case. The individual nozzle and beltline results were also available for anchoring. The PWR CRDM leak and ejection frequencies were used to determine the associated LOCA frequencies for use as a base case. A more complete description of this analysis is provided in Section 3.5.2.3 and in Appendix I.

Table 4.4 summarizes the base case LOCA frequencies for the BWR beltline weld and feedwater nozzle subcomponents as well as the combined LOCA frequencies for the BWR RPV. Table 4.5 presents the PWR base case LOCA frequencies due to CRDM ejection. LOCA frequencies are provided at 25, 40, and 60 years of plant operation in both tables. The PWR CRDM and combined BWR vessel LOCA frequencies are graphically summarized in Figures 4.18 and 4.19, respectively. The predicted 25-year PWR CRDM Category 1 LOCA frequency (Figure 4.18) is relatively high and is consistent with cracking experience within the plants. The CRDM LOCA frequencies decrease by less than an order of magnitude between the LOCA Categories 1 and 2. Category 1 LOCAs represent a partial CRDM break while the Category 2 LOCAs result from a complete ejection. The relatively small decrease between LOCA Categories 1 and 2 is a reflection of the rapid crack growth rates that are possible due to PWSCC. The Category 3 CRDM LOCAs are assumed in this analysis to result from simultaneous ejections of independent nozzles. Common cause CRDM ruptures are not considered in this base case analysis. However, the panel was asked to consider the likelihood of common cause failures in the elicitation. These base case Category 3 LOCA estimates were available for possible anchoring during this common cause consideration.

Periodic inspection over the next fifteen years is expected to decrease the LOCA frequencies by less than an order of magnitude (Figure 4.18). No further reduction in LOCA frequencies is predicted in subsequent inspections between 40 and 60 years of life. The implication is that all the cracking which previously initiated prior to the onset of periodic inspections will have been discovered by 40 years of service. After 40 years, steady state is reached and additional inspections are only likely to find new cracking indications.

Not surprisingly, the predicted BWR vessel LOCA frequencies (Table 4.4 and Figure 4.19) are much lower than the CRDM LOCA frequencies for LOCA Categories 1 and 2. This reflects the diminished concern of BWR vessel crack initiation and growth due to active degradation mechanisms, as well as the robustness of the vessel design and operation. However, the Category 3 base case BWR and PWR non-piping LOCA frequencies are actually similar. This difference is simply a function of the relative component sizes. A single PWR CRDM ejection is limited to a Category 2 LOCA while the combined BWR feedwater and vessel ruptures can contribute to all LOCA categories.

Feedwater nozzle failures provide the dominate contribution for the BWR LOCA Categories 1 – 4, but the nozzle size precludes contribution to larger category LOCAs (Table 4.4). Hence, the Category 5 and 6 LOCA frequencies are comprised solely by the beltline failure contribution. The beltline failure frequencies continue to increase with time due to the additional irradiation embrittlement while the feedwater LOCA frequencies are relatively insensitive to operating time. This time insensitivity stems from both the positive affect of mitigation procedures employed to address this degradation and the cutoff frequencies associated with the PFM analysis. More information is provided in Appendix I. These subcomponent trends explain why the total BWR vessel LOCA frequencies exhibit little time sensitivity for LOCA Categories 1 – 4 while the Category 5 and 6 LOCA frequencies continue to increase with operating time because they are governed by the beltline failures. It is interesting that after 60 years of operation, the Category 5 and 6 LOCA frequencies due to beltline failures are nearly identical to the Category 4 feedwater nozzle frequency.

It is also interesting to note (Table 4.4) that although the beltline LOCA frequencies are smaller, they decrease more gradually than the feedwater frequencies as a function of LOCA size. This trend reflects the expectation that the partial failure of the large, BWR vessel is not much more likely than a complete failure. This trend becomes even stronger after 60 years of service due to radiation embrittlement of the vessel material. The difference between a Category 1 and Category 6 LOCA after 60 years is approximately a factor of 5 (Table 4.4). Therefore, large and small LOCAs resulting for BWR vessel failures are almost equally likely after this service time. A similar trend is predicted in DH's PFM analysis for the BWR-1 piping base case frequencies for LOCA Categories 1 - 4 (Figure 4.5). LOCA decreases could be more gradual as both the component size increases and material brittleness increases. Similar trends are also discussed in the elicitation results (Section 6).

Table 4.4 BWR RPV LOCA Frequencies Broken Down by Major Subcomponent (Beltline and Feedwater Nozzles)

LOCA Category	LOCA Frequency (per calendar year)		
	25 Years	40 Years	60 Years
	RPV Beltline		
1	8.0E-9	2.4E-8	3.7E-8
2	1.9E-9	5.0E-9	2.3E-8
3	1.0E-9	2.5E-9	1.8E-8
4	4.0E-10	1.0E-9	1.4E-8
5	1.9E-10	4.5E-10	1.1E-8
6	7.9E-11	1.9E-10	8.0E-9
	Feedwater Nozzles		
1	8.0E-7	1.2E-6	1.0E-6
2	1.6E-7	2.4E-7	2.0E-7
3	3.2E-8	4.7E-8	4.00E-8
4	6.4E-9	9.4E-9	8.0E-9
	BWR Vessel – Combined		
1	8.1E-07	1.2E-06	1. 0E-06
2	1.6E-07	2.4E-07	2.2E-07
3	3.3E-08	5.0E-08	5.8E-08
4	6.8E-09	1.0E-08	2.2E-08
5	1.9E-10	4.5E-10	1.1E-08
6	7.9E-11	1.9E-10	8.0E-9

Table 4.5 PWR RPV (CRDM) LOCA Frequencies

LOCA Category	LOCA Frequencies (per calendar year)		
	25 Years	40 Years	60 Years
	RPV (CRDMs only)		
1	1.0E-03	2.2E-04	2.2E-04
2	2.0E-04	4.0E-05	4.0E-05
3	3.2E-08	1.6E-09	1.6E-09

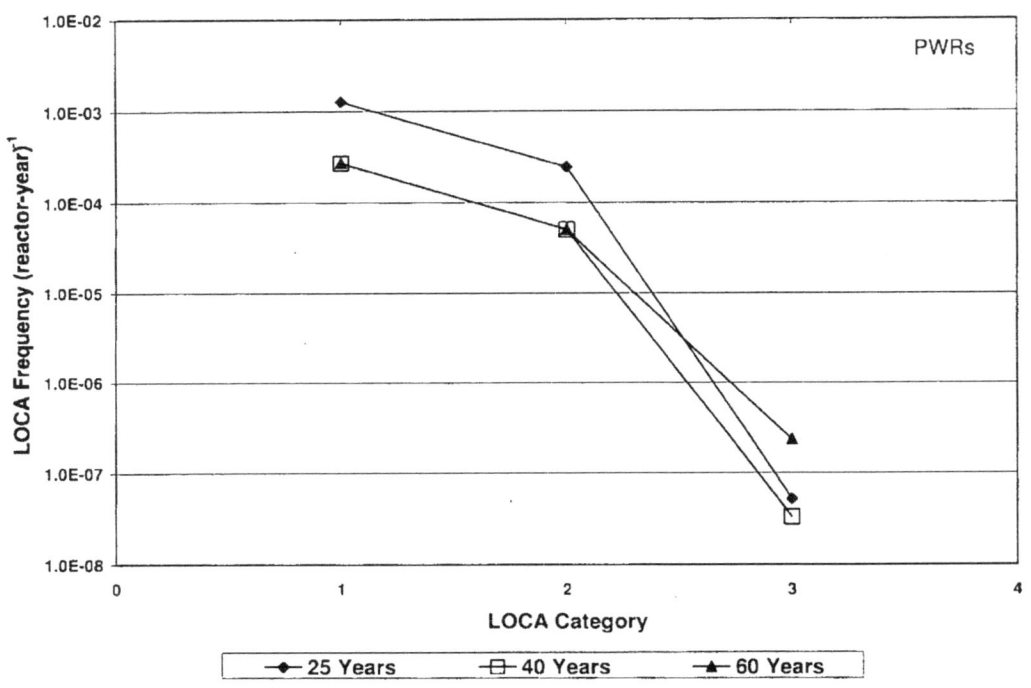

Figure 4.18 PWR CRDM LOCA Frequencies at 25, 40, and 60 Years

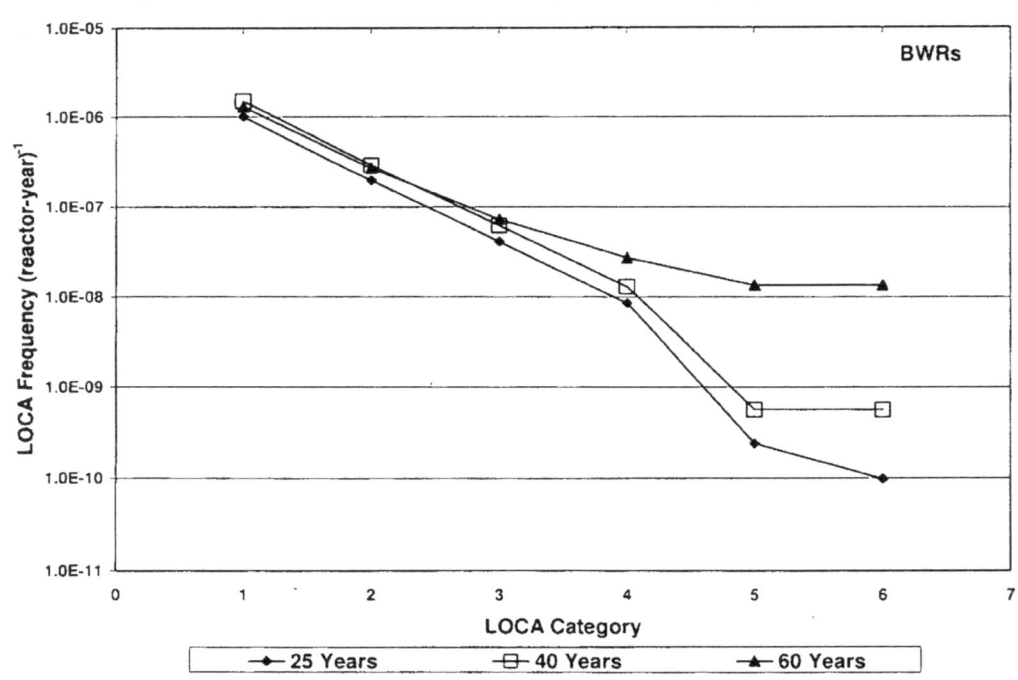

Figure 4.19 BWR Combined RPV LOCA Frequencies at 25, 40, and 60 Years

4-30

4.5 References

4.1 Poloski, J. P., et. al., "Rates of Initiating Event at U.S. Nuclear Power Plants: 1987-1995," NUREG/CR-5750, February 1999.

4.2 Beliczey, S., and Schulz, H., "Comments on Probabilities of Leaks and Breaks of Safety-Related Piping in PWR Plants," *International Journal of Pressure Vessel and Piping*, Vol. 43, pp. 219 – 227, 1990.

4.3 T. C. Chapman, et. al., "Assessment of Remedies for Degraded Piping," Electric Power Research Institute Report NP-5881-LD, 1988.

4.4 Simonen, F.A., Doctor, S.R., Schuster, G.J., and Heasler, P.G., "A Generalized Procedure for Generating Flaw-Related Inputs for FAVOR Code," NUREG/CR-6817, March 2004.

4.5 Memorandum from A.C. Thadani to S.J. Collins, Technical Basis for Revision of the Pressurized Thermal Shock (PTS) Screening Criteria in the PTS Rule (10 CFR 50.61), December 31, 2002.

5. ANALYSIS OF ELICITATION RESPONSES

The responses for each panel member were analyzed separately and then aggregated to obtain group estimates. Individual estimates of selected parameters of LOCA frequencies for BWR piping, PWR piping, BWR non-piping, and PWR non-piping failures were calculated for each panelist that provided the relevant input responses[1]. A unified response and analysis framework was used to ensure consistency and commonalty in processing the panelists' inputs. The panelists' responses were assumed to represent the median, 95th, and 5th percentiles of their subjective uncertainty distributions for each elicitation question and the analysis framework was based on the assumption that the panelist's responses could be represented by split lognormal distributions.

The final output for each panelist was a set of individual estimates for the mean, median, 5th and 95th percentiles of the total passive LOCA frequencies for BWR and/or PWR plants. The LOCA frequency parameter estimates for the individual panelists were then aggregated to obtain group LOCA frequency parameter estimates. Statistical confidence bounds were calculated to provide a measure of panel diversity for the individual LOCA frequency parameters. A number of sensitivity analyses were also conducted to examine the robustness of this analysis procedure and to identify the assumptions and techniques that most significantly affect the aggregated group estimates for each of the six LOCA categories. Details on the analysis of the panelists' elicitation responses follow.

5.1 Analysis Framework

A common analysis framework was adopted to analyze the elicitation responses as depicted in Blocks 13 and 16 in Figure 3.1. The analysis framework is "common" because the same calculation methodology is used to process the elicitation responses for each panelist. This allows the results to be compared across panelists and permits meaningful measures of panel diversity to be developed. The analysis framework is consistent with the structure of the elicitation questions (Section 3.8) which were developed to assess fundamental piping and non-piping LOCA contributing factors. The framework was also chosen to be consistent with certain principles that are subsequently discussed.

As described in Section 3.8, panel members were asked to supply three numbers for each elicitation question: a MV, an UB, and a LB. As described in Section 3.3.2, these numbers were assumed to correspond to the median, 95th percentile, and 5th percentile, respectively, of the panelist's subjective uncertainty distribution. The goal was to estimate LOCA frequencies for four plant-type combinations (BWR piping, BWR non-piping, PWR piping and PWR non-piping) and combine these to estimate total BWR and PWR passive LOCA frequencies. These bottom-line estimates are all in the form of selected parameters of the LOCA frequency distribution. The parameters used are the mean, median, 95th percentile, and 5th percentile. These parameters are considered sufficient for regulatory applications.

There are two basic approaches to aggregating the individual panelist responses to obtain group estimates of the bottom-line parameters. The first approach is to propagate the responses separately for each panelist to obtain individual total LOCA frequency estimates and then combine them to obtain group estimates of the bottom-line parameters. The second approach is to combine the responses for each question to obtain a group estimate for each elicitation question and then propagate these group estimates to determine the total LOCA frequency parameters. Because the propagation and aggregation algorithms are highly nonlinear, these two approaches may lead to significantly different results. The second approach, in effect, replaces all the panel members' responses with those of a quasi-panelist (in this case, a 13th panel member) who is informed by the collective panel wisdom. However, the strategies that individual panelists used to respond to each question varied a great deal and not all panelists answered each elicitation question. Therefore, this second approach would result in significant inconsistencies in the quasi-panelist's responses which would make it

---

[1] Not all panelists provided input responses for both BWR and PWR plants.

impractical to either interpret or apply the bottom-line estimates. In contrast, because careful attention was paid in the elicitation sessions to eliminating inconsistencies within an individual panelist's responses, the first approach should yield self-consistent individual results. Consequently, there are fewer difficulties in interpreting and applying the bottom-line estimates. Therefore, the first approach was used to aggregate the panelists' responses.

A guiding principle in the analysis framework is to make as few assumptions as are necessary to estimate the four bottom-line LOCA frequency parameters. Although the approach adopted can be extended to estimate the entire LOCA frequency distribution for each LOCA category, this was not done. Such an extension would involve fitting an entire distribution to estimates of the four bottom-line parameters and would require assumptions about the form of this distribution. Instead, the approach adopted in this study was to focus on the selected bottom-line parameters and estimate each of them separately. To go beyond this approach would necessitate assuming more structure to the panelists' responses, and more assumptions about the LOCA frequency distributions than is necessary or warranted.

It is important that the bottom-line LOCA frequency parameter estimates reflect both individual uncertainty and panel diversity. Individual uncertainty stems from the uncertainties in each panel member's responses, as embodied in the UBs and LBs for each elicited quantity. These uncertainties are propagated to obtain estimates of the means, medians, 5th and 95th percentiles of the LOCA frequency distributions. Panel diversity stems from differences among the individual panel member estimates. Because of the lack of data and the variety of approaches used by individual panel members, it is not surprising that there were large differences in their responses and in the resultant bottom-line estimates. In this sense, panel diversity is simply a reflection of the existing scientific uncertainty about the LOCA frequencies being estimated. Confidence intervals for the estimated bottom-line parameters were developed to reflect panel diversity.

The analysis framework is summarized in Figure 5.1 using a flowchart. The piping contribution is outlined in the left-hand branch of Figure 5.1. The individual responses (Section 5.2) for the anchoring frequencies (Block 1.2) and adjustment ratios (Block 1.3) are assumed to be independent split lognormal distributions (Section 5.3.1) and are multiplied as described in Section 5.3.3 to determine the piping system frequencies (Block 1.4). These system frequencies are then summed for all contributing piping systems to obtain the un-normalized piping contribution (Block 1.5) as described in Section 5.3.4. The un-normalized piping contribution is then multiplied (Section 5.3.3) by the piping contribution factor (Block 1.6) to determine the piping contribution (Block 1.7).

The non-piping calculations (right-hand branch of Figure 5.1) are analogous except for an additional step required to determine the non-piping component from the subcomponent frequencies. The individual responses for the anchoring frequencies (Block 1.9) and adjustment ratios (Block 1.10) are assumed to be independent split lognormal distributions and multiplied (Section 5.3.3) to determine the non-piping subcomponent frequencies (Block 1.11). These subcomponent frequencies are then summed (Section 5.3.4) for all contributing non-piping subcomponents to obtain the un-normalized non-piping component frequencies (Block 1.12). These frequencies are then multiplied (Section 5.3.3) by the component contribution factor (Block 1.13) to determine the non-piping component frequencies (Block 1.14). The non-piping component frequencies are then summed (Section 5.3.4) to calculate the un-normalized non-piping contribution (Block 1.15) which is then multiplied (Section 5.3.3) by the non-piping contribution factor (Block 1.16). The end result is the non-piping contribution (Block 1.17). The piping and non-piping contributions are then summed (Section 5.3.4) to develop individual estimates of the bottom-line parameters (mean, median, 95th and 5th percentiles) of the total LOCA frequencies (Section 5.3.5; Block 1.18). Finally, the individual estimates of the bottom-line parameters are combined (Section 5.4) to develop the group estimates and associated confidence intervals (Block 1.19).

Details of the analysis framework outlined in Figure 5.1 and the calculation methodology to implement this framework are discussed in the sections below.

Piping   Non-Piping

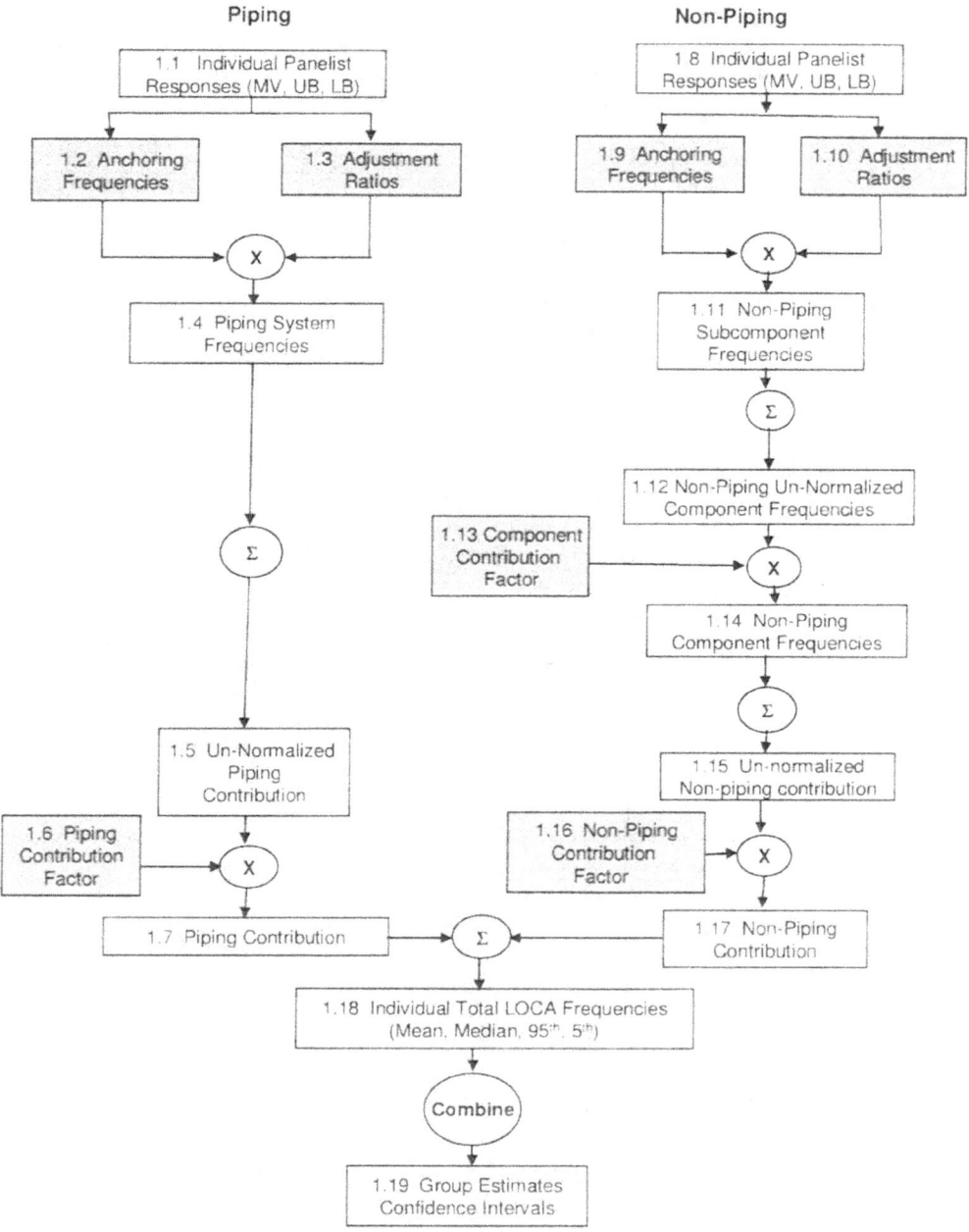

Figure 5.1 Flowchart of the Analysis Framework

5.2 Elicitation Responses

A common response format was provided to each panelist prior to their individual elicitation sessions for recording their responses. However, panelists were allowed to provide responses in a format consistent with their individual approaches. While several panelists utilized the common response format, many panelists did not. In general, the elicitation responses and associated formats varied significantly among the panelists. The input responses required from the panelists are indicated by the shaded blocks in Figure 5.1.

Anchoring frequencies were initially selected by each panelist (Blocks 1.2 and 1.9 in Figure 5.1) for anchoring subsequent responses. Often, these anchoring frequencies and conditions were adopted from the piping and non-piping base cases (Sections 3.5 and 4). The base case anchoring frequencies utilized in the common analysis framework pertain to each LOCA size category and are associated with 25 years of plant operation for the variable conditions defined for each base case (Section 3.5). However, several panelists chose not to utilize the base cases for anchoring. Some of these panelists developed absolute LOCA frequencies for each piping system/non-piping subcomponent, and some developed alternative anchoring conditions and frequencies than those described in Section 3.5.

Next, after choosing appropriate anchoring frequencies, each panelist chose adjustment ratios to relate the anchoring frequencies to the associated LOCA frequencies for that piping system/non-piping subcomponent after 25, 40, and 60 years of plant operation. Adjustment ratios were chosen for each time period and each LOCA size category to account for differences between the base case conditions and the assessed contributors to the LOCA frequencies (Blocks 1.3 and 1.10). Panelists that developed absolute LOCA frequencies did not need to supply adjustment ratios.

All anchoring frequencies and adjustment ratios had associated MVs, or medians, as well as bounds to represent 90% coverage intervals. The MVs and coverage intervals for the adjustment ratios were obtained directly from the elicitation responses. The anchoring frequency coverage intervals were based on the source of the anchoring frequency estimates. A factor of three was generally used to quantify the ratio between the coverage bounds (UB and LB) and the MVs for the piping and non-piping base case frequency estimates (Sections 3.5 and 4) when the base case results were chosen as anchoring frequencies. The factor of three was based on a consideration of operating experience uncertainties, but panelists could adjust this factor if desired in their elicitation responses (Section 3.10).

It was generally not necessary to develop anchoring frequency coverage intervals for panelists who independently developed system-based frequencies or who chose alternative anchoring frequencies. The uncertainties for the system-based frequencies were provided directly by the panelists while the uncertainty associated with any alternative anchoring frequencies was simply reflected in the adjustment ratio coverage intervals. In practice, the distinction between uncertainties in the anchoring frequencies and adjustment ratios is somewhat artificial since these uncertainties are combined as discussed in Section 5.3.

The only other inputs supplied by the panelists were the contribution factors (Blocks 1.6, 1.13, and 1.16). As discussed in Section 3.8, the panelists were not required to provide information on all piping system or non-piping subcomponents, but only on those whose total contributions were thought to represent at least 80 percent of the piping and non-piping LOCA frequencies (Blocks 1.5 and 1.15). The contribution factors were used to normalize each panelist's input to determine total LOCA frequency estimates. The piping contribution factor (Block 1.6) is the reciprocal of a panelist's MV estimate of the total percentage contribution of the quantified piping systems to the LOCA frequency for all piping systems. Each panelist also provided a coverage interval for the piping contribution factor. For example, if a panelist assessed all of the piping systems, the MV contribution factor as well as the UBs and LBs would be 1 (or 100 percent) and the contribution factor coverage interval is a single point. As another example, suppose a panelist estimated that the assessed

5-4

piping systems represent 85 percent of the total piping LOCA frequency with a LB and UB of 75 and 95 percent. The piping contribution factor is then 1.18 (1/0.85) with a LB of 1.05 (1/0.95) and a UB of 1.33 (1/0.75). The corresponding contribution factor coverage interval is (1.05, 1.33).

Similarly, panelists provided one of two non-piping contribution factors (Blocks 1.13 and 1.16). The analysis method chosen by each panelist determined which non-piping contribution factor was required. The two possible analysis methods are mutually exclusive. Therefore, the contribution factor associated with the approach not selected is a single point value of 1 (or 100%) by definition.

If a panelist assessed each non-piping component separately (i.e., pumps, valves, pressure vessel, etc.), then a component contribution factor (Block 1.13) was required for each non-piping component assessed. The component contribution factor is the total percentage contribution (> 80%) of all quantified non-piping sub-component failure frequencies to the LOCA frequency associated with that component. For example, if the panelist assessed RPV LOCA contributions and believes that vessel rupture contributes 90 percent of the frequency associated with RPV LOCAs, then the component contribution factor for the RPV is 1.11 (1/0.9). Similarly, contribution factors were then required for the remaining non-piping components (i.e., steam generator, pressurizer, pumps, valves, etc.) based on the subcomponents assessed for each component. The panelist also provided a coverage interval for each non-piping component contribution factor.

For panelists who considered specific failure scenarios without regard to the applicable component's contribution to the non-piping LOCA frequency, a non-piping contribution factor (Block 1.16) was required. Here the contribution factor is the total percentage contribution of the specific failure scenarios that were quantified to the total non-piping LOCA frequency. For example, for Category 2 PWR LOCAs, if the panelist believes that SGTRs, CRDM ejections and pressurizer heater sleeve failures together account for 95 percent of the total non-piping LOCA frequency, then the non-piping contribution factor is 1.05 (1/0.95). The panelist also provided a coverage interval for each non-piping contribution factor.

As discussed in Section 3.10, it was occasionally necessary for the facilitation team to impute missing elicitation response inputs so that complete LOCA frequency results could be developed for each panelist. Almost always, the missing inputs pertained to coverage intervals corresponding to MV responses that were provided. Missing inputs were imputed in a manner consistent with the rest of the panelist's responses. Panelists were required to either verify imputed inputs or directly provide this information to ensure that they agreed with any imputed input values.

A spreadsheet was developed for each panelist containing the input responses for each piping system and non-piping subcomponent LOCA frequency contribution. The spreadsheet structure uses the common response format described earlier and allows processing using the analysis framework (Section 5.1) with the calculation formulas (Section 5.3) to determine individual total LOCA frequency estimates. The spreadsheets are all identical to ensure consistent processing and foster comparisons of the panelists' responses.

There is one situation where this approach did not necessarily reflect the panelist's opinion about the contribution of individual piping systems to the total piping LOCA frequency contributions. One panelist, provided global LOCA piping estimates and uncertainties for PWR plants based on plant-specific studies associated with the development of risk-informed ISI methodology procedures (Appendix K, page K-2). However, to utilize the response and analysis framework discussed in Sections 5.1 and 5.2, this panelist's piping responses were apportioned among the LOCA-sensitive systems based on the percentage of welds in each system. This panelist did not confirm the validity or accuracy of this apportionment. While the final LOCA frequency estimates and uncertainties matched his elicitation input, this analysis should not be construed to reflect his beliefs about the contribution of individual piping system LOCA frequencies. This panelist did provide non-piping responses for individual failure scenarios that were consistent with the existing framework. Hence, a similar apportionment of his non-piping responses was not required. No other analysis of a panelist's responses deviated from the analysis framework described in Sections 5.1 and 5.2.

5.3 Individual Estimates

Each panelist's responses (i.e., MVs, UBs, and LBs) were propagated to estimate the bottom-line parameters (i.e., the means, medians, 95th and 5th percentiles) of the LOCA frequency distributions for each of the four plant-type combinations (BWR piping, BWR non-piping, PWR piping, PWR non-piping) that the panelist quantified. The BWR piping and non-piping estimates and the PWR piping and non-piping estimates are then summed to obtain individual estimates of the total BWR passive LOCA frequencies and the total PWR passive LOCA frequencies, respectively.

Figure 5.2 is a flow chart outlining the assumptions and calculations required to implement the analysis framework presented in Figure 5.1. The assumptions and equations associated with each specific aspect of the analysis are provided in the subsections below. Flowchart block numbers from Figures 5.1 and 5.2 are referenced within the text as appropriate using the same numbers illustrated in these figures. For example, a reference to Block 2.1 refers to block 2.1 in Figure 5.2.

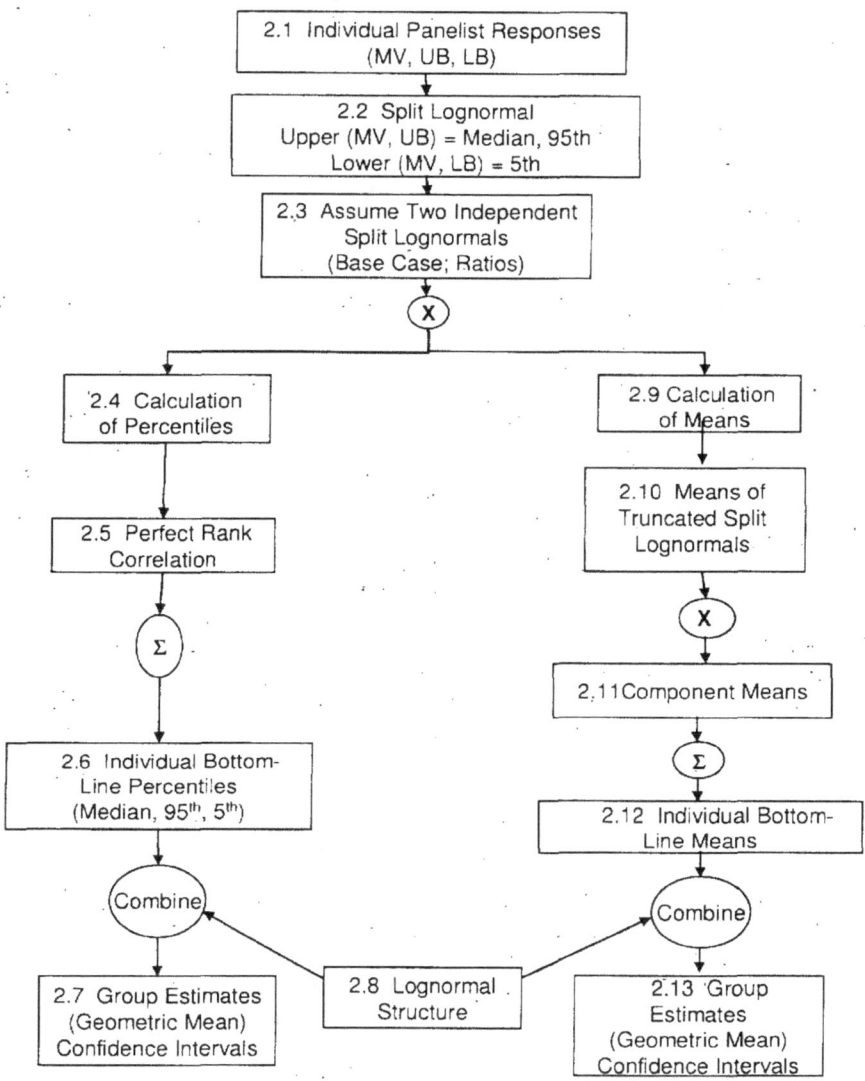

Figure 5.2 Flow Chart of Required Calculations

5.3.1 Lognormal Structure

In order to propagate the elicitation responses, it is necessary to specify a probabilistic structure for them. The assumed probabilistic structure is based on the lognormal distribution, which was chosen to reflect the structure of the elicitation questions. Because all of the questions called for a ratio as a response, the lognormal distribution is a natural choice. The loguniform or the logtriangular distributions are possible alternatives. Although no sensitivity analyses were performed to evaluate the lognormal assumption, variations in the individual LOCA frequency estimates resulting from replacing the lognormal with other plausible distributional forms are expected to be significantly smaller than the differences between the individual estimates. Accordingly, because of the very large diversity in the individual estimates (see Section 7.5), it is not expected that the particular choice of the distributional forms would result in group estimates that are significantly different than those based on the lognormal assumption.

Initially, for each question, the piping (Blocks 1.2, 1.3, 1.6) and non-piping (Blocks 1.9, 1.10, 1.13, 1.16) responses are assumed to be the corresponding percentiles of a lognormal distribution. Thus, the MV is assumed to be the median, the UB is assumed to be the 95th percentile, and the LB is assumed to be the 5th percentile of a lognormal distribution. A complication arises if the UB and LB are not multiplicatively symmetric about the MV, as often happens. In such cases, there is no lognormal distribution determined by the given MV, UB and LB which satisfies the assumed percentile assignments. Therefore, a split lognormal distribution (Block 2.2) is assumed as the probabilistic structure to address this situation. This probabilistic structure was endorsed by a peer reviewer as reasonable and appropriate [5.1].

A split lognormal distribution is a combination of two halves of separate lognormal distributions. Because a lognormal distribution can be determined by its median and one other percentile, two lognormal distributions can be determined from the response to each elicitation question -- one by the MV and LB and another by the MV and UB. For values less than the median, the split lognormal is the lower half (i.e., the part less than its median of the lognormal distribution determined by the MV and LB). For values greater than the median, the split lognormal is the upper half (i.e., the part greater than its median) of the lognormal distribution determined by the MV and UB. Clearly, if the UB and LB are symmetric about the MV, then the split lognormal is identical to the lognormal distribution determined by the MV, UB, and LB. Accordingly, the split lognormal distribution is assumed for both the symmetric and non-symmetric cases. With respect to the three bottom-line percentiles in Block 2.6, note that the median of a split lognormal is equal to the common median of its upper and lower halves, the 95th percentile is equal to the 95th percentile of its upper half and the 5th percentile is equal to the 5th percentile of its lower half.

The LOCA frequency for each of the four plant-type combinations is a sum of contributions from piping systems and non-piping subcomponents (Block 1.18). Each of these contributions is a product of an anchoring frequency from a base case (Block 1.2 or 1.9) and an adjustment ratio (Block 1.3 or 1.10). As noted above, the anchoring frequencies and adjustment ratios are each modeled by a split lognormal distribution. Thus, the inputs to Blocks 1.4 and 1.11 are products of two split lognormal distributions. Each panelist's bottom-line LOCA frequency estimate for the four plant-type combinations is a normalized sum of these products over all piping systems (Block 1.7) and non-piping components (Block 1.17). Thus the bottom-line LOCA frequencies are sums of products of the split lognormal distributions which represent the panelist's responses.

From Figure 5.2, the calculation of the individual bottom-line parameters is carried out in two parallel paths – one for the percentiles and the other for the means. Section 5.3.2 provides the basic lognormal formulas, Section 5.3.3 provides formulas for the product of two independent split lognormal distributions and Section 5.3.4 shows these results are propagated to obtain the bottom-line LOCA frequency parameters.

5.3.2 Basic Lognormal Formulas
By definition, Y has a lognormal distribution if the distribution of $\ln(Y)$ is normal. A lognormal distribution is denoted by $LN(\mu, \sigma)$, where $\ln(Y)$ is normal with a mean μ and variance σ^2 (or standard deviation σ). Note that $\ln(Y)$ is sometimes referred to as the underlying normal of the lognormal Y. Denote the median of Y by $m(Y)$, its expected value by $E(Y)$, its variance by $V(Y)$, its standard deviation by $SD(Y)$ and its pth percentile by $b_p(Y)$. These lognormal parameters can be written as follows [5.2].

$$m(Y) = \exp(\mu) \tag{5.1}$$

$$E(Y) = \exp(\mu + \sigma^2/2) \tag{5.2}$$

$$V(Y) = \exp(2\mu + \sigma^2) \cdot [\exp(\sigma^2) - 1] \qquad (5.3)$$

$$SD(Y) = \exp(\mu + \sigma^2/2) \cdot [\exp(\sigma^2) - 1]^{1/2} \qquad (5.4)$$

$$b_p(Y) = \exp(\mu + k_p\sigma) \qquad (5.5)$$

where k_p is the pth percentile of the standard normal distribution ($\mu = 0$ and $\sigma^2 = 1$).

Some other useful relations are the following.

$$E(Y) = m(Y) \exp(\sigma^2/2) \qquad (5.6)$$

$$V(Y) = [E(Y)]^4 / [m(Y)]^2 - [E(Y)]^2 \qquad (5.7)$$

$$m(Y) = E(Y) [1 + V(Y)/[E(Y)]^2]^{-1/2} \qquad (5.8)$$

Solving Equation 5.6 for σ yields

$$\sigma = [2 \ln\{E(Y)/m(Y)\}]^{1/2} \qquad (5.9)$$

A commonly used parameter is the error factor, defined as the ratio of the 95th percentile to the median. Because the 5th and 95th percentiles of a lognormal are multiplicatively symmetric about the median, the error factor is also equal to the ratio of the median to the 5th percentile. Setting $k_{95} = 1.645$, it follows from Equations 5.1 and 5.5 that the error factor of Y is

$$EF(Y) = b_{95}(Y)/m(Y) = m(Y)/b_5(Y) = \exp(1.645\sigma) \qquad (5.10)$$

5.3.3 Piping System and Non-piping Subcomponent Frequencies

From Figure 5.1, each piping system frequency (Block 1.4) or non-piping subcomponent frequency (Block1.11) is a product of an anchoring frequency from a base case, or a suitable alternative as described in Section 5.2 (Block 1.2 or 1.9), and an adjustment ratio (Block 1.3 or 1.10). As indicated in Block 2.2 and described in Section 5.3.1, the anchoring frequencies and adjustment ratios are each modeled by a split lognormal distribution. Thus, the piping and non-piping frequencies are products of split lognormal distributions.

In order to evaluate the product of any two distributions, it is necessary to specify their correlation structure. The statistical characteristics of the base case anchoring frequency and adjustment ratio distributions depend on the statistical characteristics of the uncertainty bounds provided by the panelists. Because the anchoring frequencies and adjustment ratios are generally provided by different panel members and because they have different structures (one is an absolute number and the other is a ratio), their associated uncertainty bounds are assumed to be statistically independent. Accordingly, the inputs to Blocks 1.4 and 1.11 are each assumed to be the product of two independent split lognormal distributions (Block 2.3). The calculations of the percentiles (Block 2.4) and the means (Block 2.9) of the product of two independent split lognormal distributions are discussed below.

5.3.3.1 Calculation of Percentiles - The required percentiles are calculated by the following algorithm. First, the median of the product of two independent split lognormal distributions is calculated by multiplying the medians of the two split lognormals. Next, the 95th percentile of the product is set equal to the 95th percentile of the product of the two lognormals which are the upper halves of the anchoring frequency and adjustment ratio split lognormal distributions. Finally, the 5th percentile of the product is set equal to the 5th percentile of the product of the two lognormals which are the lower halves of the split lognormals.

Although this algorithm for determining the required percentiles of the product of two independent split lognormals is not exactly equivalent to multiplying the split lognormals as two-part distributions, the results are a good approximation to those obtained through exact multiplication. The algorithm is used for computational convenience, and also because the median and 95th percentiles of each product are largely determined by the contributions from both upper halves of the split lognormal responses. The median and 95th percentiles are the most important percentiles in this study.

To implement the algorithm, a formula for the percentiles of the product of two independent lognormal distributions is required. As shown below, the product of two independent lognormals is a lognormal. Because the percentiles of a lognormal are determined by the mean and standard deviation of its underlying normal distribution (Equation 5.5), formulas are required to calculate the mean and standard deviation of the underlying normal of the lognormal which is the product of two independent lognormals.

Denote the two independent lognormals by Z_1 and Z_2, where $Z_i = LN(\mu_i, \sigma_i)$ for i = 1, 2. The lognormals Z_1 and Z_2 represent either the upper or lower halves of the corresponding anchoring frequency and adjustment ratio split lognormals. Recall that the MV, UB, and LB values supplied by a panelist in response to each question are assumed to correspond to the median, 95th and 5th percentiles, respectively, of a lognormal distribution. Denote the panelist responses corresponding to Z_i by MV_i, UB_i and LB_i and denote the median of Z_i by m_i. Then $m_i = MV_i$ and, from Equation 5.1,

$$\mu_i = \ln(m_i) = \ln(MV_i), \ i = 1, 2. \tag{5.11}$$

It is convenient to convert the responses supplied by a panelist into error factors of the corresponding lognormal distributions in order to determine the σ_i. Denote the error factor of Z_i by EF_i. Then, for Z_i corresponding to the upper half of a split lognormal, $EF_i = UB_i/MV_i$ and, from Equation 5.10,

$$\sigma_i = \ln(EF_i)/1.645 = \ln(UB_i/MV_i)/1.645, \ i = 1, 2 \tag{5.12}$$

Let $Z = Z_1 Z_2$ be the product of the two independent lognormals, where the variable Z represents either the piping system (Block 1.4) or non-piping subcomponent (Block 1.11) frequencies. Because the sum of any number of normal distributions is always normal, the product of two lognormals is always a lognormal, whether or not they are independent. Hence $Z = LN(\mu, \sigma)$ is lognormal. Because multiplying two lognormals is equivalent to adding their underlying normal distributions, it follows from the independence of Z_1 and Z_2 that $\mu = \mu_1 + \mu_2$ and $\sigma = [(\sigma_1)^2 + (\sigma_2)^2]^{1/2}$. In terms of the panelist responses, it follows from Equations 5.11 and 5.12 that

$$\mu = \ln(MV_1) + \ln(MV_2) = \ln[(MV_1)(MV_2)] \tag{5.13}$$

$$\sigma = [(\ln(UB_1/MV_1))^2 + (\ln(UB_2/MV_2))^2]^{1/2} /1.645 \tag{5.14}$$

From Equation 5.5, the median, 95th and 5th percentiles of Z are given by

$$b_{50}(Z) = \exp(\mu) = (MV_1)(MV_2) = m_1 m_2 \tag{5.15}$$

$$b_{95}(Z) = \exp(\mu + 1.645\sigma) \tag{5.16}$$

$$b_5(Z) = \exp(\mu - 1.645\sigma) \tag{5.17}$$

The formulas in Equations 5.13 through 5.17 apply when Z corresponds to the upper halves of the split lognormals. When Z corresponds to the lower halves, the error factor of Z_i is defined by $EF_i = MV_i/LB_i$ and the formulas in Equations 5.12 through 5.17 apply with UB_i/MV_i replaced by MV_i/LB_i for i = 1, 2. Note that Equation 5.15 demonstrates that the median of the product of two independent split lognormal distributions is the product of the medians of the two split lognormals, as stated above.

For reference, from Equations 5.2 and 5.3, the mean and variance of Z are given by

$$E(Z) = (MV_1)(MV_2) \exp[\{(\ln(UB_1/MV_1))^2 + (\ln(UB_2/MV_2))^2\}/5.412] \qquad (5.18)$$

$$V(Z) = [E(Z)]^2 \{\exp[\{(\ln(UB_1/MV_1))^2 + (\ln(UB_2/MV_2))^2\}/2.706] - 1\} \qquad (5.19)$$

5.3.3.2 Calculation of Means - The mean of the product of two independent distributions is always equal to the product of the means of the two distributions. Accordingly, to calculate the means in Block 2.9, a formula for the mean of a split lognormal distribution is needed. The mean of a lognormal distribution is more strongly influenced by the values in the upper half of the distribution as a result its skewed shape. Additionally, as can be seen from Equation 5.2, the mean of a lognormal distribution is proportional to an exponential function of the variance of its underlying normal. These features imply that the mean of a lognormal is a sensitive function of the upper percentiles (i.e., > 95th) of the underlying normal distribution. Consequently, the mean of a split lognormal is also a sensitive function of the upper percentiles of the distribution.

The upper percentiles and consequently the mean of the elicitation response largely are defined by the ratio of the UB to the MV. For some panelists, this ratio can be as large as 1,000. In such a case, the ratio of the mean to the median for a lognormal distribution is 6747 (see Equation 5.6). This mean to median ratio implies that the mean corresponds to a much higher percentile than the 95th percentile which is the UB elicited from the panelist. In other words, the mean is extrapolated far beyond the elicitation responses in situations where the error factor is large.

To remedy this problem, the split lognormals in Block 2.2 that are used to calculate the means are replaced by split lognormals truncated at the 99.9th percentile. The remaining 0.1 percent of each distribution is concentrated at the 99.9th percentile. This truncation point was chosen to be reasonably conservative, yet ensure that the mean is not dominated by the extreme upper tail of the distribution. Furthermore, because the panelists provided no information beyond the 95th percentile, any truncation point beyond the 95th percentile is consistent with their responses. The truncated split lognormal is used only to calculate the means because truncation at this percentile minimally affects the 5th, median, or 95th percentile estimates.

The mean of a truncated split lognormal distribution, W, is given by

$$E(W) = m(W)\left\{\exp\left(\frac{\sigma_L^2}{2}\right)\Phi(-\sigma_L) + \exp\left(\frac{\sigma_U^2}{2}\right)\left[\Phi(\sigma_u) - \Phi(\sigma_u - t_p)\right] + \left(1 - \Phi(t_p)\right)\exp\left(t_p\sigma_u\right)\right\} \quad (5.20)$$

where

       $m(W)$ = median of W
       Φ_L^2 = variance of lower half of $\ln(W)$
       Φ_U^2 = variance of upper half of $\ln(W)$
       M = cumulative distribution function of the standard normal distribution
       t_p = truncation point corresponding to the pth percentile

The means of the truncated split lognormals corresponding to the anchoring frequencies and adjustment ratios (Blocks 1.2, 1.3, 1.9, 1.10) are multiplied to obtain the means in Blocks 1.4 and 1.11. Denote the distribution of the frequency in Block 1.4 or Block 1.11 by Z. Then

$$Z = U V \qquad (5.21)$$

where U is a truncated split lognormal corresponding to an anchoring frequency and V is a truncated split lognormal corresponding to an adjustment ratio. Based on the assumed independence of U and V

$$E(Z) = E(U) E(V) \qquad (5.22)$$

where E(U) and E(V) are given by Equation 5.20.

5.3.4 Piping and Non-piping Contributions

Section 5.3.3 provides formulas for calculating the piping system and non-piping subcomponent percentiles and means for the four bottom-line frequency parameters (Blocks 1.4 and 1.11). This section describes the methodology used to calculate the percentiles and means of the piping and non-piping contributions (Blocks 1.7 and 1.17).

5.3.4.1 Calculation of Percentiles – From Figure 5.1, the piping and non-piping contributions depend on sums and products of frequencies (i.e., Blocks 1.5, 1.12, 1.15 for the sums and Blocks 1.6, 1.13, 1.16 for the products).

Denote any of the sums of frequencies by X with its jth term denoted by Z_j. Then

$$X = \Sigma Z_j \tag{5.23}$$

To calculate the percentiles of a sum, it is necessary to specify the correlation structure of its individual terms. The Z_j in Equation 5.23 express the uncertainties of the panelists about their responses as provided by the elicited UBs and LBs. The correlation structure is determined by the correlations between all pairs of the individual Z_j terms. Although the exact correlation structure is unknown, this study assumes that the Z_j are positively correlated. This means that if one Z_j is high, then the other Z_j's will tend to be high. This is a reasonable assumption because most panelists noted that degradation mechanisms that significantly contribute to the LOCA frequencies affect many systems similarly. Therefore, if conditions lead to higher failure probabilities in one piping system due to a specific degradation mechanism, then the failure probabilities of other susceptible systems will also increase. This positive correlation also should extend to the uncertainties. In other words, if a panelist expressed large (or small) uncertainties for one system, then other systems will also have large (or small) uncertainties if these measures are positively correlated. It was observed in the elicitation responses that the uncertainties were indeed positively correlated which implies that the susceptibility of each system is also positively correlated.

For positively correlated Z_j, the correlation structure is bounded by the independent case (all pair-wise correlations are zero) and the perfect rank correlation case (all pair-wise correlations are maximal). These correlation bounds can be translated into bounds on the variance V(X) of X. To calculate V(X), let ρ_{jk} be the correlation coefficient of Z_j and Z_k and let $SD_j = [V(Z_j)]^{1/2}$ be the standard deviation of Z_j. The variance of X is given by

$$V(X) = \Sigma V(Z_j) + \Sigma_{j \neq k} \rho_{jk} (SD_j)(SD_k) \tag{5.24}$$

The LB on the variance is attained when the components are independent and $\rho_{jk} = 0$. For this case, the variance of the sum in Equation 5.23 is the sum of the variances, and the LB on the variance of X is

$$V_L(X) = \Sigma V(Z_j) \tag{5.25}$$

Clearly, an UB on the variance of X is given by Equation 5.24 with $\rho_{jk} = 1$ for all $j \neq k$ (perfect correlation) and is equal to

$$V_U(X) = \Sigma V(Z_j) + \Sigma_{j \neq k} (SD_j)(SD_k) = [\Sigma (SD_j)]^2 \tag{5.26}$$

However, this UB is not typically achieved because it requires that the correlation coefficients for all pairs (Z_j, Z_k) be simultaneously equal to 1. For this to happen, each pair (Z_j, Z_k) has to be linearly related or, in other words, each Z_j must be a linear function of Z_1. Because this constraint is almost never satisfied, the UB on the variance is achieved by assuming a perfect rank correlation structure where the correlation between all pairs (Z_j, Z_k) is maximal instead of being equal to one. This condition is achieved if Z_j and Z_k, are functionally related by some monotonically increasing function F, i.e., $Z_j = F(Z_k)\}$.

To see why this condition implies maximal correlation, consider the lognormal case. Let Z_i be lognormal with underlying parameters μ_i and σ_i. Therefore, $\ln(Z_i)$ is normal with mean μ_i and standard deviation σ_i. and $Z_i = \exp(\mu_i + \sigma_i W_i)$ where W_i is a standard normal with mean $\mu_i = 0$ and standard deviation $\sigma_i = 1$. Then Z_1 and Z_2 will have maximal correlation if $W_1 = W_2 = W$, i.e., if the same standard normal distribution is used in the expressions for Z_1 and Z_2. Therefore, the sum $X = Z_1 + Z_2$ is a monotonic function of W and the percentiles of X correspond directly to the percentiles of W. It follows that $b_p(X) = \exp[\mu_1 + \sigma_1 b_p(W)] + \exp[\mu_2 + \sigma_2 b_p(W)] = b_p(Z_1) + b_p(Z_2)$, where $b_p(X)$ is the pth percentile of X. In particular, the median $m(X) = m(Z_1) + m(Z_2)$, and the 95th percentile of the sum equals the sum of the 95th percentiles. The assumption of perfect rank correlation generalizes to the sum of any number of components by setting $Z_j = \exp(\mu_j + \sigma_j W)$. Thus, the pth percentile of the sum of any number of lognormals with perfect rank correlation is the sum of the pth percentiles of the components.

This conclusion for the lognormal distribution can be generalized when summing arbitrary distributions, provided that the distributions satisfy the perfect rank correlation assumption. From Equation 5.23, if $b_p(X)$ is the pth percentile of X, then

$$b_p(X) = \sum b_p(Z_j) \qquad (5.27)$$

The multiplication algorithm is used to calculate the percentiles of the product of two independent split lognormals for Blocks 1.6, 1.13, and 1.16 described previously in Section 5.3.3.1. Let

$$Z = Z_1 Z_2 \qquad (5.28)$$

where Z = piping (Block 1.7) or non-piping (Block 1.17) contribution
   Z_1 = un-normalized piping (Block 1.5) or non-piping (Block 1.15) contribution
   Z_2 = piping (Block 1.6) or non-piping (Block 1.16) contribution factor

The non-piping analysis requires an additional product calculation to determine the non-piping component frequencies (Block 1.14). If Z represents the non-piping component frequency, then Z_1 and Z_2 represent the non-piping un-normalized component frequencies (Block 1.12) and the non-piping component contribution factor (Block 1.13), respectively.

There are two potential complications in applying the multiplication algorithm to Equation 5.28. From Section 5.3.3.1, the multiplication algorithm assumes that both Z_1 and Z_2 are split lognormals. First, because the contribution factors all lie between 1 and 2, a split lognormal with its unbounded range is at best an approximation to the distribution of a contribution factor. However, because the uncertainties in the contribution factors all have very narrow ranges, the distribution of any contribution factor and the split lognormal Z_2 representing it are essentially the same constant. Second, from Blocks 1.5 and 1.15, Z_1 is a sum of split lognormals, which does not have a split lognormal distribution. However, with the multiplication algorithm, a split lognormal is fit to the median, 95th and 5th percentiles of Z_1. Therefore, the median, 95th and 5th percentiles of the split lognormal fit to Z_1 all coincide with the respective percentiles of Z_1. Because multiplying Z_1 by Z_2 is approximately equivalent to multiplying Z_1 by a constant, the percentiles of Z are not significantly altered by the assumption that Z_1 is split lognormal. Thus, using the multiplication algorithm yields a good approximation to the percentiles of Z.

As discussed previously, the actual correlation structure is expected to be much closer to perfect rank correlation than independence. Degradation mechanisms and other LOCA contributing factors can affect many piping systems and non-piping subcomponents similarly. Hence, there is an expectation that a strong correlation exists. This correlation is, in fact, evident in many of the panelists' responses. Consequently, the baseline estimates are all calculated assuming perfect rank correlation.

To implement the calculation, a perfect rank correlation structure is assumed for all the summation blocks in Figures 5.1 and 5.2 that calculate percentiles. These include the un-normalized piping contribution (Block 1.5), the non-piping un-normalized component frequencies (Block 1.12), and the un-normalized non-piping contribution (Block 1.15), and the baseline estimates for the LOCA frequency percentiles (Blocks 1.18 and 2.6). However, in a sensitivity study, the perfect rank and independent correlation structures are compared using case studies to represent the individual elicitation responses which are used to determine these un-normalized contributions (Sections 5.6.3 and 7.6.3). The cases were chosen to span the range of features (i.e., symmetry, error factors, etc) characterizing the panelists' responses.

5.3.4.2 Calculation of Means - The calculation of the means (Block 2.12) is more straightforward than the calculation of the percentiles. As noted at the beginning of Section 5.3.4.1, the piping and non-piping contributions depend only on sums and products of frequencies. Because the contribution factors in Blocks 1.6, 1.13 and 1.16 are structurally different than the frequencies which they multiply, it is reasonable to assume that their distributions are independent of the frequency distributions. Therefore, the means of the frequency contributions in Blocks 1.7 and 1.17 are calculated by simply replacing the component frequencies by their means. Because all of these components have split lognormal distributions, the formula in Equation 5.20 is used.

5.3.5 Total LOCA Frequencies
The analysis framework described above is used with each panelist's responses to calculate the estimated bottom-line parameters of the LOCA frequencies for BWR and PWR piping and non-piping systems. The total BWR and PWR LOCA frequencies (Blocks 1.18 and 2.6) are calculated by summing the frequencies corresponding to the piping and non-piping contributions (Blocks 1.7 and 1.17, respectively) as described in Section 5.3.4. The total LOCA frequencies are calculated only for those individual panelists who have supplied sufficient quantitative input response information.

5.4 Group Estimates and Confidence Intervals

Section 5.3 describes how to calculate the bottom-line parameters (means, medians, 95th, and 5th percentiles) of total passive LOCA frequencies based on the responses of any one panelist. The final step in the analysis framework is to combine the individual estimates to obtain group estimates for the total passive BWR and PWR LOCA frequencies (Blocks 1.18, 2.7, 2.13). Once the individual bottom-line parameters have been calculated, each parameter is combined independently of all other parameters. In other words, the group mean is a function only of the individual means, the group median is a function only of the individual medians, and so on. Even though the individual parameter estimates are constrained by the split lognormal structure applied to each panelist's responses, the group parameter estimates are not constrained by this structure. This is consistent with the guiding principle discussed in Section 5.1 of making as few assumptions as necessary. In particular, no assumption is made about the form of the total LOCA frequency distribution, e.g., that it is split lognormal. Because estimated values of the four bottom-line parameters are sufficient for regulatory applications, any additional constraints on them are not required.

5.4.1 Justification for Using Group Estimates
The justification for combining the individual estimates into group estimates for the bottom-line parameters is the empirical observation that a group estimate of a quantity about which its individual members have only vague information can be surprisingly close to the "truth". Here, a group estimate is defined as a value near the center of the individual estimates. This empirical observation is

documented in [5.3], which reports on many examples of this phenomenon. Many of these quantitative examples use "almanac-type" questions, where the answers are known but about which the respondents can only guess.

Part of the elicitation training of the panelists consisted of an exercise in which the panelists were asked "almanac-type" questions about chronic disease statistics (see Section 3.3.2 and Appendix C). The training exercise questions were designed with a similar structure as the LOCA frequency elicitation questions to make them as relevant as possible. The training exercise responses were then combined into group estimates, and the results of the training exercise (Appendix C) supported the empirical observation that group estimates are more accurate than individual estimates and are relatively accurate.

A natural question which arises is the relevance of this training exercise to this study's elicitation procedure which uses an expert panel chosen for their extensive knowledge and experience with highly complex physical phenomena. On the surface, this elicitation would appear to be very different from asking an arbitrary group of people --- with no direct topical expertise --- to estimate chronic disease statistics. However, the elicitation questions asked the panelists to estimate quantities which differ in three important ways from their relevant scientific expertise. First, virtually all of the questions ask the panelists to extrapolate well beyond their knowledge and experience with the LOCA-initiating phenomena. Secondly, the questions all concern the frequencies of occurrence for LOCA-initiating phenomena, which is a very different concept than the phenomena themselves. Finally, the questions all effectively asked the panelists to estimate fractals of their subjective probability distributions for the LOCA frequencies. This is an abstract concept which is even further removed from the LOCA-initiating phenomena. For all of these reasons, the elicitation questions asked the panelists to estimate quantities about which they have no specific, direct expertise. Thus, asking questions related to LOCA frequencies of these panelists is analogous to asking almanac-type questions of an arbitrary group of people. In other words, the elicitation questions are analogous to almanac-type questions, for which the improved accuracy of group compared to individual estimates has been validated [5.3].

5.4.2 Confidence Intervals -- Lognormal Model

Let $\{u_k\}$, $k = 1, 2,..., n$, where "n" is the number of panelists, denote a set of individual estimates of any one bottom-line parameter. For example, $\{u_k\}$ could be the set of current-day estimates for the mean BWR Category 1 LOCA frequency, as calculated from the panelists' responses. A basic premise of any expert elicitation process is that the panel responses as a whole have no significant systematic bias. In other words, while individual responses can be highly uncertain and differ drastically, they do not systematically over- or underestimate the quantities of interest. The goal, based on the u_k values only, is to calculate an unbiased group estimate of each bottom-line parameter and to also obtain a measure of panel diversity associated with the group estimate of each parameter. The baseline approach (Section 5.5) used here simultaneously provides a group estimate and a measure of panel diversity. The baseline approach is motivated by the observation that $\{u_k\}$ typically spans several orders of magnitude (see Section 7.5) and is best described by a highly skewed distribution.

The assumption of *statistical* independence applies only to each panelist's individual responses (MVs, UBs, and LBs) to the elicitation questions and is used to derive each panelist's uncertainty distribution from their responses. Since the panelists' individual quantitative responses were never discussed in the group sessions, the statistical independence assumption is warranted. It should also be noted that statistical independence of two distributions describes the variability of the distributions about their means rather than the relation between their absolute values. Two distributions can have the same mean and still be statistically independent. Thus, even if the MVs for a particular question *were* correlated among panelists as a result of group discussions, there is no reason to believe that the uncertainties about their responses from the individual elicitation sessions were also correlated.

To construct confidence intervals for the group estimate (Blocks 1.19, 2.7, 2.13), a lognormal structure (Block 2.8) is assumed. Specifically, it is assumed that the $\{u_k\}$ values are a sample from a lognormal distribution, U. This assumption is made only for analytical purposes and should not be interpreted to mean that the individual panelist estimates are all generated by the same process. An analogy is a sample of heights from a defined population. For descriptive or statistical purposes, it is customary to assume that the heights in the sample are normally distributed, but this assumption does not imply that the heights are determined by the same physical process.

It is convenient to write U in the form

$$U = g \exp(W) \tag{5.29}$$

where W has a normal distribution with mean 0 and variance τ^2. The interpretation of U is that the true value of the bottom-line parameter estimated by the u_k is g and that $\exp(W)$ is a bias factor which describes the differences between the panelists' estimates and the true value of the parameter. This structure for W assumes no systematic bias in the panelists' estimates of the true value U. Therefore g is the median of U and $\ln(U)$ has a normal distribution with mean $\ln(g)$ and variance τ^2.

Define

$$w_k = \ln(u_k) \text{ for } k = 1, 2,..., n. \tag{5.30}$$

From Equation 5.29, the standard estimate for $\ln(g)$ is the sample mean of the w_k individual estimates for the bottom-line parameter of interest. Hence, the group estimate of g is the GM of the u_k. Denoting the group estimate of g by g^*, it follows that

$$g^* = [\Pi \, u_k]^{1/n} \tag{5.31}$$

A measure of panel diversity is a confidence interval for g^*. From Section B.4.2 of Reference 5.2, a two-sided $100(1-\alpha)\%$ confidence interval (where $1-\alpha$ is the desired confidence level) for $\ln(g)$ is given by

$$\ln(g^*) - t_{1-\alpha/2}(n-1)[S/n^{1/2}] \leq \ln(g) \leq \ln(g^*) + t_{1-\alpha/2}(n-1)[S/n^{1/2}] \tag{5.32}$$

where $t_{1-\alpha/2}(n-1)$ is the $100(1-\alpha/2)$ percentile of a Student's t distribution with n-1 degrees of freedom, and

$$S^2 = \Sigma \, [w_k - \ln(g^*)]^2 \,/(n-1) \tag{5.33}$$

is the sample variance of the w_k. The corresponding $100(1-\alpha)\%$ confidence interval for g is obtained from Equation 5.32 as.

$$g^* \exp\{- t_{1-\alpha/2}(n-1)[S/n^{1/2}]\} \leq g \leq g^* \exp\{ t_{1-\alpha/2}(n-1)[S/n^{1/2}]\} \tag{5.34}$$

For this analysis, α is chosen as 0.05 to calculate a 95% confidence interval for g.

5.5 Baseline LOCA Frequency Estimates

A baseline analysis was conducted to process the panelists' elicitation responses (Section 5.2). The purposes of this analysis were to develop a set of LOCA frequency estimates that could be used to investigate important LOCA frequency contributing factors (Sections 7.2 – 7.4), assess panel member uncertainty and group variability (Section 7.5), provide a reference for comparison with subsequent sensitivity analysis results (Sections 5.6 and 7.6), and serve as the basis for the summary elicitation results (Sections 7.7 and 7.8). These summary results are meant to provide appropriate group LOCA frequency estimates, taking the elicitation structure and results into account. The baseline analysis procedure and assumptions for each analysis step have been discussed in Sections 5.3 and 5.4, but they are summarized here for convenience.

There are six key elements of the baseline analysis framework:

(i) The MV, UB, and LB supplied by the panelists for each elicitation question were assumed to correspond to the median, 95th percentile, and 5th percentile, respectively, of a split lognormal distribution with the mean calculated assuming the upper tail is truncated at the 99.9th percentile.

(ii) Panelist responses are not adjusted to account for possible overconfidence in the uncertainty estimates for each elicitation response.

(iii) Split lognormal distributions were summed by assuming perfect rank correlation among the individual terms.

(iv) The individual estimates are the total LOCA frequency parameters (i.e., mean, median, 5th percentile, and 95th percentile) determined for each panelist

(v) The group estimates of the total LOCA frequency parameters are defined as the GMs of the individual estimates.

(vi) Panel diversity is characterized with two-sided 95% confidence intervals based on an assumed lognormal model for the individual estimates.

These key baseline analysis elements were selected to be consistent with the analysis framework (Section 5.1). If several consistent choices were possible, then conservative analysis choices were made. Element (i) naturally results from the lognormal analysis framework defined in Sections 5.1 and 5.3.1. Sensitivity analyses were also conducted to examine the effects of the truncation method for calculating means (Section 5.6.1). Element (ii) uses the raw responses provided by the panelists. However, panelist overconfidence is a well-known phenomenon [5.4] and possible adjustment schemes were evaluated as sensitivity analyses (Section 5.6.2). Element (iii) assumes maximal correlation between the contributions to the piping and non-piping frequencies for each panelist. Sensitivity analyses were conducted to evaluate the effects of different distributional shapes on the mean frequencies (Section 5.6.1), and when assuming an independent rank correlation structure for sample panelist responses.

A number of natural aggregation points (element iv) exist for combining the individual panelist responses: (1) after the total LOCA frequencies have been determined, (2) after the separate piping and non-piping component frequencies have been calculated, and (3) at each individual elicitation response. Aggregation point (1) was chosen for element (iv). This selection stems from the careful attention paid during the individual elicitation sessions on eliminating inconsistencies between the responses of each panelist. Also, the correlation between a panelist's piping and non-piping responses is retained by this selection. The other two aggregation points lead to inconsistencies (Section 5.1) because the strategies that individual panelist used to respond to each question varied a great deal and not all panelists answered each elicitation question. Therefore, aggregation of the individual total LOCA frequency estimates is the best choice, although sensitivity analysis (Section 5.6.4.2) was conducted to evaluate the effects of aggregation responses at points 2 and 3.

The use of the GM to aggregate the individual results (element v) and confidence intervals to capture panel diversity (element vi) results from the interpretation and structure of the individual responses. As discussed in Section 5.4, both of these elements are a consequence of the assumed lognormal structure of the individual bottom-line LOCA frequency estimates, although the use of the GM for aggregation can be justified without assuming a lognormal structure (see below). In addition to being consistent with the actual elicitation results, this approach is motivated by the fact that the estimated LOCA frequencies are based on expert elicitation responses, which are fundamentally assumed to not be systematically biased (see Section 3.3). The individual responses can be highly uncertain and they can differ drastically, but no significant systematic bias is assumed. The questionnaire structure and the elicitation training were designed to achieve this goal, and the training exercise results (Appendix C) are consistent with this fundamental assumption.

A consequence of the assumption of no systematic bias is that the group estimate should be somewhere in the middle of the individual estimates, especially if there are wide differences in the results. One obvious choice would be to calculate the group estimate using the median of the individual estimates. Other possible choices include the GM and the trimmed geometric mean (TGM) because they are also estimates of the median. The arithmetic mean (AM) is another possible choice. However, the AM of individual estimates is often not a good measure of the median group opinion when the individual estimates are widely varying. In this case, the AM is dominated by the one or two largest results and cannot be fairly described as a group estimate.

There is support for the use of the median or GM in the literature. In previous NRC applications (see References 5.5-5.8), the median was used as a group estimate when the individual estimates varied by several orders of magnitude. In Reference 5.4, Meyer and Booker state: "To overcome the influence of extreme values when forming an aggregation estimate, use the median or geometric mean" (p. 310). In Reference 5.9, von Winterfeldt and Edwards recommend averaging probabilities (p. 136). However, they conclude: "The only context in which we have any reservations about this conclusion is that of very low probabilities - - - - - for such extreme numbers, we would prefer averaging log odds to averaging probabilities." For very low probabilities, averaging log odds is equivalent to using the GM. Taking these considerations into account, and noting that a sensitivity study showed that there is little difference between using the median or the GM to aggregate the study results (see Section 7.6.4.1), the GM was chosen as the most appropriate group estimate which utilizes all the individual estimates.

Sensitivity analyses were also conducted to evaluate the ramifications of assuming a lognormal structure for the individual LOCA frequency estimates. Different aggregation schemes (element v) are examined in Section 5.6.4, including AM aggregation (Section 5.6.4.1) and mixture-distribution aggregation (Section 5.6.4.4). Estimating panel diversity through the quartiles of the distribution is examined in Section.5.6.5.

5.6 Sensitivity Analyses

A number of sensitivity analyses were conducted to examine the effects of modifying the analysis framework used to calculate the baseline LOCA frequency estimates outlined in Section 5.5. There are many possible alternatives to the baseline elements which could have been chosen to estimate the group LOCA frequency bottom-line parameters. The analyses described below were selected either to evaluate plausible alternatives or to quantify the maximum possible LOCA frequency differences which could result from using alternative calculation schemes. The sensitivity analyses apply to only one element of the baseline analysis individually; combinations of alternatives were not evaluated.

5.6.1 Calculation of Means
The baseline estimates of the mean are based on a split lognormal truncated at the 99.9th percentile. Sensitivity analyses analyzed the effects that other distributional shapes have on the calculated means for the individual elicitation responses. Several other plausible choices were evaluated: an untruncated split lognormal, a lognormal corresponding to the upper half of a split lognormal (upper tail lognormal), a split log-triangular or a normalized split lognormal truncated at the 99.9th percentile. For these later two distributions the mass beyond the truncation point is proportionately distributed within the upper half of the distribution. The effects of varying the truncation point for the split lognormal distribution were also examined. See Section 7.6.1 for the results of the sensitivity analyses for the calculation of means.

5.6.2 Overconfidence Adjustment
The baseline estimates assume that the UB and LB supplied by the panelists for each elicitation question correspond to the 95th percentile and 5th percentile, respectively, of a lognormal distribution. In other words, the baseline estimates are based on the assumption that all the (LB, UB) uncertainty intervals have 90 percent coverage, i.e., they all have a 90 percent chance of containing the true value.

However, extensive experience with such subjectively determined intervals has demonstrated that people tend to underestimate the uncertainty in their answers, i.e., they tend to be overconfident. Instead of having about a 90% chance of containing the true value, nominal 90% coverage intervals have been found to only have about 30% to 70% coverage of the true value [5.4].

Accordingly, sensitivity analyses were performed to evaluate the effects of assuming that the panelists' uncertainty intervals did not always correspond to the stated 90% coverage interval. The MVs supplied by the panelists were not adjusted, i.e., they always correspond to the median of the assumed lognormal distributions. Two types of overconfidence adjustments were investigated. The first type adjusted the coverage intervals directly for each individual elicitation response. For example, one sensitivity analysis assumed that the 90% intervals supplied by the panelists correspond to only 50% coverage, i.e., the associated UBs and LBs were assumed to correspond to the 75th percentile and 25th percentiles, respectively, of their associated lognormal distributions. The second type of overconfidence adjustment was applied to the error factors associated with the bottom-line LOCA frequency estimates and not the individual responses as before. Because the error factor of any estimated LOCA frequency parameter depends on the uncertainty intervals on which the estimate is based, increasing the error factor is equivalent to an indirect overconfidence adjustment of the underlying uncertainty intervals. Both blanket and target overconfidence adjustments were applied. In a blanket adjustment, all panelists were adjusted by the same amount. In a targeted adjustment, panelists with relatively small uncertainties were adjusted more than panelists with larger uncertainties. The rationale for a targeted adjustment is that panelists with relatively small uncertainties are more likely to be overconfident. See Section 7.6.2 for the results of the overconfidence adjustment analyses.

5.6.2.1 Coverage Interval Adjustment - For any UB and LB supplied by a panelist, formulas were developed to calculate the results of any level of overconfidence adjustment. The overconfidence adjustment is accounted for by defining an adjusted UB and LB corresponding to the 95th and 5th percentiles, respectively, of the panelist's distribution. This allows the results of the overconfidence adjustment to be calculated by simply substituting the adjusted UB and LB into the formulas developed for the unadjusted UB and LB. As stated previously, the MV supplied by the panelist were not altered by the overconfidence adjustment.

The formula for adjusting a UB is presented first. Let b be the value of a UB supplied by a panelist and let p be the value of the percentile of the panelist's assumed lognormal distribution corresponding to b. Define $UB_p(b)$ as the adjusted value of the UB, i.e., $UB_p(b)$ corresponds to the 95th percentile of the panelist's adjusted distribution. (For the case of no adjustment, $p = 95$ and $UB_{95}(b) = b$.) Let m be the value of the associated MV supplied by the panelist. Denote the error factor of the panelist's adjusted lognormal distribution by r_p. From Equation 5.10,

$$r_p = UB_p(b)/m \qquad (5.35)$$

Before adjustment, the error factor, r, is equal to b/m. Hence, the overconfidence adjustment is equivalent to changing the error factor from r to r_p for each elicitation response.

Referring to Section 5.3.2, let σ^2 be the variance of the underlying normal of the panelist's lognormal distribution. From Equation 5.10,

$$r_p = \exp(1.645\sigma) \qquad (5.36)$$

From Equation 5.1, Equation 5.5 and the relation $r = b/m$,

$$r = \exp(k_p\sigma) \qquad (5.37)$$

It follows from Equations 5.36 and 5.37 that

$$r_p = r^{c(p)}, \tag{5.38}$$

where $c(p) = 1.645/k_p$.

From Equations 5.35 and 5.38, and the relation $r = b/m$,

$$UB_p(b) = b^{c(p)} \cdot m^{1-c(p)}, \tag{5.39}$$

where b is the UB and m is the MV supplied by the panelist. This is the desired formula for the overconfidence adjustment to the UB.

Because the UB and LB are not necessarily symmetric about the MV, the LB must be adjusted independently of the UB. As for the UB adjustment, let b' be the value of a LB supplied by a panelist and let p' be the value of the percentile of the panelist's assumed lognormal distribution corresponding to b'. Define $LB_p(b')$ as the adjusted value of the LB, i.e., $LB_p(b')$ corresponds to the 5th percentile of the panelist's distribution. (For the case of no adjustment, $p' = 5$ and $LB_5(b') = b'$.)

The derivation of the formula for $LB_p(b')$ is analogous to the derivation of the formula for $UB_p(b)$. The only difference is that the error factor is equal to the MV divided by the LB, instead of the UB divided by the MV. The adjusted error factor ($r_{p'}$) is defined then defined as in Equation 5.35 as

$$r_{p'} = m / LB_{p'}(b'), \tag{5.40}$$

where m is the MV supplied by the panelist. Equation 5.36 remains unchanged for $r_{p'}$, but Equation 5.37 for the unadjusted error factor (r') becomes

$$r' = \exp(-k_{p'}\sigma) \tag{5.41}$$

where $k_{p'}$ is the pth percentile of the standard normal distribution associated with the LB and MV response.

Note that, because p' is the percentile corresponding to a LB, $p' < 50$ and $k_{p'}$ is negative. From the symmetry of the normal distribution, $k_{p'} = -k_{100-p'}$. Hence Equation 5.38 for the LB adjustment is

$$r'_{p'} = [r']^{c'(p')}, \tag{5.42}$$

where $c'(p') = 1.645/k_{100-p'}$. Then Equation 5.39 becomes

$$LB_p(b') = [b']^{c'(p')} \cdot m^{1-c'(p')}, \tag{5.43}$$

where b' is the LB and m is the MV supplied by the panelist. This is the desired formula for the overconfidence adjustment to the LB. Equations 5.39 and 5.43 express similar forms for the adjusted UB and LB responses.

Because an overconfidence adjustment implies that the adjusted percentile is always closer to 50% than the nominal percentile, $p < 95$ and $p' > 5$. Hence, $c(p)$ and $c'(p')$ are always > 1, and the adjusted bounds are always further away from the median than the unadjusted bounds. Therefore, the panelist's adjusted distribution is always broader than the unadjusted distribution. In other words, an overconfidence adjustment always increases the uncertainty in the panelist's response.

Five separate overconfidence adjustments were evaluated to examine their effects on the LOCA frequency estimates: two blanket adjustments and three targeted adjustments. As previously stated, the blanket adjustments adjust the responses for all panelists by the same amount. The targeted adjustments use a two-level adjustment. Those panelists expressing larger uncertainties were adjusted by smaller amounts, while those panelists with smaller uncertainties were adjusted by larger amounts.

The various overconfidence adjustment schemes are summarized in Table 5.1. The Blanket 1 (B1) adjustment is the most severe adjustment because the responses which nominally represent 90% coverage intervals are assumed to represent 50% coverage intervals. Each successive adjustment scheme is less severe. Thus, the Blanket 2 (B2) adjustment assumes that the nominal 90% coverage intervals actually have only 60% coverage. The targeted adjustment schemes are based on a division of panelist responses into three error-factor bins: less than 10 (less uncertainty), between 10 and 50, and greater than 50 (greater uncertainty). The Targeted 1 (T1) adjustment does not adjust the 3 or 4 panelists with error factors greater than 50 (coverage interval 1) and adjusts the coverage intervals to 50% for the remaining panelists (coverage interval 2). The Targeted 2 (T2) adjustment uses the same criteria to represent coverage intervals 1 and 2, but adjusts the more uncertain responses to 80% coverage while the remaining panelists are adjusted to 60% coverage. The Targeted 3 (T3) adjustment adjusts only the panelists with error factors less than 10 (coverage interval 2) from 90% to 60% coverage. Results of this analysis are presented in Section 7.6.2.1.

Table 5.1 Elicitation Response Adjustment Schemes

Overconfidence Adjustment	Coverage Interval 1	Population	Coverage Interval 2	Population
Blanket 1	50%	All	NA	NA
Blanket 2	60%	All	NA	NA
Targeted 1	90%	4	50%	remaining
Targeted 2	80%	4	60%	remaining
Targeted 3	90%	4	60%	remaining

5.6.2.2 Error-Factor Adjustment - The error-factor adjustment is applied to the one-sided error factors associated with a panelist's 95th and 5th percentile estimates. For the present discussion, a one-sided error factor is defined as either the ratio of the 95th percentile to the median or the ratio of the median to the 5th percentile. The median is unadjusted in this analysis. Therefore, the error-factor adjustment is equivalent to adjusting the 95th or 5th percentile of a panelist's LOCA frequency distribution. The adjustment can be performed by either changing the value of the 95th or 5th percentile or by changing the value of the error factor. Changing the percentile value is similar to the confidence level adjustment in the previous section except that it applies to the bottom-line 95th or 5th percentile, rather than the LB and UB elicitation responses.

As for the coverage interval adjustment (Section 5.6.2.1), the effective coverage interval of the original responses can be determined to quantify the magnitude of the overconfidence adjustment. For any one-sided error factor which is adjusted, let EF_0 be its original value and let EF_a be its adjusted value. Given EF_a and EF_0, define p_a as the percentile in the adjusted distribution corresponding to the original error factor. From Equation 5.10, the underlying standard deviation of the adjusted lognormal distribution (σ_a) is given by

$$\sigma_a = \ln(EF_a)/1.645 \tag{5.44}$$

since EF_a now is assumed to represent the true percentile.

From Equation 5.5,

$$EF_0 = \exp(k_a \, \sigma_a), \tag{5.45}$$

where $= k_a$ is the p_a percentile of the standard normal distribution. Solving Equation 5.45 for k_a and using Equation 5.44 yields

$$k_a = 1.645 \, [\ln(EF_0)/\ln(EF_a)] \qquad\qquad (5.46)$$

Then k_a can be calculated from Equation 5.46 and p_a can be determined from tables of the normal distribution. For the 95th percentile, $EF_a > EF_0$, $k_a < 1.645$ and therefore $p_a < 95$. This result is to be expected, because increasing the error factor broadens the distribution. A similar result holds for the 5th percentile.

The error-factor adjustment described above is applied to the error factors associated with the 95th and 5th percentiles of the panelists' various total LOCA frequency estimates (Section 7.6.2). Separately for each percentile and LOCA estimate, each panelist's error factor is compared with the GM of all the panelists' error factors. Error factors falling below the GM are adjusted up to the GM. Error factors above the GM are not adjusted. For each panelist, the 95th and 5th percentiles are recalculated using the formulas in Section 5.3.2 with the adjusted error factors, assuming the median remains unchanged. The mean is calculated using the adjusted error factors for the 95th percentile. This results in a variable adjustment as a function of plant type, LOCA category, and operating time. See Section 7.6.2.2 for additional information and the results of this analysis.

5.6.3 Correlation Structure

The baseline estimates of the percentiles (Section 5.5) are calculated assuming perfect rank correlation between the summed component distributions (Block 2.6). As discussed in Section 5.3.4, this baseline assumption maximizes the pairwise correlations between the distributions. Because the correlations are assumed to be positive, they are minimized by assuming the distributions are independent such that all pairwise correlations are equal to zero. These bounding assumptions lead to the largest differences in the summed component distributions attributable to the correlation structure. Accordingly, as a sensitivity analysis, the effect of assuming that the summed distributions are independent is evaluated.

An evaluation of the effects of an assumed independent correlation structure is only possible through Monte Carlo simulation. Because of the very large number of individual distributions that comprise the elicitation responses, the bottom line estimates were calculated only for selected panelist responses. Accordingly, ten simulation trials were selected to span several important variables as summarized in Table 5.2. First, both the panelists and the range of LOCA plant type combinations (i.e., PWR piping, PWR non-piping, BWR piping, and BWR non-piping) and time periods were sampled. Second, a number of distributions representing contributing piping components (or non-piping subcomponents) which must be summed to develop the bottom-line estimates were used. Third, and most important, the Monte Carlo trials were selected to span the range of distinguishing characteristics representative of the elicitation responses. These distinguishing characteristics (last column in Table 5.2) indicate whether the elicitation responses are generally symmetric (S in Table 5.2) or asymmetric (U in Table 5.2), and indicate the relative magnitude (small, moderate, large) of the upper or lower error factors (UEF and LEF, respectively) in Table 5.2. The characteristics are representative of the variability among the entire population of elicitation responses. Additional details concerning the Monte Carlo simulation and the results are contained in Section 7.6.3.

Table 5.2 Summary of Monte Carlo Trials

Trial Number	Number of Distributions	Panelist	LOCA Plant Type/Time Period	Distinguishing Characteristics
1	12	A	BWR-1 Piping @ 25 yrs	S, small LEF/UEF
2	12	A	BWR-2 Piping @ 25 yrs	U, small UEF
3	2	C	PWR-6 Piping @ 25 yrs	U, large UEF
4	4	C	BWR-3 Piping @ 25 yrs	S, moderate LEF/UEF
5	14	G	PWR-5 Non-Piping @ 60 yrs	U, large LEF
6	5	C	BWR-3 Non-Piping @ 25 yrs	U, large LEF
7	8	J	PWR-5 Non-Piping @ 25 yrs	S, large LEF/UEF
8	7	I	PWR-4 Piping @ 25 yrs	S, moderate LEF/UEF
9	4	E	BWR-4 Non-Piping @ 25 yrs	U, large LEF
10	9	B	PWR-3 Non-Piping @ 25 yrs	S, small LEF/UEF

5.6.4 Aggregation

As discussed in Section 5.4, the individual total LOCA frequencies were combined to obtain group estimates of the total BWR and PWR LOCA frequencies. The baseline estimates were calculated using the GM of the individual estimates (see Equation 5.31). However, as discussed in Section 5.4), there are several other ways to construct group estimates. A number of sensitivity analyses were performed to examine other methods for aggregating the panelists' responses. Many different methods were considered which use various measures of group opinion, combine panelist responses at different stages of the analysis, and employ different methods of calculating the bottom-line LOCA frequency parameters.

5.6.4.1 Group Estimate – As discussed in Section 5.4, the principal requirement for a group estimate based on expert opinion is that it be somewhere in the middle of the group. Consequently, a group estimate should not be overly dependent on outliers or extreme values. This requirement implies that a group estimate should be a measure of central tendency, and the GM is a plausible candidate (Section 5.4), especially since the estimated frequencies often differ by several orders of magnitude. However, because the individual estimates based on the panelists' responses were usually highly variable, the GM could conceivably be driven by one or more outliers. An alternative group estimate is the TGM, which is the GM of the values remaining after the largest and smallest values have been omitted from the set. Still another alternative is to use the median of the distribution of individual estimates. While the median is potentially much less dependent on outliers than the GM or TGM, it literally depends on only the center of the distribution and does not reflect any other aspect of the distribution. It is of interest to note that the median has been used as a group estimate in previous NRC applications (see References 5.5 - 5.8).

Another measure which could potentially be used as a group estimate is the AM of the panelists' bottom-line parameter estimates. The AM is always larger than the GM, except in the trivial case when the distribution is a constant. Therefore AM aggregation always yields more conservative LOCA frequency estimates. However, when the individual estimates span one or more orders of magnitude, as they do in this study, the AM largely reflects the highest estimates and contributions from estimates below the median value are negligible. Therefore, for the results of this study, the AM is not a good measure of central group opinion.

This study uses the GM to aggregate the individual estimates to obtain the baseline group estimates (Section 5.5). Sensitivity analyses were conducted to evaluate the differences resulting from using the TGM, median, or AM as aggregation schemes. Results of this analysis are contained in Section 7.6.4.1. Section 7.6.4.1 also includes a more in-depth comparison of the AM to the GM as a group estimate for the results in this study.

5.6.4.2 Aggregation Point - The aggregation point is the point in the analysis framework where the individual responses are combined to construct a group response (Sections 5.4 and 5.5). The baseline analysis used the last possible aggregation point and combined the individual total LOCA frequencies. This choice is most consistent with the elicitation structure and responses as discussed in Section 5.5. However, one alternative is to combine the individual piping and non-piping LOCA frequencies to construct group piping and non-piping LOCA frequencies, respectively. The two group estimates are then added as described in Section 5.3.4 to construct group estimates of total LOCA frequencies.

A second alternative is to combine at the earliest possible aggregation point, i.e., to combine the individual responses to each of the elicitation questions. Both MV and bounding responses are developed for the responses, and the responses are analyzed using the calculation procedure in Section 5.3 to determine total LOCA frequency estimates for BWR and PWR plant-types. This is equivalent to synthesizing responses for a 13th panelist from the responses of the other 12 panelists (Section 5.2). Section 7.6.4.2 provides more information on these calculations and summarizes the differences in the baseline results attributable to these two alternative aggregation points.

5.6.4.3 Aggregation Parameters - In the baseline methodology, estimates for the mean and other bottom-line parameters (i.e., median, 5th and 95th percentiles) were aggregated directly from the individual panelist responses (Section 5.5). An alternative approach is to calculate the bottom-line mean estimates from the aggregated percentile estimates by assuming a distributional relationship for the percentile estimates. The method used for this sensitivity analysis (called the MEF method) calculates the group 5th and 95th percentile estimates from the aggregated median and error factors instead of directly aggregating each percentile as in the baseline methodology. The bottom-line mean is then calculated from the aggregated median, 5th and 95th percentile values by assuming that the underlying distribution is a split lognormal which is truncated at the 99.9[th] percentile. This is the same distributional form assumed for calculating the mean estimates from the individual elicitation responses in the baseline methodology (Section 5.5)

The 5th and 95th percentile estimates determined by the MEF approach are identical, by definition, to the estimates determined by direct aggregation since all the individual panelist estimates can be equivalently represented by the 5th, 50th, or 95th percentiles (baseline methodology) or by the median, UEF and LEF (MEF approach). However, the means can be determined at several points in the analysis and differences do result from this choice. The baseline approach calculates the means at the earliest possible point in the analysis from the individual distributions while the MEF approach calculates the means as the last step in the analysis after aggregating all individual responses. The means could also be calculated after the bottom-line percentile estimates for each individual panelist are determined. It is postulated, however, that the means from the MEF approach will differ most from the baseline approach because the characteristics of the aggregated distributions are likely to be the most different from the individual elicitation response distributions due to the intermediate processing steps (Section 5). It should be noted that calculation of the group mean in this way also requires that the distributional form of group LOCA frequency estimates be assumed, which is counter to the objective (Section 5.4) of minimizing assumptions related to the bottom line parameters. Additional details and results of this sensitivity analysis are provided in Section 7.6.4.3.

5.6.4.4 Mixture-Distribution Aggregation –This is a method of expert opinion aggregation for probability distributions used in the NUREG-1150 study [5.10] which uses the arithmetic average of the panelists' probability distributions as the aggregate probability distribution. The method is described briefly on pages 3-6 of Reference 5.11. It is stated there that one of the principles used in the expert judgment process was that the aggregation of judgments from various experts should preserve the uncertainty that exists among alternative points of view.

The mixture-distribution approach assumes that the expert panel is a random sample from the population of all experts and that the goal is to obtain an unbiased estimate of the aggregate distribution function of the LOCA frequencies averaged over the population of all experts. For a panel of N experts, denote the cumulative distribution function for expert i by $F_i(x)$ and the group cumulative distribution function for the LOCA frequency by $G(x)$. Then

$$G(x) = \frac{1}{N} \sum_{i=1}^{N} F_i(x) \tag{5.47}$$

Equation 5.47 has the form of a mixture distribution where $F_i(x)$ is the probability that expert i would assign to the statement that the LOCA frequency X is less than x. $G(x)$ is the group estimate of the probability that the LOCA frequency X is less than x. The corresponding density functions are obtained by differentiating Equation 5.47 to obtain

$$g(x) = \frac{1}{N} \sum_{i=1}^{N} f_i(x) \tag{5.48}$$

Integrating Equation 5.48, the mean LOCA frequency for a given set of conditions (e.g., current-day, BWR, Category 1 LOCA frequency) is given by

$$\bar{x} = \int x g(x) dx = \frac{1}{N} \sum_{i=1}^{N} \int x f_i(x) dx = \frac{1}{N} \sum \bar{x}_i \tag{5.49}$$

As noted above, the panel of experts is considered a random sample of experts from the population of experts. Denoting the number of experts in the population by N_{all}, the population aggregate distribution is given by

$$G_{pop}(x) = \frac{1}{N_{all}} \sum_{j=1}^{N_{all}} F_j(x) \tag{5.50}$$

If the panel of N experts is a random sample from the population of N_{all} experts, $G(x)$ from Equation 5.47 is an unbiased estimate of $G_{pop}(x)$.

Note that no matter how large a sample is taken, there will always be large variability in the expert opinions. Mixture-distribution aggregation is a technique that attempts to capture the uncertainty expressed by this variability (see page 9.7 of Reference 5.12).

The mixture-distribution aggregation scheme will always result in higher mean and 95th percentile estimates and lower 5th percentile estimates than the other aggregation schemes (see Section 7.6.4.4). Consequently, the mixture-distribution scheme exhibits the greatest difference between the 5th and 95th percentiles. However, the mean and 95th percentile estimates are often dominated by the maximum individual estimates while the 5th percentile is often dominated by the minimum estimate. These characteristics imply that the extreme individual estimates will often dominate the mean, 5th, and 95th percentile estimates. Therefore, the mixture distribution may not represent central estimates of group opinion when the differences among the individual estimates are large.

Obtaining central estimates of group opinion was a principal objective of this elicitation (Section 2). Even if conservative central estimates of LOCA frequencies are required, the mixture-distribution scheme may not be appropriate. In such cases, either high confidence bounds (Section 5.4) or the quartile method (Section 5.6.5) could be used to provide conservative estimates of group opinion.

5.6.5 Panel Diversity

The baseline approach (Section 5.5) for quantifying panel diversity is to construct two-sided confidence intervals for the group estimates of the bottom-line LOCA frequencies parameters. As discussed in Section 5.4, this calculation assumes that the individual estimates are characterized by a lognormal structure. The baseline measures of panel diversity are 90% confidence intervals centered on the GM of the individual estimates.

5-25

An alternative approach is a non-parametric one, where no assumptions are made about the probabilistic structure of the individual estimates. For each set of individual panelist estimates, a quartile interval is constructed from the upper and lower quartiles of the individual estimates. The lower and upper quartiles represent the 25th and 75th percentiles, respectively, of the individual panelist estimates. In this quartile method, the median (or 50th percentile estimate) is an obvious measure of central group opinion. A sensitivity analysis comparing the confidence interval and quartile method is presented in Section 7.6.5. In Section 7.6.4.1, it is shown that the GM and median provide similar measures of central group opinion. Thus, for these elicitation results, a quartile interval is a measure of group diversity about the median just as the confidence interval measures diversity about the GM for the assumed lognormal structure.

5.7 References

5.1 Atwood, C. L., "Review of Draft Report on LOCA Frequency Estimates by Expert Elicitation," NRC ADAMS Accession No. ML051430327, May 2005

5.2 Atwood, C.L., LaChance, J.L., Martz, H.F., Anderson, D.J., Englehardt, M., Whitehead, D., and Wheeler, T., "Handbook of Parameter Estimation for Probabilistic Risk Assessment," NUREG/CR-6823, U.S. Nuclear Regulatory Commission, Washington, DC, September 2003.

5.3 Surowiecki, J., "The Wisdom of Crowds", Doubleday, New York, NY, 2004.

5.4 Meyer, M.A. and Booker, J.M., "Eliciting and Analyzing Expert Judgment: A Practical Guide," NUREG/CR-5424, U.S. Nuclear Regulatory Commission, January 1990.

5.5 Vo, T.V., Heasler, P.G., Doctor, S.R., Simenon, F.A., and Gore, B.F., "Estimates of Rupture Probabilities for Nuclear Power Plant Components: Expert Judgment Elicitation," *Nuclear Technology*, Vol. 96, pp. 259-271, 1991.

5.6 Vo, T.V., Simenon, F.A., Gore, B.F., and Livingston, J.V., "Expert Judgment Elicitation on Component Rupture Probabilities for Five PWR Systems," *Reliability and Risk in Pressure Vessels and Piping*, PVP-Vol. 251, pp. 127-140, American Society of Mechanical Engineers, 1993.

5.7 USNRC, "A Pilot Application of Risk-Informed Methods to Establish Inservice Inspection Priorities for Nuclear Components at Surry Unit 1 Nuclear Power Station ," NUREG/CR-1681, Rev.1, PNNL-9020, Rev.1, U.S. Nuclear Regulatory Commission, February 1997.

5.8 USNRC, "Technical Elements of Risk-Informed Inservice Inspection Programs for Piping," Draft Report, NUREG-1661, U.S. Nuclear Regulatory Commission, January 1999.

5.9 Von Winterfeldt, D. and Edwards. W., "Decision Analysis and Behavioral Research," Cambridge University Press, New York, NY, 1986.

5.10 "Severe Accident Risks: An Assessment for Five U.S. Nuclear Power Plants," NUREG-1150, U.S. Nuclear Regulatory Commission, December 1990.

5.11 USNRC, "Analysis of Core Damage Frequency from Internal Events: Expert Judgment Elicitation," NUREG/CR-4550, Vol. 2, U.S. Nuclear Regulatory Commission, April 1989.

5.12 Gorman, E.D., Breeding, R.J., Helton, J.C., Brown, T.D., Murfin, W.B., Harper, F.T., and Hora, S.C., "Evaluation of Severe Accident Risks: Methodology for the Containment, Source Term, Consequence, and Risk Integration Analyses," NUREG/CR-4551, Vol. 1, Rev. 1, U.S. Nuclear Regulatory Commission, December 1993.

6. QUALITATIVE RESULTS AND DISCUSSION

The detailed results from the elicitation exercise are presented in the next two sections of this report. In Section 6, the qualitative results (i.e., rationale and insights) are presented. In Section 7 the quantitative (i.e., numerical LOCA frequency estimates) are presented. The qualitative results presented herein focus on recurring insights for important technical issues which contribute to passive system LOCAs. These insights usually summarize individual responses, and do not necessarily represent a panel consensus or majority opinion. However, most insights were generally raised by more than two panel members, unless explicitly noted. Individual insights expressed by less than two of the participants are typically not presented. However, some of the more interesting individual opinions are presented with their associated quantitative results in Appendix L, where the detailed results are discussed. The bulk of the qualitative rationale contained in this section was obtained during the individual elicitations, although insights obtained from the three group panel meetings are also included. The insights from the individual elicitation sessions have been mined from the associated minutes, handouts, and written responses. Minutes were recorded during each interview session. In addition, the participants often provided handouts with their preliminary responses prior to their elicitations. Then, after each session, most participants provided more complete written rationale to accompany their final elicitation responses.

6.1 Base Case Evaluations

There were two basic approaches conducted to evaluate the base case conditions: review and assessment of operating experience and PFM analyses. Some broad opinions surfaced regarding each approach. The advantages of using an operating-experience-based approach are that it naturally captures the effects of all observed degradation mechanisms as well as including the actual mitigation, loading, and plant operating conditions experienced in the plants. As such, operating-experience analysis provides an accurate characterization of past precursor events (e.g., leaks). The perceived difficulties include the fact that it is difficult to extrapolate precursor event data to predictions of LOCA frequencies, especially for future operating periods.

A number of the panelists commented that they believe that the surface history predictions are accurate for assessing LOCA frequencies over the first 25 years of plant operation, but are somewhat skeptical of future operating period estimates. Future predictions can be difficult using operating experience because emerging and future degradation mechanisms are not contained within the passive-system failure data. Also, the use of operating experience can bias the predictions if data related to past problems which have already been successfully mitigated (e.g., IGSCC in BWRs) and infant mortality rates have not been adequately screened. The data must be carefully evaluated to avoid these types of analysis pitfalls. At least one panelist also commented that operating experience is not effective for inferring larger LOCA sizes since the events largely consists of partial cracking and low-leak rate events. Finally, a number of the panelists commented that only the reporting of actual leaking cracks is actually comprehensive within the database. The reporting of the non-leaking precursor events is likely underreported.

There are several perceived advantages of the PFM approaches. First, they are capable of predicting the future damage for both currently existing and newly occurring degradation mechanisms that have been adequately modeled. This is valuable in assessing the required action necessary to mitigate emerging mechanisms before their impact is fully evident in the passive-system failure data. Also, PFM is capable of predicting the possible evolution of leaking and non-leaking cracks to a larger LOCA sizes.

There are several disadvantages that panelists expressed when using PFM to predict LOCA frequencies. At least one panelist indicated that only a few degradation mechanisms are realistically modeled. Specifically, the accuracy of the models for predicting crack initiation and crack linking from multiple initiated cracks was questioned for SCC-type mechanisms. This potential limitation can lead to underestimates of service cracking frequency and predicted crack lengths and result in unconservative LOCA frequency estimates. However, several panelists countered that conservative design stresses and transients are usually applied in the PFM analyses that are not representative of actual service stress values. These stresses result in conservative LOCA frequency estimates. Finally, it was felt by some panelists that the existing PFM models are too simplistic. The input variables, assumptions, and models are usually idealizations whereas the variables which are most important to LOCA susceptibility are often very complex. While the intent is to make conservative idealizations in the analysis, this contention is sometimes difficult to prove.

As a result of these beliefs, most panelists chose to anchor their elicitation responses to one or both of the operating-experience-based base case analyses (Sections 3.5 and 4) for making current-day (i.e., 25 year) LOCA frequency estimates. For future year projections (i.e., 40 or 60 year estimates), a number of the participants augmented the operating-experience-based assessments with results from the PFM studies and associated sensitivity analysis to develop their responses. Several panelists also relied primarily on the results from PFM analysis for all their predictions, but benchmarked them in some way by the passive-system failure data. A more complete description of anchoring philosophy chosen by each panelist is found in Appendix K.

6.2 Safety Culture Effects

There were two basic principals underlying the panelists' safety culture evaluations. First, most panelists assumed and expected that regulatory oversight policies and procedures will continue to be used to identify and mitigate risk associated with plants with deficient safety practices. It was also anticipated that current regulatory oversight practices would continue to evaluate aging management and mitigation strategies with the objective of reasonably assuring that future plant operation and maintenance is sufficient.

As discussed in Section 3.4.2, there are several competing issues that can either positively or negatively affect the future utility and regulatory safety cultures that were considered. The safety culture elicitation question was structured so that the change in the future safety culture resulting from each issue was quantitatively assessed relative to the existing safety culture. Panelists were also asked to qualitatively identify the strongest factors. Many panelists also provided qualitative opinions on the existing US safety culture. The panelists agreed that the utility and regulatory safety cultures are highly correlated and that there are no real distinctions between regulatory and utility safety culture. The panel members also generally believe that the future overall safety culture will not differ dramatically from the current culture. In fact, most participants expected a small improvement due primarily to continued nuclear experience and technology advances. Some of the qualitative rationale supporting this weak trend follows.

There are several positive safety culture influences that are expected. Several panelists believe that deregulation will have a positive effect on safety culture by requiring the industry to adopt an asset management strategy to optimize continued plant operation. These panelists believe that favorable economic incentives are in place today that bode well for the future of nuclear power. Plants are therefore being viewed as assets that need to be maintained to maximize their investment potential. As a result, management realizes that proactive and periodic maintenance is preferable to the risk of an extended outage. Another positive factor for improving future safety is the large experience and knowledge base developed during the previous 25 years of operation. This information allows plant operators to focus on actual problems, and not just perceived or postulated problems. Utilities and regulators also now have risk-informed methods to use as a basis for identifying important areas of concern. However, one panelist was disturbed by his perception that some utilities do not want to invoke risk-informed practices unless there is an immediate economic payback. However, this was not a majority opinion.

Some panelists also expect the safety culture to improve over the next 15 years due to the increased awareness of the consequences of passive system degradation. Recent high-profile examples of vessel head degradation at Davis-Besse and CRDM cracking at a number of PWR plants have illustrated the potential economic ramifications associated with extended plant outage. Also, there is a realization that problems at one plant can affect the public's perception of the entire nuclear industry. One panelist commented that another high visibility event, similar to the Three Mile Island (TMI) accident, could result in a loss of public trust that would be difficult to restore.

There are also several negative safety culture influences that may temper these predicted improvements. Several panelists cautioned that focusing on economic performance can also be detrimental. Economic pressures to keep plants operating, especially at high capacity factors, may create a poor safety environment. Utilities could be tempted to forego maintenance and inspections unless there are near-term economic incentives. Another potential negative safety culture factor is the inevitable decommissioning of plants. As plants approach their decommissioning date, safety may suffer as the cost benefit for plant maintenance decreases. There are also a number of personnel issues associated with decommissioning that could negatively affect safety culture. The morale of the plant operating personnel may suffer due to job uncertainty. Furthermore, the most competent plant personnel may migrate to other industries first and new workers may be difficult to entice to the field. The resultant workforce may then be less competent, overworked, and unmotivated during the latter years of plant operations.

A number of panel members raised concerns about the currently aging nuclear workforce. Obviously, there is a need to attract young, energized new staff into the field. However, a potential stumbling block to satisfying this goal is the fact that the number of the universities with nuclear engineering programs has decreased. Another negative influence mentioned by several panelists is complacency. With a few notable exceptions, the nuclear industry has an excellent safety record over the past 25 years. Unfortunately, this good record may breed a false sense of security. These panelists are concerned that complacency could lead to an overall erosion of safety unless proper vigilance is maintained.

Some safety culture concerns mentioned during the elicitations are due to the influence of politics and economics on the regulatory environment. Some panelists commented that the safety culture could be adversely affected by the country's energy demands. If demand continues to increase, and supplies become stretched, there may be pressure to continue plant operations at the detriment of the safety culture. A few panelists also expressed concerns that political pressure may be applied to keep plants operating and people employed at the risk of safety. Research and plant oversight can also be affected by future budget cuts at the NRC.

In addition to these positive and negative safety culture influences, the panelists believe that some factors exhibit compensating features. One example is the role of industry in ASME code activities. A few panelists believe that strong industry participation benefits safety culture outright. However, others believe that this participation can be detrimental if regulatory opinions are underrepresented on important issues. This possible weakness is compensated by the NRC's ability to not endorse aspects of the code where important disagreements exist. The design and construction of new plants also has compensating factors. As new plants, with improved designs and materials are developed, the knowledge gained could benefit maintenance, repair, and replacement activities in the existing fleet. Conversely, resources could shift to the newer, more efficient plants, which could harm safety at the older, less efficient plants. Multiple versus single plant sites are also associated with compensating factors. The detriment of a multiple plant utility is that there is typically less oversight per plant. However, some potential advantages that the multiple plant utility have is that there is often more people reviewing important decisions, and there is an increased opportunity for reassignment of personnel as plants are decommissioned. This somewhat offsets the previously discussed concerns about deteriorating morale as a result of decommissioning.

Many panelists discussed issues related to current safety culture differences among plants as a function of design-type, U.S. attitudes, and plant management philosophy. For instance, a number of panelists expect that the plant-type will influences safety culture. Specifically, the BWR plant inspectors and operators are believed to be more experienced in identifying degradation and developing successful mitigation strategies than their PWR counterparts. This opinion results from the BWR experience in mitigating IGSCC in the 1980's. The implication is that the current BWR safety culture may be better than the PWR safety culture. However, many of the same panelists also believe that the PWR experience with emerging PSWCC issues may equalize the future safety culture.

Additionally, panelists with the most extensive international experience tend to view the current U.S. safety culture less favorably than the other panelists. Specifically, a number of these panelists believe that the current U.S. safety culture is weaker than in other countries. A concern is that common U.S. practice strives for the most expedient solution without addressing the root cause of the problem. A cited example of this practice was the U.S. industry's perceived delay in adopting the lessons-learned from earlier French CRDM cracking experience. These and other possible, absolute differences among the panelists are not reflected in the safety culture results since the elicitation questions assessed the future safety culture relative to the current safety culture. Hence, the elicitation results normalize any initial panelist differences. However, the effects of the current-day safety culture were both implicitly and explicitly assessed by the panelists to develop the current-day LOCA frequency estimates.

Many panelists also expressed the opinion that safety culture is a highly plant-specific factor. Most panelists believe that the industry is generally acting in a consistent, safety conscious manner. However, individual plants can deviate markedly from general industry philosophies and practices. The vessel head degradation at Davis-Besse was a commonly cited example. Some panelists argued that the LOCA frequencies at these less safety conscious plants could be greatly elevated (i.e., as much as 1 to 2 orders of magnitude higher for plants with particularly egregious safety culture) compared with the remaining population. However, a possible poor safety culture at a few plants would not affect either the median LOCA frequency or safety culture estimates, but it could affect the UBs of both distributions. Some panelists increased their UB estimates on the effect of safety culture on LOCA frequencies (Section 7.1) to account for this effect. It should be noted however that this concern is an indication of the possible effects and it does not imply that these conditions are in existence. Panelists provided no particular examples of existing egregious safety cultures. There is also an expectation, as mentioned previously, that regulatory actions using existing enforcement measures would diminish both the possibility and impact of safety culture deficiencies at particular plants. These UB safety culture estimates therefore serve as more of a warning for the agency to remain vigilant in addressing plant safety culture deficiencies.

A few panelists also commented that safety culture has a cyclical component as well. In their opinion, when passive system degradation is first identified, the safety culture improves somewhat as the industry identifies and then develops mitigation strategies. Then, once effective mitigation has been employed, the focus on safety wanes. None of the panelists expressing this believe were able to quantify the cyclic variations about the median safety culture. They also expressed that this effect is much smaller than the variability associated with the safety culture at individual plants.

The panelists were also asked to comment on the relationship between safety culture and LOCA size category. Generally, the panel members expressed the opinion that safety culture is only very weakly correlated with LOCA size, if at all. Small leaks tend to impact plant availability, so the plants have an economic incentive to guard against such events. However, a few panel members believe that plant safety is focused on preventing failure in the largest pipes due to their increased risk significance.

6.3 Passive System Failure Insights

In the sections that follow, some broad passive system failure insights are discussed. The insights are first segregated into issues associated uniquely with piping and non-piping LOCA frequency contributions. Then, insights related to general issues that contribute to the underlying LOCA frequencies are presented including aging mechanisms; the effect of component size; mitigation and maintenance; the effect of operating time; and estimation uncertainties.

6.3.1 Piping and Non-Piping LOCA Frequency Contributions

Table 6.1 indicates the major piping systems and non-piping subcomponents that provide the greatest LOCA frequency contribution for each LOCA size category. A LOCA of a given size can occur by either a complete break of the smallest component supporting that LOCA size, or a partial break of a larger component. Most panelists believe that complete failure of a smaller component is more likely than partial failure of a bigger component. Therefore most panelists expect that the smallest diameter piping system or subcomponent that could support a particular LOCA size or category is the dominant LOCA frequency contributor. This belief is usually apparent in Table 6.1.

The major BWR and PWR piping Category 1 and 2 LOCA contributors are the smaller diameter instrument and drain lines (Table 6.1). These lines are susceptible to a number of failure scenarios that the larger diameter lines are not susceptible to, e.g., mechanical fatigue of socket weld fittings. In addition, a flaw of a given size in the smaller diameter pipe is a larger percentage of the pipe circumference than an equivalent size flaw in a larger diameter pipe. Therefore, this flaw is closer to the failure flaw size in the smaller diameter pipe. Finally, these smaller diameter pipes are not subject to the same level of ISI as their larger counterparts, if they are inspected at all.

Table 6.1 Major Piping and Non-Piping Contributors to the Various Size LOCA Categories

LOCA Category	BWRs		PWRs	
	Major Piping Contributors	Major Non-Piping Contributors	Major Piping Contributors	Major Non-Piping Contributors
1	Small diameter instrument and drain lines	CRDM (i.e. stub tube) penetrations	Small diameter instrument and drain lines	Steam generator tubes, pressurizer heater sleeves, CRDM penetrations
2	Small diameter instrument and drain lines	CRDM (i.e. stub tube) penetrations	Small diameter instrument and drain lines	CRDM penetrations
3	Primary: Recirculation, Secondary: feedwater, SRV, RWCU, RHR, Core Spray	RPV nozzles, pump and valve bodies	CVCS, RHR, SIS (DVI), PSL, surge line	Nozzles and component bodies
4	Primary: Recirculation, Secondary: feedwater, SRV, RWCU, RHR, Core Spray	RPV nozzles, pump and valve bodies	Surge Line, SIS (Accumulator), RHR, Hot leg, SRV	Nozzles and component bodies
5	Recirculation, RHR	RPV, pump, and valve bodies	Hot Leg, Surge line, RHR	Manways and component bodies
6	N/A	RPV body	Hot Leg	Component bodies

CRDM = Control rod drive mechanism
RPV = Reactor pressure vessel
SRV = Safety relief valve line
RWCU = Reactor water cleanup system
CVCS = Chemical volume control system
RHR = Residual heat removal
SIS (DVI) ≈ Safety injection system (direct volume injection)
SIS (Accumulator) = Safety injection system (accumulator lines)
PSL = Pressurizer spray lines

The important non-piping contributors for Category 1 and 2 LOCAs are CRDM penetrations in both PWR and BWR plants. Additionally, in PWR plants, steam generator tubes and pressurizer heater sleeves are important contributors. The pressurizer and CRDM concerns stem from relatively recent emergence of SCC in Alloy 600 base and associated weld materials. This concern is currently at a peak because the mechanism is prevalent, yet mitigation strategies have yet to be fully implemented. The impact of PWSCC for PWR small-diameter non-piping components is expected to decrease over the next 10 to 15 years as effective mitigation strategies are developed and implemented. Steam generator tube failure is also an expected dominant contributor based on the historically high failure rates and the decreased degradation tolerance associated with these components since the design safety factors for these tubes are less than for small bore piping. However, many panelists believe that future failure rates for steam generator tubes will decrease due to improved inspection programs, steam generator replacement initiatives, and improved secondary side water chemistry.

There is relatively good agreement that the recirculation lines for Category 3 and 4 LOCAs in BWR piping are the dominant LOCA frequency contributor. There are lingering concerns about IGSCC in these larger diameter recirculation system lines, especially for those lines which have not been replaced. This selection is counter to the general trend that the smallest lines supporting a LOCA category are the biggest contributors. The safety relief valve (SRV) lines, reactor water cleanup (RWCU) systems, residual heat removal (RHR) lines, and core spray lines all are smaller than the recirculation lines, but they were considered to be of secondary importance for this BWR LOCA category. The feedwater line is relatively large as well, and its inclusion was driven by FAC concerns.

In contrast to the BWR piping assessment, there was no clear consensus expressed about the major contributing PWR piping systems and non-piping subcomponents for Category 3 and 4 LOCAs. This uncertainty stems from the large number of piping systems and non-piping subcomponents that contribute to these LOCAs, and the lack of clear operational differences and data suggesting that any one system or subcomponent could dominate. The nozzle and component body references in Table 6.1 refer to all nozzles (RPV, steam generator, and pressurizer nozzles) and/or all component bodies (RPV, steam generator, pressurizer, pumps, and valve bodies). Hence, the dominant non-piping contributors cannot be additionally refined.

Nozzle failures are a concern because system and transient stresses can be highest at these locations. Additionally, past degradation has been experienced in these locations. Valve and pump component bodies are a concern because they typically are fabricated from cast stainless steel materials which are notoriously difficult to inspect. These materials are also subject to thermal aging which reduces the fracture toughness of the material. Fortunately, no known cracking mechanism exists for these materials as of this date. Reactor pressure vessel concerns were due to either PTS for PWR plants or LTOP in BWR plants.

There are only a few BWR and PWR piping systems that can support Category 5 and 6 LOCAs, and there is subsequently greater consensus about the dominant systems. The surge line and RHR line are thought to be important based on past degradation and the relatively large transients that are possible during normal operation. The hot leg was also typically a greater concern than the cold leg for Category 6 LOCAs because of its higher operating temperatures. For Category 5 and 6 PWR non-piping contributors, there remains little agreement on the dominant subcomponent failure mechanisms. There is also no consensus about Category 5 BWR subcomponent failures and only BWR vessel failures lead to Category 6 LOCAs. Hence, by default, BWR vessel failures are the dominant Category 6 LOCA contributor.

It was almost universally expressed that the contribution to the overall LOCA frequencies is greater for the non-piping components than for piping for the smaller category LOCAs in PWR plants. Specifically, steam generator tube, CRDM, and pressurizer heater sleeve failures are expected to be the most important Category 1 and 2 total LOCA frequency contributors. These expected higher failure rates also result in the expectation that PWR LOCA frequencies are higher than BWR LOCA frequencies for Category 1 and 2 LOCAs. However, the non-piping component contributions to the larger (Category 3 - 5) LOCA frequencies decreases dramatically due to the robustness (e.g., increased design margin) of these non-piping components compared to piping. Several panelists expect that non-piping contributions become significant again for the Category 6 LOCAs because the operating margins are similar to the largest piping systems.

Most panelists also believe that the quantitative non-piping LOCA frequency assessments are more challenging than the piping assessments. For non-piping, there are multiple components to consider, each with different operating requirements and characteristics. The associated design margins, materials, and inspection considerations also vary widely. Furthermore, there is little precursor data available on non-piping component degradation due to the historical focus on piping reliability.

6.3.2 Important Aging Mechanisms

Most panelists believe that precursor events (e.g., cracks and leaks) are a good barometer of LOCA susceptibility even if it is not straightforward to extrapolate precursor events to larger LOCA sizes. Operational experience provides an indication of the historical precursor frequency associated with degradation mechanisms in the operating fleet. Therefore, almost all panelists anchored their responses against available passive-system failure data. In addition, a number of the participants used the weld census data to determine the relative LOCA contributions for piping systems with similar operating characteristics. In this approach, the LOCA frequency ratio of two systems is identical to their ratio of LOCA-susceptible welds. For example, assume that a panelist believes that PWSCC is an important degradation mechanism and PWSCC susceptibility is equivalent for both the hot leg and surge line. Also assume that the surge line has a single PWSCC-susceptible weld while the hot leg has 7 PWSCC-susceptible welds. The hot leg PWSCC contribution is therefore a factor 7 higher than the surge line PWSCC contribution. This application of weld census information is a natural approach because many mechanisms, including fatigue and SCC, preferentially attack welds instead of the base metal. Reasons for this preferential attack are related to the weld metallurgy and chemistry, the relatively high residual stresses in the weld vicinity, and the higher weld defect density.

A discussion of the most significant aging mechanism identified by the panelists in BWR and PWR plants follows. Most participants identified thermal fatigue, FAC, IGSCC, and mechanical fatigue as the important degradation mechanisms to consider in BWR piping. The BWR plants are expected to be more prone to thermal fatigue problems compared with the primary side of PWR plants because they experience greater temperature fluctuations during the normal operating cycle. In BWR plants, thermal fatigue remains an important contributor for the feedwater lines and the RHR system. There was a rash of feedwater nozzle cracks reported in the 1970 to early 1980 time period in BWRs. Plant and system modifications were implemented after a detailed study of the problem and augmented inspections are being conducted based on NUREG-0619 [6.1] requirements. These mitigation measures have proven effective as no new thermal fatigue cracks have been discovered in these BWR feedwater nozzles over the last 20 years. Additionally, the U.S. BWR plants have been less susceptible to thermal fatigue damage than some of the foreign designs. However, thermal fatigue is an aging mechanism that could lead to a large LOCA because it does not manifest itself as a single crack, but as a family of cracks over a wide area. Thermal fatigue cracks also tend to propagate rapidly, and since it is not material sensitive (i.e., it can attack a number of materials), it is difficult to prioritize critical areas for inspections. These reasons explain why thermal fatigue is still regarded as an important LOCA contributor by many panelists.

The carbon-steel feedwater piping system in BWRs is the most susceptible to FAC of all the primary side (i.e., LOCA-sensitive) components (Section 3.4.5). The main steam line is the other major carbon-steel primary side system which experiences constant fluid flow. However, it is not as susceptible to FAC as the feedwater system because the erosion rates associated with two-phase flow are much less. While FAC caused a serious accident in the secondary side piping at Surry some 15 years ago, most panel members believe that the industry has inspection programs in place today to prevent the reoccurrence of such an event, especially in the primary side piping systems. However, one panel member expressed the concern that the water chemistry improvements (hydrogen water chemistry – HWC) that mitigate IGSCC could lead to FAC in unanticipated locations that are not monitored as part of these inspection programs.

Most BWRs that have implemented HWC routinely inject oxygen into the feedwater line to mitigate this possibility. In fact there has been virtually no FAC in BWR feedwater piping since the advent of the feedwater chemistry specifications with elevated oxygen levels. FAC and elevated oxygen do not coexist – the piping maintains a protective oxide coating, thus protecting the steel below. Therefore, FAC degradation can only result from interruptions in oxygen injection or excessive hydrogen injection could potentially result in the onset of FAC. Current U.S. instances of FAC in BWR piping have been confined to Class 3 service water systems and do not directly affect the LOCA frequencies calculated herein. Thus, many panelists believe that a properly designed and monitored HWC program should not result in higher FAC susceptibility, and their LOCA concerns due to FAC are minimal. Only one panelist remains concerned about the LOCA susceptibility due to FAC degradation.

The panel consensus is that the susceptibility of BWR piping systems to IGSCC is greatly reduced compared to what it was in the past. Measures such as improved hydrogen water chemistry, weld overlay repairs, stress relief, and pipe replacement with more crack resistant materials have led to this reduction. Inspection quality has also improved such that the probability of crack detection is much better than in the past. However, as indicated earlier, there remains concern about the failure likelihood of the large recirculation piping and the RHR lines that have not been replaced. The original piping materials are much more susceptible to IGSCC and many lines retain preexisting cracks that initiated and grew before HWC was adopted. Furthermore, at least one panelist is also concerned that the more IGSCC-resistant replacement piping materials may still crack under service conditions. This panelist cited the German plant experience with cracking in Type 347 stainless steel. Also, it is possible that cold work (e.g. due to grinding) could increase the IGSCC susceptibility of the low carbon (L grade) stainless steel that has been used as a replacement material in many U.S. plants. However, the U.S. BWR experience with L grade stainless steel piping has been very good thus far. For these reasons, many panelists believe that continued vigilance is required through the augmented inspection requirements in Generic Letter 88-01 [6.2] and NUREG-0313 [6.3].

Another aging mechanism of concern in BWR plants is mechanical fatigue. This is primarily a problem in smaller diameter piping, especially those with socket welds, and is caused by an adjacent vibration source. It was noted that locations susceptible to mechanical fatigue damage are not always obvious and that it is impossible to eliminate all plant vibrations. Another concern is that plant configuration changes can result in newly susceptible areas. This mechanism is also a prevalent root failure cause of small diameter piping in the passive-system failure data.

Lastly, it was noted by some panelists that certain BWR piping systems (e.g., relief valve lines) may have an increased likelihood of operating transients (e.g., water hammer) compared with similar PWR systems. This is primarily due to increased valve openings during plant operation. These transients, combined with existing degradation, could make the BWR plants more susceptible to LOCAs. However, this is not a universal sentiment. At least one panelist strongly disagrees with this contention and believes that water hammer is a design-specific issue. This panelist believes that while some BWR RPV head cooling layout designs could be susceptible to water hammer, it is primarily a concern with secondary side systems and some secondary side standby safety systems. Furthermore, this panelist argues that even though relief valve lines in the main steam systems do get exposed to transient loads, they are designed for these loads.

Most participants identified thermal fatigue, mechanical fatigue, and PWSCC, as the important degradation mechanisms to consider in PWR piping. The concerns associated with thermal and mechanical fatigue in PWR plants are similar to those previously discussed for BWR plants. The locations susceptible to thermal fatigue in PWR plants include the surge line which is subject to cyclic thermal stratification stresses. Also, as identified in Table 6.1, concerns were expressed about the DVI and chemical volume and control system (CVCS) lines due to periodic testing of these lines which imposes additional thermal cycles. The DVI and accumulator safety injection system (SIS) lines have also experienced some thermal fatigue cracking due to cold water leakage past the check valves.

Stress corrosion cracking of nickel-based alloys is a relatively new mechanism that has manifested itself in U.S. plants over the last 5 years. It is a temperature dependent mechanism that attacks Alloy 600 type base materials and Inconel 82/182 welds. It has many similar characteristics to previous IGSCC cracking in sensitized stainless steels in BWR reactors. Many panel members believe that the PWSCC problems in PWRs will be resolved (i.e., mitigated) over the next 15 years. Therefore, its contribution to the overall LOCA frequencies may peak sometime over the next several years, but then decrease after that. To date, instances of PWSCC in piping systems have been observed in surge lines at the surge line-to-pressurizer weld in the United States at Three Mile Island, as well as in plants in Belgium and Japan. The hot legs have also experienced PWSCC at the hot leg-to-reactor pressure vessel weld in the United States at the V.C. Summer plant and in Sweden at the Ringhals plant. Other piping systems where PWSCC could surface in the future are the cold leg and the pressurizer spray lines in PWRs. However, since the cold leg, and many of the BWR systems, operate at slightly lower temperatures than the hot legs and surge lines in PWRs. Problems in these lower-temperature systems may not materialize until later in the operating life of the plant.

Many of the same degradation mechanisms existing for piping are important for non-piping components as well. For example, SCC of nickel-based alloys, as described above for piping systems, is prevalent in smaller Alloy 600 components, such as steam generator tubes, CRDMs and other penetrations. Similarly, thermal fatigue is an important consideration at nozzle inlets and other locations which experience thermal stratification. However, there are also several degradation mechanisms that are only a concern in certain non-piping components. These include radiation embrittlement which reduces the fracture toughness of the RPV, especially in PWRs where there is less shielding. In addition, steam generator tubes, which were generally cited as a major small break LOCA contributor in PWRs, are susceptible to a variety of unique degradation mechanisms, including fretting and wear and denting from secondary side contamination.

There was also more concern expressed about the possibility of common cause failures in non-piping components. Multiple steam generator tube or CRDM failures could result from the failure of a single component due to the proximity of other components. Multiple steam generator tubes can also fail due to a sudden secondary side pressure drop if multiple tubes are sufficiently degraded. Also, bolting failures are only expected to lead to a LOCA if multiple bolts fail due to common causes, such as improper installation and inspection, or the emergence of degradation mechanisms such as steam cutting or boric acid corrosion which affect multiple bolts.

The panelists were also asked to assess the impact of future degradation mechanisms that may not have been apparent in the operating experience. A number of the panelists indicated that if a new degradation mechanism arises in the future, the LOCA frequency may temporarily increase until the industry has developed an effective mitigation strategy. After this mitigation strategy has been fully implemented, the LOCA frequencies will return to values applicable before the manifestation of the new degradation mechanism. The severity of any increase, and the time necessary to implement effective mitigation was often benchmarked by the panelists against experience with known degradation mechanisms (e.g., IGSCC or FAC) using either passive-system failure data, PFM modeling, or both.

6.3.3 Effect of Component Size on LOCA Frequencies
The panelists generally believe that smaller LOCAs are more likely than larger LOCAs in piping because small piping has a greater failure propensity. As previously mentioned (Section 6.3.1), the panelists generally believe that complete rupture of a smaller pipe is more likely than an equivalent size opening in a larger pipe. A given size crack is a larger percentage of the pipe circumference in the smaller diameter line, but cracking length and likelihood is not expected to vary with piping size. Smaller piping is also often subject to fabrication flaws which exacerbates this decreased failure margin. Additionally, smaller diameter lines are often fabricated from socket welded pipe which has a history of mechanical fatigue damage from plant vibrations and is also susceptible to external failure mechanisms arising from human error (e.g., damage from equipment). Finally, small piping is typically more difficult to inspect and ISI is

not routinely performed on these lines. In contrast, the larger diameter lines are inspected more rigorously and routinely and quality control/quality assurance programs are more stringent as piping size increases. One outcome of the increased failure propensity of smaller diameter lines is that the industry is experienced in dealing with failures, repair, and replacement of the smallest diameter lines such that the mitigation of small pipe damage is relatively routine.

A number of the panelists expressed the similar opinions that the smallest applicable non-piping component will contribute the greatest risk to each LOCA category and that smaller non-piping LOCAs are much more likely. As discussed previously, there are a number of LOCA-sensitive small diameter non-piping PWR components, i.e., steam generator tubes, pressurizer heater sleeves, and CRDMs, that have all experienced either damage or failure in service. Therefore, these components should provide the largest PWR Category 1 and 2 LOCA frequency contributions. In contrast, some larger non-piping components (e.g., vessels) are fabricated with more stringent controls while other larger non-piping components (e.g., valve and pump bodies) have bigger design margins than pipes. Hence non-piping failures are generally expected to be less likely than piping failures for the larger LOCA Categories. The main residual concern with large non-piping components is that inspections are difficult and not usually performed on these components. Therefore, there is more uncertainty about the level of degradation in these components.

A number of panelists believe that aging degradation may have the greatest effect on intermediate diameter (6 to 14-inch diameter) piping systems because of the small and large component trends previously discussed. As discussed, the industry has a greater tendency to replace the smallest diameter piping as problems arise and the failure rates associated with these components are likely to be unchanged in the future. Conversely, the largest piping has increased leak-before-break margin, and is subject to the highest quality inspection programs due to the high failure consequences associated with this piping. Furthermore, there is a large number of piping systems containing intermediate size piping with an associated large number of degradation-sensitive welds. Therefore, the greater future LOCA concerns are often associated with intermediate diameter piping.

6.3.4 Influence of Mitigation and Maintenance Actions

Timely and proper maintenance and mitigation practices were almost always considered by the panelists to result in decreased LOCA susceptibility. Practices such as improved ISI techniques (e.g., eddy current inspection programs for steam generator tubes), greater use of risk-informed ISI, new materials, and pipe and component replacement programs were all mentioned as having a positive influence on the estimated LOCA frequencies. Both steam generator and RPV head replacement programs were cited as effective strategies that the industry is undertaking for reducing LOCA susceptibility. Also, secondary side water chemistry improvement programs were mentioned as positive measures for decreasing steam generator tube degradation.

However, there are several potential detrimental aspects associated with any mitigation or maintenance practices. For one, the frequent opening and closing of systems for inspections can cause problems. There is also the increased likelihood for human error if components are reassembled improperly or artifacts (e.g., tools and loose debris) are left within the component after servicing. Also improper maintenance activities associated with active system components (such as a valve) could cause increased loading in a passive system component, and thus greater failure likelihood.

Although ISI programs are clearly beneficial, a number of participants expressed reservations about the viability of ISI programs for small diameter piping systems, for piping systems and components fabricated from cast stainless steels, and for welds with limited accessibility for inspections. Inspection of smaller piping may not be as viable because degradation can initiate and lead to failure in some instances even within inspection cycles. The concern with cast stainless steel components (i.e., pump and valve bodies) is that there still is no proven technique for inspecting these materials. In addition, these materials are subject to thermal aging which can reduce the fracture toughness. Fortunately, no service-induced crack initiation mechanism has been identified for these materials or else this issue would become much more serious. Limited accessibility locations are a concern because inspection quality may suffer or an inspection may not even be possible. An example of a limited accessibility location is the surge line-to-pressurizer bimetallic weld in Westinghouse PWRs. This particular weld also has experienced service cracking and is particularly susceptible to PWSCC due to the high operating temperature of this location.

As the industry moves toward greater risk-informed inspection (RI-ISI) criteria, many panelists expressed caution with this approach. The philosophy of inspecting risk-important locations is generally regarded to be sound. However, inspection locations are largely based on experience. As a result, it is imperative that early signs of degradation in new locations are aggressively addressed and incorporated into RI-ISI programs. Some panelists are concerned about relying too heavily on risk-informed regulation and eschewing good engineering practices (i.e., proactive maintenance, inspection, and material replacement programs) if the associated risk are expected to be small.

There were a few maintenance and mitigation issues discussed that are unique to non-piping components. There was some concern expressed that maintenance and inspection programs of the large component bodies (pressurizer, steam generator, RPV) is not as rigorous as for piping systems. For smaller non-piping components such as steam generator tubes and CRDMs, improved inspection methods and mitigation programs are expected to reduce future failure frequencies. For steam generators, improved eddy current techniques have been incorporated in inspection programs and more degradation resistant materials are replacing older materials. Also, as mentioned above, better water chemistry control is possible for the secondary side of the steam generator. Similarly for CRDMs, ongoing head replacement programs with more resistant Alloy 690 materials and better and periodic inspection programs are being followed to minimize the occurrence of further PWSCC cracking.

Mitigation and maintenance practices are also generally expected to help uncover the emergence of new degradation mechanisms before the risk impact becomes severe. Additionally, as mentioned in Section 6.3.2, a number of the panelists expect that targeted mitigation strategies will be developed and implemented once the mechanism is identified. The LOCA frequencies may be higher than historical values in the time period between the discovery and mitigation of the degradation mechanism. After mitigation is fully implemented, the same panelists expect the frequencies to return to values applicable before the manifestation of the new degradation mechanism.

6.3.5 Effect of Operating Time on LOCA Frequencies

Unabated aging mechanisms typically lead LOCA frequency increases with continued operating time. However, nearly all panelists expect that the NRC and industry will aggressively respond to emerging mechanisms. Some historical examples cited to support this contention include the IGSCC cracking issue in BWR plants in the late 1970s and early 1980s and the PWSCC issue currently being addressed in PWR plants. Overall, the participants believe that maintenance and mitigation will offset the natural tendency for LOCA frequencies to increase due to aging. Most of the participants think that the various compensating factors will balance such that the LOCA frequencies will remain relatively constant over the future operating period, up to 35 years from today.

Furthermore, there are several factors which will promote LOCA frequency decreases in the future. A number of the panelists expect that inspection techniques will continue to improve and result in greater knowledge of existing plant degradation and allow degradation evolution to be more accurately monitored. This will allow the effectiveness of mitigation strategies to be more effectively evaluated. Other mitigation strategies are expected to improve and mature as well. In addition, the infant mortality rates typically associated with new systems and technologies have already occurred. Thus, the fabrication-related defect failure rates should continually decrease in the future.

Conversely, the panelists identified several factors associated with degradation mechanisms which promote future LOCA frequency increases. Thermal fatigue, SCC, and FAC challenges are all expected to increase. Thermal fatigue issues will increase near the end of the plant license-renewal period from the high usage factors at many plant locations. One panelist expressed concern about future IGSCC failure rates from potential sulfate increases resulting from infrequent ion-exchange filter replacement. Also, over the next 10 to 15 years, a number of the participants believe that PWR LOCA frequencies will increase until appropriate PWSCC mitigation strategies have been developed and implemented. Once these strategies are in place, then the associated LOCA frequencies would naturally diminish. Panelists also tempered their expectations about possible future LOCA decreases based on the possibility of new degradation mechanisms emerging. While no specific mechanisms were identified, operating experience has shown that previously unseen mechanisms have periodically surfaced in the fleet. The expectation of many panelists is that new mechanisms will continue to arise periodically and present new LOCA challenges.

6.3.6 Uncertainty in LOCA Frequency Estimates

The panel members generally expressed greater uncertainty in their predictions as the LOCA size (or category) increases and as the future operating time period increases. These trends are natural because assessment of larger and more distant LOCAs requires greater extrapolation of existing passive-system failure data. There was also generally more uncertainty associated with the PWR LOCA frequency assessments. One reason is that a number of panelists believe that the uncertainty associated with BWR piping is less than with PWR piping. The important BWR LOCA issues largely remain IGSCC in the future, but the BWR plants have more experience mitigating these degradation effects. Similar experience does not exist for PWR plants, which leads to greater uncertainty about the possible effects of degradation. However, there is also an expectation that PWR plants will gain experience from addressing current PWSCC piping concerns. At this point in time, the associated PWR and BWR uncertainties may again be similar. Another reason for increased PWR uncertainty is that PWR plants contain more LOCA-sensitive non-piping components. Therefore, the uncertainty associated with assessing their LOCA frequency contributions is naturally greater.

6.4 References

6.1 Snaider, R., "BWR Feedwater Nozzle and Control Rod Drive Return Line Nozzle Cracking," NUREG-0619, April 1980.

6.2 Generic Letter 88-01, "NRC Position on IGSCC in BWR Austenitic Stainless Steel Piping," January 25, 1988.

6.3 Hazelton, W. S., "Technical Report on Material Selection and Processing Guidelines for BWR Cooling Pressure Boundary Piping," NUREG-0313, Rev. 2, June 1986.

7. QUANTITATIVE RESULTS

Each panelist provided quantitative answers to the elicitation questions using the framework described in Section 3. The general approach and philosophy used by each panelist in developing their responses is provided in Appendix K (General Approach and Philosophy of Each Panelist). Each panelist's quantitative elicitation responses can be found through the "Electronic Reading Room" link on the NRC's public website (http://www.nrc.gov/) using the Agencywide Documents Access and Management System (ADAMS). The document is found in ADAMS using the following accession number: ML080560005.

The analysis procedures of Section 5 were then applied to process the results from each panelist and aggregate the individual results to generate LOCA frequency estimates which reflect the estimates of the entire panel. As described in Section 5.3.1, each panelist's input (Section 5.2) is processed using the assumed split lognormal structure of each response to obtain frequency estimates for each contributing piping system and non-piping sub-component. These individual contributors are then summed (Section 5.3.5) to obtain total LOCA estimates for each panelist corresponding to BWR and/or PWR plant types as appropriate. The individual estimates are then aggregated as described in Section 5.4 to estimate central group opinion and evaluate panel diversity. This analysis produces the baseline LOCA frequency estimates as described in Section 5.5. A number of sensitivity analyses are then performed as described in Section 5.6 to examine the robustness of these results.

This section provides the principal LOCA frequency estimates resulting from the expert elicitation process and the analysis procedures described in Section 5. Appendix L (Detailed Results) contains additional detailed results and discussion, including the LOCA frequency estimates for each panelist. These estimates are also supported by the qualitative technical insights in Section 6. These qualitative insights are generally not replicated in this section, although it is necessary to understand them to evaluate the quantitative results and trends. For example, the underlying damage mechanisms (Section 6.3.2) are fundamental to the quantitative BWR and PWR LOCA frequencies. The locations of other supporting qualitative rationales in Section 6 are referenced whenever possible in this section.

The results have been partitioned to examine the effects of key variables in various subsections. The effects of safety culture on the LOCA frequencies are described in Section 7.1. The total LOCA frequencies are documented along with the relative piping and non-piping contributions and the effect of operating time in Sections 7.2 to 7.4. While the elicitation structure defined the LOCA sizes by leak rate categories, the LOCA categories are correlated to break sizes as described in Section 3.7 and the results are presented with respect to break sizes in Section 7.2. Correlating the LOCA flow rate categories to physically equivalent break diameters is helpful for identifying major system and component contributors. Nevertheless, the results for each applicable LOCA category include contributions for all the variously-sized systems and components. Estimates of SGTR frequencies are provided in Section 7.8 and compared with results from prior studies.

Panelist uncertainties and variability among the panelists' responses are discussed in Section 7.5. In addition, numerous sensitivity analyses were performed to examine the robustness of the analysis techniques and to understand the impact of various analysis choices on the calculated LOCA frequencies. The results of these analyses are presented in Section 7.6. Summary results that are consistent with the elicitation objectives and structure and are reasonably representative of the panelists' quantitative judgments are presented both for the total LOCA frequencies (Section 7.7) and the PWR SB LOCA frequencies without SGTR contributions (Section 7.8). The summary results are compared to the LOCA frequency estimates developed in previous studies (Section 7.9) and the SB LOCA estimates are compared to operating-experience estimates in Section 7.10.

Note that many figures in this section use straight lines to connect the plotted points associated with adjacent LOCA categories. This graphical aid was added merely for visual clarity and not to imply a linear relationship between adjacent LOCA categories. The elicitation estimated LOCA frequencies only at each of the six defined LOCA categories (Section 3.4.1). Interpolation of frequencies between LOCA category sizes can be done at the user's discretion depending on the conservatism required by the application. Some common interpolation schemes are linear, multi-point nonlinear and cubic spline. A step-wise or stair-step interpolation between two categories where the frequency for the lower category size is used for all flow rates or corresponding break sizes between the two categories provides the most conservative interpolation scheme. Note that any interpolation scheme does not reflect the uncertainty in the interpolated frequencies.

7.1 Safety Culture

There were two basic tenets that guided the panelists' safety culture quantitative evaluations. First, most panelists assumed and expected that regulatory oversight policies and procedures will continue to be used to identify and mitigate risk associated with plants having deficient safety practices. It was also anticipated that current regulatory oversight practices would continue to evaluate aging management and mitigation strategies with the objective of assuring that future plant operation and maintenance are sufficient. Therefore, the analysis and evaluation of safety culture required the assessment of the overall, future effectiveness of regulatory practices in determining MV estimates of safety culture trends. The UB estimates typically represent the effect of inadequate safety culture and regulatory oversight at a few (or possibly one) specific plants.

Figures 7.1 and 7.2 illustrate the effects of the industry and regulatory safety cultures, respectively, on the ratios of the LOCA frequency in the future to the LOCA frequency at 25 years (i.e., LOCA Ratio). The Category 1 LOCA ratio results are shown using box and whisker plots[1]. Similarly, Figure 7.3 shows the effect of the industry safety culture on the LOCA ratio for Category 4 LOCAs. Only the Category 4 LOCA results are shown because they are representative of the results for all larger LOCA categories. Qualitative information provided by the panelists supporting the results in these figures is provided in Section 6.2.

In Figures 7.1 to 7.3, LOCA ratios less than 1.0 indicate an expected LOCA frequency reduction as a result of safety culture improvements. The median ratios in all three figures are equal to 1.0 and the interquartile ranges (represented by the shaded boxes) are all bounded by 0.5 and approximately 1.0. This result implies that the panel members generally expect that the safety culture will either improve slightly, or remain essentially constant up to the end of the plant license-renewal period. Also, the majority of the panelists expect that any future LOCA frequency reduction will be less than a factor of 2. Only some panelists expect that the smaller LOCA categories will significantly benefit from an improved safety culture in the future. However, there is nearly universal agreement that larger LOCA categories will not be significantly affected (within a factor of 2) by future safety culture changes. Nearly all panelists also expressed the opinion (Section 6.2) that the industry and regulatory safety cultures are highly positively correlated and will remain this way in the future.

---

[1] For a discussion of how to interpret the box and whisker plots in Figures 7.1 through 7.3, the reader is referred to Appendix L.

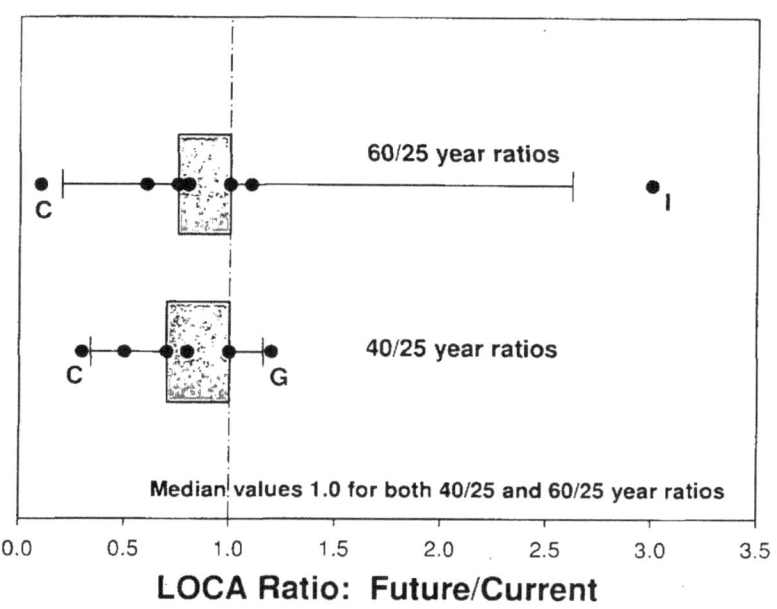

Figure 7.1 Effect of Industry Safety Culture on Category 1 LOCAs

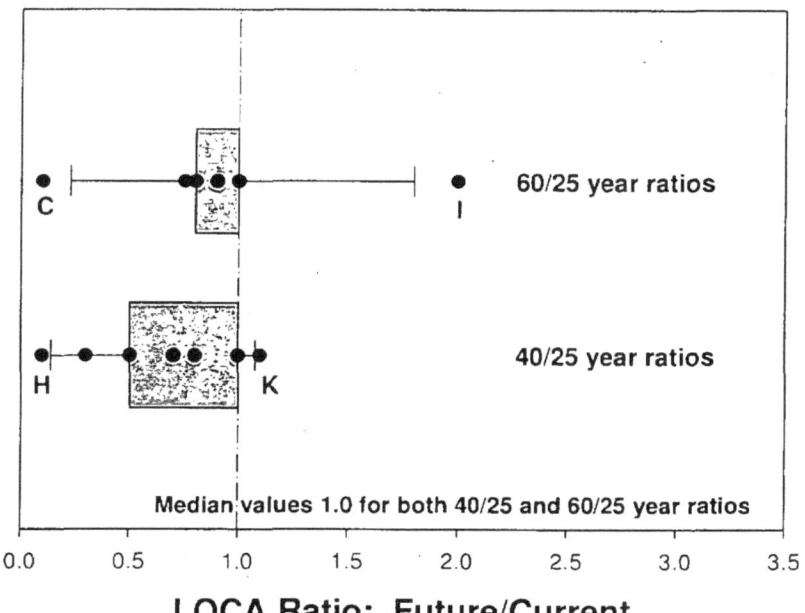

Figure 7.2 Effect of Regulatory Safety Culture on Category 1 LOCAs

7-3

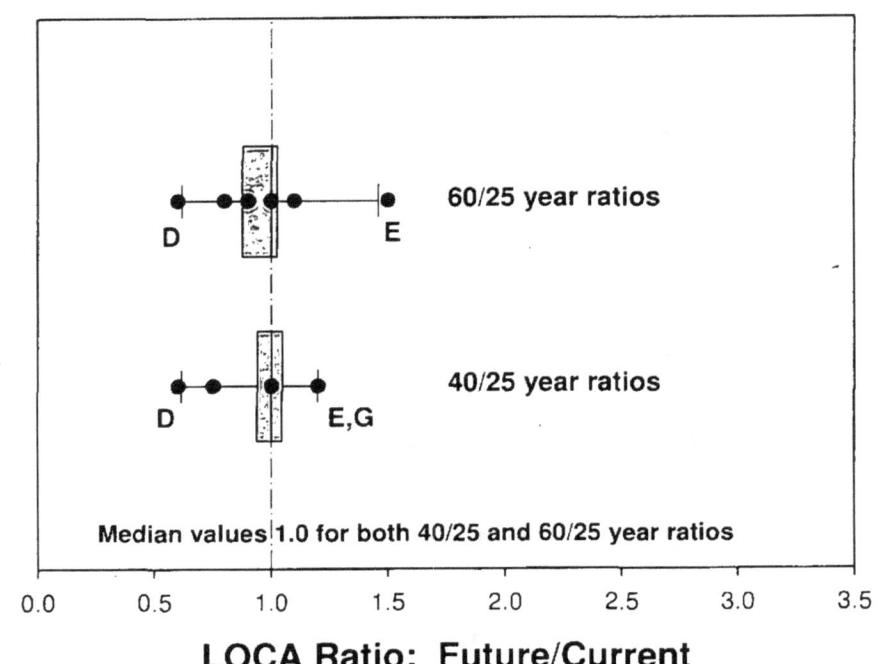

Figure 7.3 Effect of Industry Safety Culture on Category 4 LOCAs

There are several positive and negative factors that counteract to result in the expected stability in the regulatory and industry safety cultures. Factors that the panelists generally believe improve safety culture include deregulation; an increased awareness of the consequences of passive system degradation; additional industry nuclear experience; and improved technology. The positive factors are largely balanced by countervailing panelist concerns that could increase the LOCA frequencies. These concerns include a possible focus on economic performance instead of plant safety; the eventual decommissioning of plants; the aging nuclear workforce; complacency resulting from the generally good nuclear safety record; and economic and political pressures to maintain operations. Other issues evaluated have both positive and negative aspects that tend to cancel. More discussion on these issues is contained in Section 6.2. While the median group opinion was that safety culture would slightly improve, some panelists were more pessimistic and predicted higher LOCA ratios near the end of the plant license-renewal period (i.e., the 60/25 year ratios) when compared with the 40/25 year ratios in Figures 7.1 through 7.3.

The most serious general concern is that LOCA frequencies at a safety-deficient plant could be significantly increased by its operating philosophy (Section 6.2). The vessel head degradation that occurred at the Davis-Besse plant was frequently cited as an example of this effect (Section 6.2). Several panelists expressed the view that safety culture deficiencies at a plant could increase the LOCA frequencies at that plant by a factor of 10 or more. Such safety culture deficiencies at only a small number of plants would not affect the median, or generic, LOCA frequency estimates but could affect the upper tails of the distributions. The UB safety culture effects provided by a few panelists did reflect this possible increase. As discussed in Section 6.2, there is an expectation that regulatory actions using existing enforcement measures would lessen both the possibility and the impact of safety culture deficiencies at particular plants. However, the panelists recommended continued vigilance to ensure that deficient plant safety cultures do not develop.

In summary, the three principal conclusions from the safety culture elicitation questions are that (1) safety culture effects on the future generic LOCA frequencies are expected to be minimal; (2) industry and regulatory safety cultures are highly correlated; and (3) plant-specific safety culture can significantly affect that plant's LOCA frequencies and therefore vigilant regulatory oversight should be maintained. **Because of these findings, no modification or adjustment was applied when calculating the** *generic* **LOCA frequencies presented in this study.** This decision was endorsed by the elicitation panel. However, it is recognized that these generic results are only applicable provided the plants continue to operate within the context of existing regulations and their licensing basis.

7.2 Total BWR and PWR LOCA Frequencies (Baseline Estimates)

The total (piping plus non-piping) BWR and PWR passive system LOCA frequencies are provided in Table 7.1. These frequencies are calculated using the baseline analysis framework described in Section 5.5. The medians, means, 5th (5th Per.) and 95th percentiles (95th Per.) are the GMs of the panel members' individual total BWR and PWR LOCA frequency estimates as described in Section 5.4. Estimates are provided for the current-day frequencies (25 year average plant life) and end-of-plant-license frequencies (40 year average plant life). Generally, the 95th percentiles for both the BWR and PWR plants are between a factor of 2 to 4 higher than the mean values. The LOCA flow rate threshold associated with each LOCA category is also provided in Table 7.1.Recall that, as described in Section 3.4.1, the LOCA categories are defined in terms of cumulative thresholds rather than binned intervals because the thresholds are more consistent with the various PFM and operating-experience analyses that the panelists performed.

Figure 7.4 depicts the mean and 95th percentile, current-day (25 years of plant operations) BWR and PWR LOCA frequency estimates from Table 7.1. As illustrated, the current-day PWR Category 1 results are approximately an order of magnitude higher than the BWR Category 1 results. This reflects the contribution of the SGTRs which is the dominant LOCA contributor for PWR Category 1 LOCAs (Section 6.3.1). Conversely, for the current-day Category 5 LOCAs, the BWR results are approximately an order of magnitude higher than the PWR results. This reflects the continued risk associated with IGSCC for the BWR recirculation system (Section 6.3.2). The BWR frequencies decrease precipitously for the Category 6 LOCAs because only a catastrophic rupture of the RPV in BWR plants can result in a primary LOCA of this size[2] For the other LOCA categories, the total LOCA BWR and PWR frequencies are generally comparable and differ by a factor of 5 or less.

---

[2] As noted in Section 3, breaks can occur in either LOCA sensitivity piping or non-piping systems and components. No single failure within a BWR primary piping system could cause a Category 6 LOCA.

Table 7.1 Total BWR and PWR LOCA Frequencies

Plant Type	LOCA Size (gpm)	Eff. Break Size (inch)	Current-Day Estimate (per cal. year) (25 years fleet average operation)				End-of-Plant-License Estimate (per cal. year) (40 years fleet average operation)			
			5th Per.	Median	Mean	95th Per.	5th Per.	Median	Mean	95th Per.
BWR	>100	½	4.3E-05	3.0E-04	5.6E-04	1.7E-03	3.5E-05	2.6E-04	5.3E-04	1.7E-03
	>1,500	1 7/8	4.0E-06	5.0E-05	1.1E-04	3.7E-04	3.3E-06	4.5E-05	1.0E-04	3.6E-04
	>5,000	3 ¼	8.0E-07	9.7E-06	2.4E-05	8.5E-05	7.3E-07	9.8E-06	2.7E-05	9.5E-05
	>25K	7	1.3E-07	2.2E-06	6.2E-06	2.2E-05	1.2E-07	2.3E-06	7.6E-06	2.7E-05
	>100K	18	1.3E-08	2.9E-07	1.1E-06	3.9E-06	1.2E-08	3.1E-07	1.5E-06	5.0E-06
	>500K	41	1.1E-11	2.9E-10	3.3E-09	8.5E-09	1.3E-11	4.0E-10	5.0E-09	1.3E-08
PWR	>100	½	9.1E-04	3.9E-03	6.7E-03	1.8E-02	5.5E-04	2.6E-03	4.9E-03	1.4E-02
	>1,500	1 5/8	1.1E-05	1.4E-04	4.7E-04	1.5E-03	1.1E-05	1.6E-04	5.8E-04	1.9E-03
	>5,000	3	2.7E-07	3.4E-06	1.3E-05	4.4E-05	6.2E-07	7.6E-06	2.8E-05	9.8E-05
	>25K	7	2.0E-08	3.1E-07	1.2E-06	4.1E-06	4.0E-08	6.6E-07	2.8E-06	9.5E-06
	>100K	14	5.7E-10	1.2E-08	1.1E-07	3.0E-07	1.4E-09	2.8E-08	2.7E-07	7.4E-07
	>500K	31	4.7E-11	1.2E-09	1.5E-08	4.0E-08	1.2E-10	2.9E-09	3.9E-08	9.9E-08

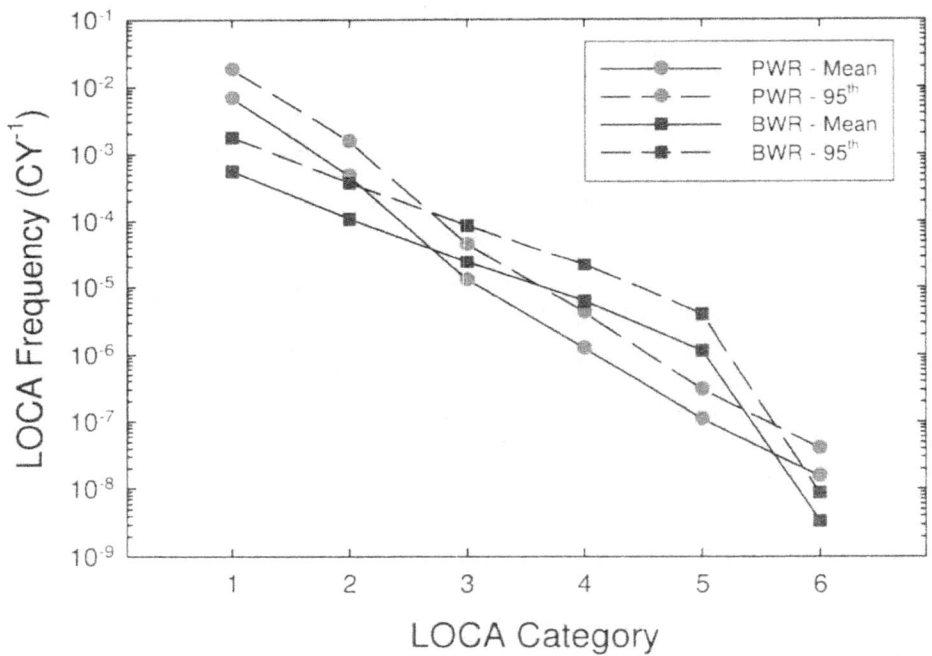

Figure 7.4 Total BWR and PWR LOCA Frequencies (Means and 95th Percentile Values) as a Function of LOCA Category at 25 Years of Plant Operations

Table 7.1 also lists the minimum effective break size corresponding to each LOCA flow rate category using the correlations described in Section 3.7. As described in Section 3.7, the break size is the equivalent diameter of the associated break area required for each of the six threshold flow rates. However, these break sizes do not represent actual pipe diameters. Specifically, there is no 41 inch diameter (Table 7.1) piping in BWR plants. Only the PWR and BWR liquid correlations (Section 3.7) were used to develop the Table 7.1 break size equivalents because the major BWR contributions come from the liquid lines. The main steam line is generally not expected to be a significant contributor to the overall LOCA frequencies. It is worth stressing that most panel members did utilize the effective break size to assess failure rates for each LOCA category instead of the flow rate definitions. The LOCA frequencies are plotted as a function of threshold break diameter in Figure 7.5 for both the PWR and BWR mean and 95th percentile estimates. The break diameter and LOCA category relationship is approximately logarithmic and there is approximately a factor of 2 difference in each successive break size. These relationships explain why Figures 7.4 and 7.5 appear to be nearly identical.

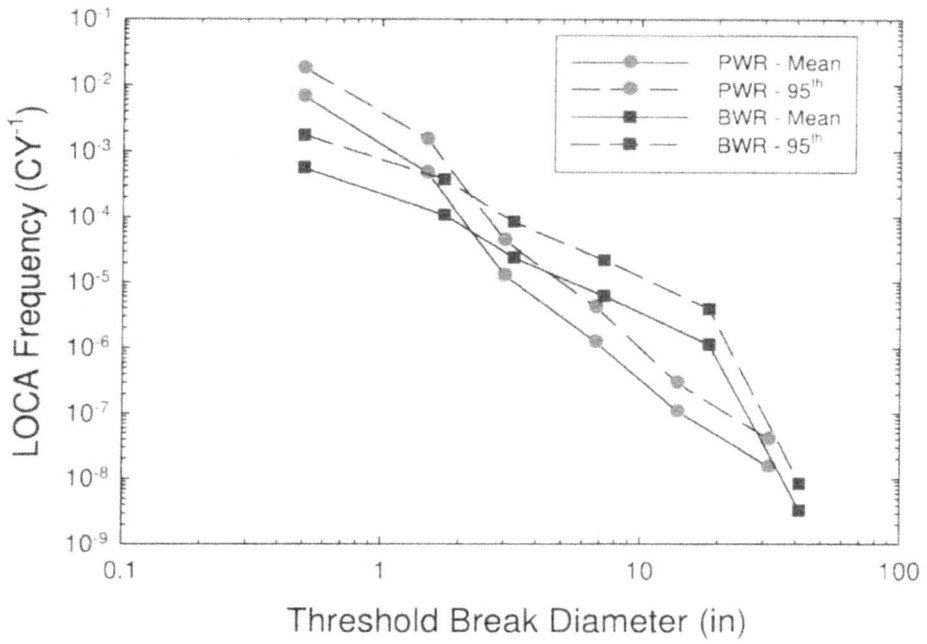

Figure 7.5 Total BWR and PWR LOCA Frequencies (Means and 95th Percentile Values) as a Function of Threshold Break Diameter at 25 Years of Plant Operations

It is worth reiterating that the elicitation focused solely on determining event frequencies that initiate by unisolable primary system side failures that can be exacerbated by material degradation with age. Therefore, active system failures (e.g., stuck open valve, pump seals, interfacing system LOCAs), consequential primary pressure boundary failures due to either secondary side failures, or failures of other plant structures (e.g., crane drops) are not reflected in the LOCA frequency estimates in Table 7.1. In addition, only normal plant operational cycles and loading histories that occur over the extended licensing period (i.e., frequency approximately 0.01 or greater per calendar year) were explicitly addressed. Rare event loading from seismic, severe water hammer, and other sources was not considered in this generic evaluation because of their strong dependence on plant specific factors. As with other consequential LOCAs, the LOCA frequency contribution from rare event loading may be an important consideration when evaluating total plant risk and determining total LOCA risk contributions.

As indicated in Section 2, a separate program was conducted to assess seismic LOCA risk [7.1]. Both unflawed and flawed seismic piping analyses were conducted in order to ascertain the magnitude of any potential adjustments to the LOCA frequency estimates provided in this section. The unflawed piping analyses showed that the seismic-induced failure probabilities for unflawed piping are significantly lower than the passive LOCA frequencies (Table 7.1) for large LOCAs (i.e., initiating event frequencies greater than 1.0E-5 per calendar year).

The flawed piping seismic analysis demonstrated that even for very long circumferential cracks, the surface-flaw depth must be greater than 40 percent of the wall thickness in order to lead to failure for a seismic event with a 1.0E-5 annual probability of exceedance. A flaw this size is large enough so that periodic nondestructive ISI coupled with required leak detection should ensure that such a flaw would be detected well in advance of failure in service. Hence, the failure frequencies associated with large LOCAs (> 1.0E-5 per calendar year) due to seismic loading of flawed piping will be significantly less than the passive system seismic LOCA frequencies if appropriate ISI programs are in place for large primary system piping and non-piping components (i.e., those that can lead to a LOCA size corresponding to a failure frequency greater than 1.0E-5 per calendar year).

7.3 Comparison of Piping and Non-Piping Contributions to Total LOCA Frequencies

The piping and non-piping current-day mean LOCA frequency estimates are compared in Figures 7.6 and 7.7 for BWR and PWR plants, respectively. The quantitative differences between the piping and non-piping contributions are summarized in Table 7.2. For the BWR plants, the piping contributions to the Category 1 and 2 LOCA frequencies are a factor of 2 to 4 higher than the contribution from the non-piping components. The BWR primary piping system contributors to these smaller category LOCAs are the small diameter instrument and drain lines (Section 6.3.1). The main BWR non-piping contributor to these smaller category LOCAs is from vessel penetrations, especially the lower head control rod drive (CRD) housings (Section 6.3.1), due to current PWSCC concerns (Section 6.3.2).

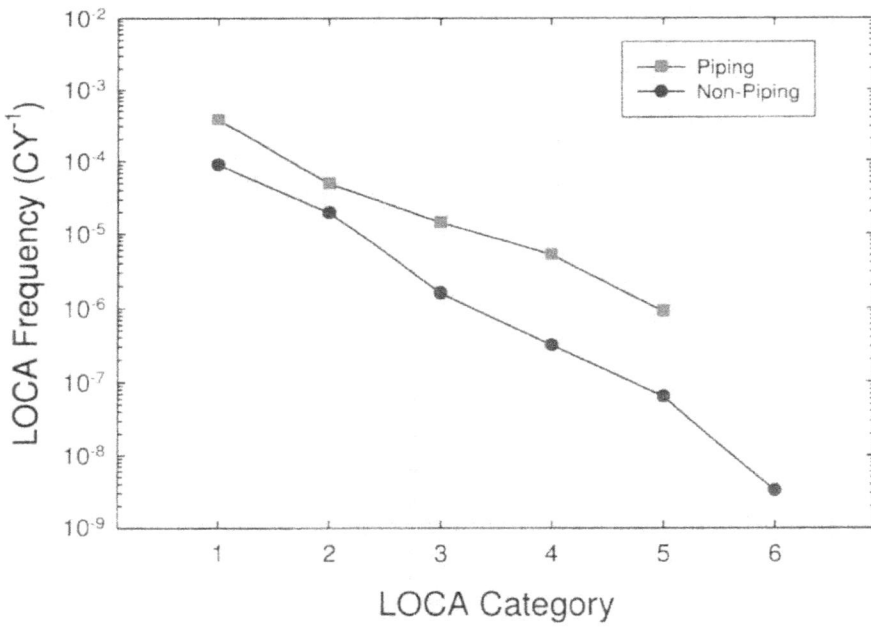

Figure 7.6 BWR Piping and Non-Piping LOCA Frequencies (Current-day Estimates)

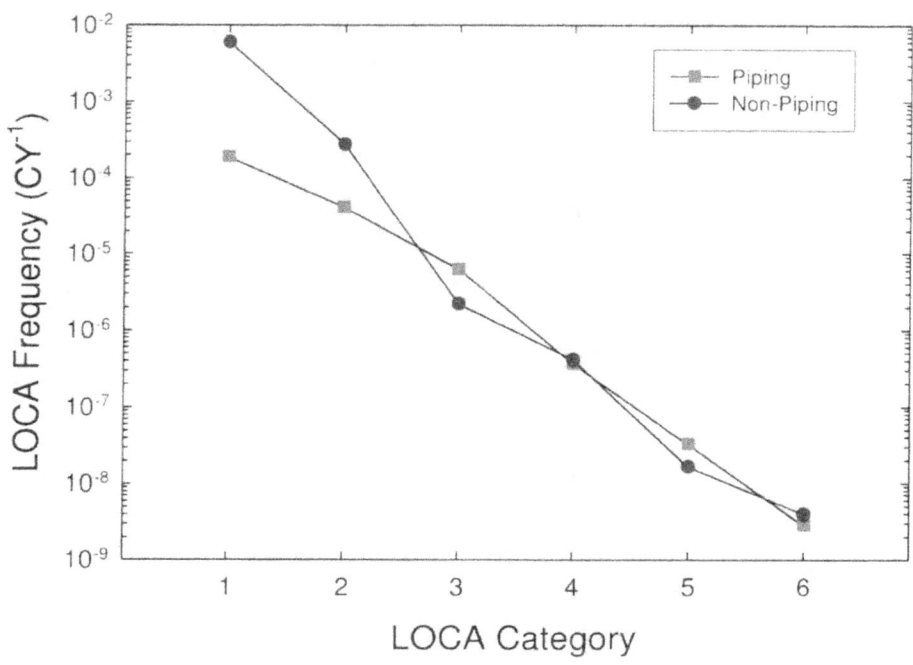

Figure 7.7 PWR Piping and Non-Piping LOCA Frequencies (Current-day Estimates)

Table 7.2 Ratio of BWR and PWR Non-Piping Contribution to Piping Contribution

LOCA Category	Ratio of BWR Non-Piping Contribution to Piping Contribution	Ratio of PWR Non-Piping Contribution to Piping Contribution
1	0.24	32
2	0.40	6.7
3	0.11	0.36
4	0.06	1.1
5	0.07	0.50
6	NA	1.4

The principal BWR piping system contributor to LOCA Categories 3 to 5 is the recirculation system piping (Section 6.3.1) where IGSCC concerns remain (Section 6.3.2). The panel consensus is that the susceptibility to IGSCC is greatly reduced compared to the past (Section 6.3.2). Mitigation measures such as improved HWC, weld overlay repairs, post-weld heat treatment (PWHT), and pipe replacement using more resistant materials have effectively reduced the likelihood of IGSCC. However, there is still residual concern about the failure likelihood of the large recirculation piping material that has not been replaced. The main BWR non-piping contributors to LOCA Categories 3 to 5 are the RPV nozzles and the component bodies (Section 6.3.1). However, the BWR non-piping contributions to the total LOCA frequencies are substantially less than the piping contributions for these LOCA sizes due to the comparative robustness of the non-piping components. The relative non-piping contribution for all the LOCA categories is generally expected to remain relatively constant in the future, although mitigation of vessel penetration cracking may decrease this specific contribution over the next fifteen years (Section 6.3.4).

The PWR LOCA frequencies (Figure 7.7 and Table 7.2) are dominated by the non-piping contributions for LOCA Categories 1 and 2. The mean non-piping LOCA frequencies are 32 and 6.7 times greater than the piping contribution for Category 1 and 2 LOCAs, respectively. The major piping contributors for PWR Category 1 and 2 LOCAs are, as was the case for the BWR plants, the instrument and drain lines. The concerns associated with these PWR systems are analogous to those for BWR plants (Section 6.3.1). However, the relative non-piping contributions are significantly higher for PWR plants because of plant design differences and the increased population of non-piping primary pressure boundary components. The PWR plants operate at higher temperatures and several non-piping components (e.g. pressurizer, steam generator) have experienced service degradation due to PWSCC or actual rupture (e.g. steam generator tubes). For LOCA Category 1, the large non-piping contribution is provided by steam generator tubes, CRDM penetrations, and pressurizer heater sleeves (Section 6.3.1). For LOCA Category 2, the principal contribution is from the CRDM penetrations (Section 6.3.1). The non-piping contributions for LOCA Categories 1 and 2 are expected to decrease somewhat in the future as PWSCC degradation is mitigated and as steam generator replacement programs proceed.

The relative non-piping contribution for PWR plants drops precipitously after LOCA Category 2, and the piping and non-piping contributions are nearly the same (less than a factor of 3 difference) for LOCA Categories 3 to 6. The steam generator tube, CRDM, and pressurizer heater sleeve components do not substantially contribute anymore due to size limitations. The major PWR piping system contributions include various auxiliary piping systems for LOCA Category 4; the hot leg, surge line and RHR lines for Category 5; and the hot leg for Category 6 (Section 6.3.1). The major PWR non-piping contributors are nozzles and component bodies for LOCA Categories 3 and 4; the manways and component bodies for Category 5; and the component bodies for Category 6.

7.4 Effect of Plant Operating Time on Total LOCA Frequencies

The future LOCA initiating event frequencies after 40 years of operation are also provided in Table 7.1. As indicated in Section 2, the elicitation responses did not explicitly require the determination of either average frequencies over the time period or instantaneous frequencies at the end of this time period. It is reasonable to assume, however, that most of the panelists did provide average values. Note that these estimates include failure contributions from continued degradation due to mechanisms identified in the first 25 years of operating experience as well as possible new mechanisms that could surface over the next 15 years. Current operating experience contains information about the response rate and the effectiveness of mitigation measures in addressing several known degradation mechanisms (e.g., IGSCC and FAC). Several panelists used this operating experience as a basis for assessing the impact of possible new degradation mechanisms. Alternatively, PFM models can be used to assess the impact of new degradation mechanisms by comparison with the failure frequencies associated with known degradation mechanisms. Other panelists chose this approach.

Despite these considerations and the additional uncertainty associated with predicting future events, the differences between the 25 and 40 year estimates (Table 7.1) are not large. The 40-year PWR median and mean estimates are less than a factor of 3 greater than the 25-year estimates for all LOCA categories. The LOCA Category 1 mean frequencies are expected to actually decrease slightly (by ~30%) due to expected improved mitigation practices for SGTRs and CRDM cracking (Section 6.3.4). The modest PWR LOCA frequency increases are largely attributed to continuing PWSCC concerns in other components. The 40-year BWR LOCA mean frequencies are even more stable, and differ by less than 35% from the 25 year estimates for LOCA Categories 1 to 5. This stability reflects the opinion that IGSCC (Section 6.3.2) susceptibility is expected to remain constant in the near term. The BWR mean Category 6 LOCA frequency only increases by 50% over the next fifteen years. This minor increase results from an increased RPV failure rupture frequency due to continued radiation embrittlement (Section 6.3.1). Because of the predicted stability in these estimates over the near-term, it is recommended that the 25-year results be used to estimate the average LOCA frequencies over the next 15 years of fleet operation.

The panelists expect somewhat larger LOCA frequency differences between the 25 and 60 year estimates. The 25 and 60 year mean frequency estimates are graphed in Figure 7.8 for both BWR and PWR plants. Except for LOCA Category 6, the increases in the BWR estimates are relatively modest. In fact, for LOCA Categories 1 to 5, the mean BWR LOCA frequency estimates for 25 and 60 years differ by less than a factor of 3. The BWR mean frequency increases are largely due to uncertainty about the future, which is reflected in general increases in the 95th percentile estimates (not shown in Figure 7.8),and the concern that new mechanisms could arise in the operating fleet (Section 6.3.4). The only significant mean frequency increase is predicted to result from continued RPV aging and only affects the BWR LOCA Category 6 frequency (Section 6.3.1). However, even this increase is less than a factor of 10 over the current-day Category 6 LOCA estimate.

From Figure 7.8, the increases in the PWR estimates in the next 35 years are typically much larger than the BWR increases. The PWR mean frequencies increase by a factor of less than 3 for LOCA Categories 1 and 2, factors of 8 and 10 for LOCA Categories 3 and 4, and factors of 10 and 18 for LOCA Categories 5 and 6, respectively. The larger increases in the higher LOCA categories are driven by the expectation that they will be impacted more severely by possible future degradation mechanisms. Because, as discussed above, the differences between the 25 and 40 year estimates are not significant, the PWR frequencies increase more than the BWR frequencies after 40 years of operation. This increase stems from uncertainty about the long-term mitigation effectiveness of PWSCC and the effect of future degradation mechanisms on PWR LOCA frequencies. While no specific new degradation mechanism has been identified for PWR plants, many panelists expect that they are more susceptible to future degradation than BWR plants. The net effect of the relative changes after 40 years of operation is that at 60 years, the PWR frequencies are expected to be greater than the BWR frequencies for LOCA Categories 1, 2, 3, and 6 and just slightly less than the BWR frequencies for LOCA Categories 4 and 5.

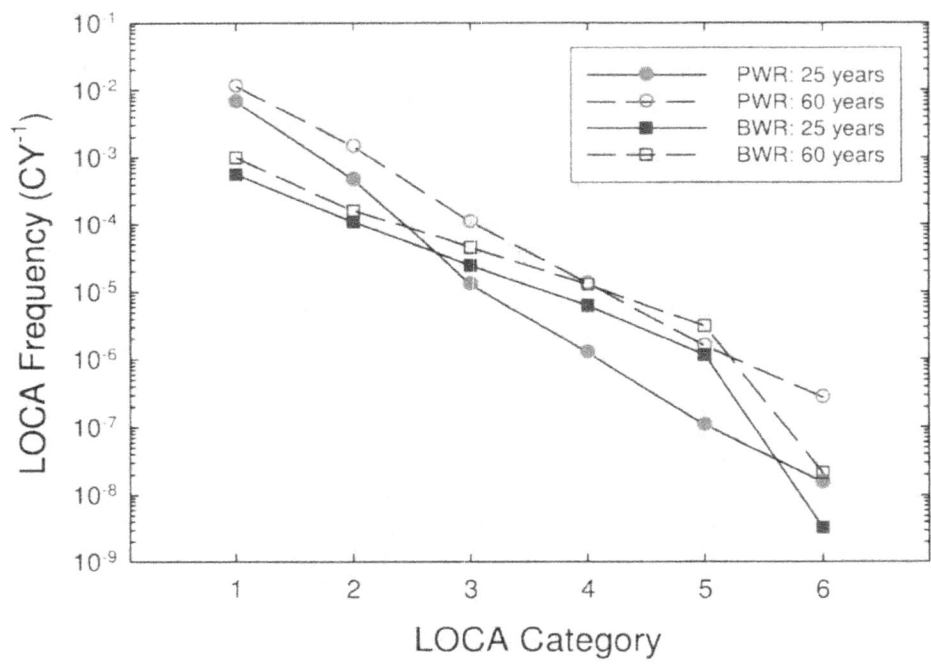

Figure 7.8 BWR and PWR Mean Total LOCA Frequencies at 25 and 60 Years

7.5 Individual Uncertainty and Panel Diversity

7.5.1 Individual Uncertainty

The BWR Category 1 and Category 3 LOCA frequency percentiles are plotted in Figures 7.9 and 7.10, respectively, for the various panelists who answered the BWR elicitation questions. Figures 7.11 and 7.12 are similar plots for the PWR Category 1 and Category 6 LOCA frequency percentiles, respectively. In each of these plots, the panelists' medians and 5th and 95th percentile estimates are shown. The letter designators for the horizontal lines in these plots are the reference letters which identify each panelist. For the BWR plants (Figures 7.9 and 7.10), the uncertainties associated with most of the panelists' responses (i.e., the differences between their 5th and 95th percentiles) are comparable for their Category 1 and Category 3 responses. The exceptions to this are Panelists A, E, and F, who expressed almost twice as much uncertainty in their Category 3 responses as they did in their Category 1 responses. The uncertainties in the larger LOCAs, Categories 4 to 6, increase with successively larger LOCA sizes. This increase is reflected by the individual UEFs which are defined as the UB/MV responses (Sections 5.6.2.2 and 7.6.2.2) and is another measure of panelist uncertainty. The GM of the individual UEFs also increases from a factor of about 7 for LOCA Category 1 to 28 for Category 6. However, the differences in the relative uncertainties of the panelists are consistent from LOCA Categories 3 to 6. That is, the ratios of any two panelists' UEFs (e.g., see Figure 7.10 for BWR LOCA Category 3) are relatively constant for LOCA Categories 3 to 6.

The PWR LOCA frequency estimates (Figures 7.11 and 7.12) exhibit many similar features to the BWR estimates. As above, the panelists express more uncertainty as the LOCA size increases. For the PWR Category 1 responses (Figure 7.11), the panelists' uncertainties varied from less than one order of magnitude for Panelists A, B, and L to two orders of magnitude for Panelist H. This relatively low level of uncertainty for LOCA Category 1 for PWRs reflects the fact that steam generator rupture is a significant contributor to this LOCA category and there is actual rupture data to support the estimates. Conversely, for the PWR Category 6 LOCAs, the level of uncertainty ranges from approximately two orders of magnitude for Panelists A, B, H, and L to greater than four orders of magnitude for Panelists C, E, and J. This level of uncertainty is a reflection of the fact that more uncertainty is naturally related to very low frequency events, such as a Category 6 LOCA (Section 6.3.6). The UEFs vary from approximately 5 to 30 for Category 1 and Category 6 respectively (see Section 7.6.2.2). This is comparable to the increase for the BWR results. The largest increases are for Panelists C and J, who expressed three orders of magnitude more uncertainty in their Category 6 responses than they did in their Category 1 responses.

It is of interest to note that almost all of the coverage intervals in Figures 7.9 to 7.12 are quite symmetric about the medians. Because the frequency scale in the figures is logarithmic, this reflects the general logarithmic symmetry in the panelists' responses to the elicitation questions. However, a few of the panelists' responses exhibit significant asymmetries for the higher LOCA categories. For instance, the median estimates for Panelists A and E are relatively closer to their 95th percentiles for both BWR Category 3 (Figure 7.10) and PWR Category 6 (Figure 7.12) LOCAs. This implies that these panelists expect that the actual LOCA frequencies could only be a little higher, but significantly lower then their median estimates. Conversely, the Category 6 PWR LOCA responses for Participants C and J have median values closer to the 5th percentile and substantial LOCA frequency increases above the medians would not be surprising to these panelists.

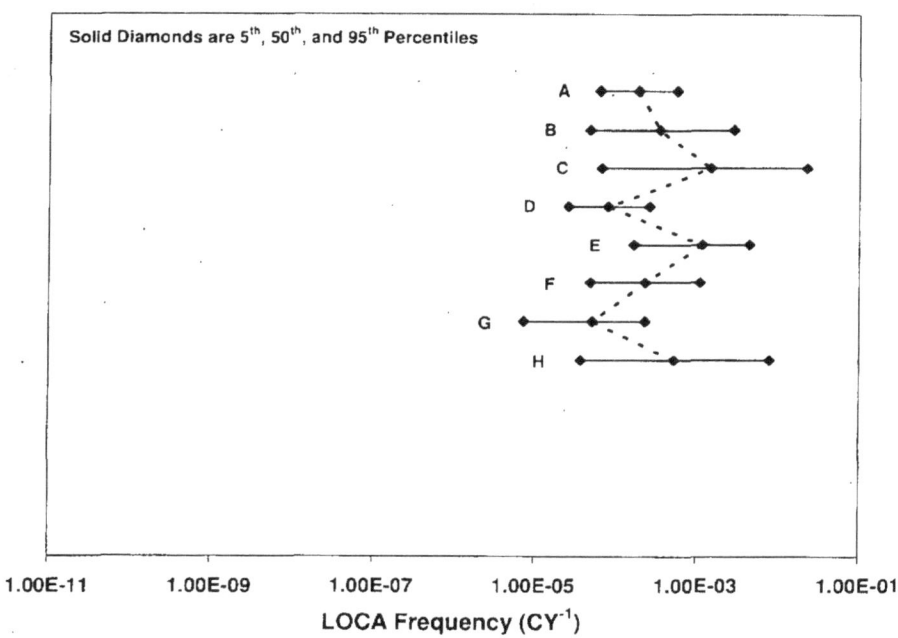

Figure 7.9 Individual Estimates of BWR Category 1 LOCA Frequency Percentiles

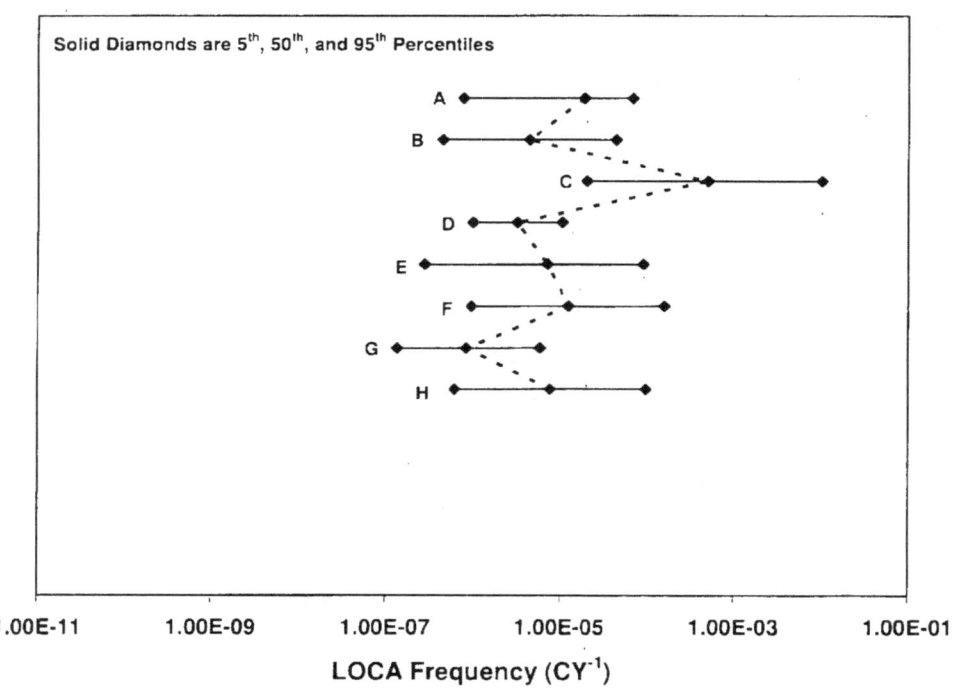

Figure 7.10 Individual Estimates of BWR Category 3 LOCA Frequency Percentiles

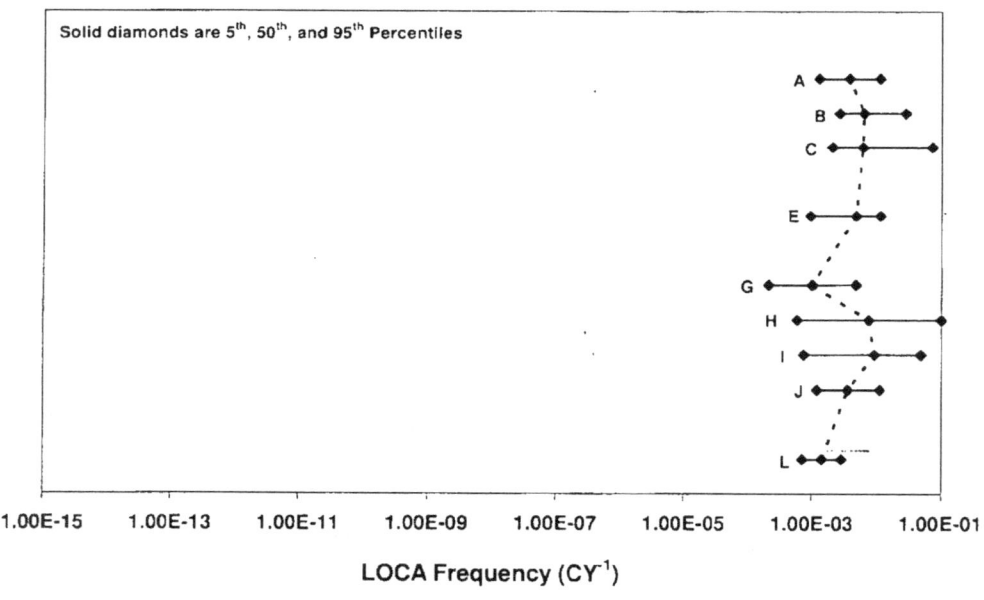

Figure 7.11 Individual Estimates of PWR Category 1 LOCA Frequency Percentiles

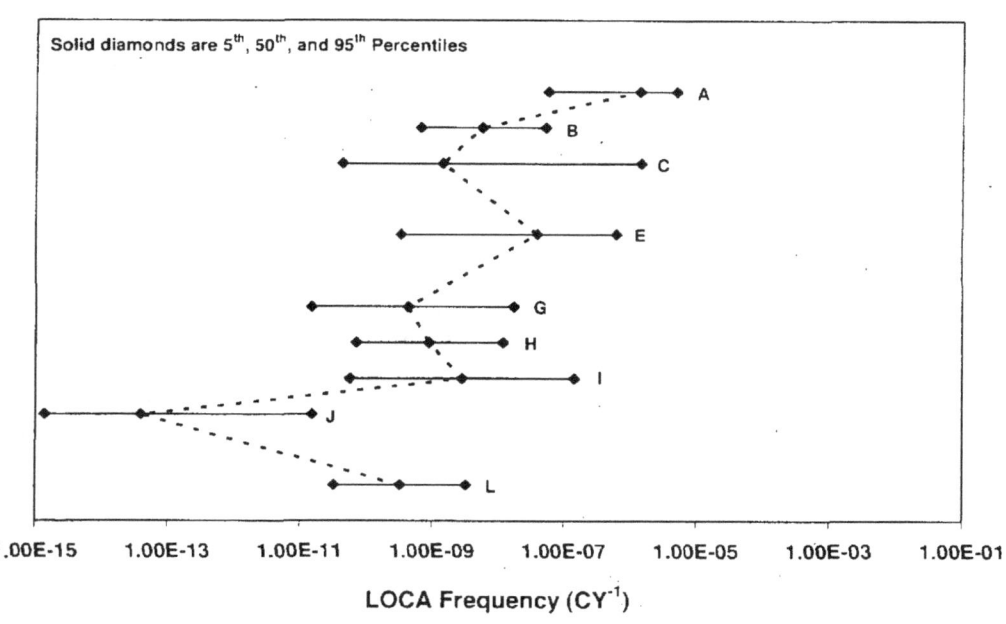

Figure 7.12 Individual Estimates of PWR Category 6 LOCA Frequency Percentiles

7.5.2 Panel Diversity

There is considerable diversity in the panelists' opinions. For example, there is approximately a factor of 30 difference between the lowest and highest BWR Category 1 median frequency estimates (Panelists G and C in Figure 7.9). This difference is relatively small compared to the factor of 600 difference between the medians for these two panelists for BWR LOCA Category 3 (Figure 7.10). Panelist C also provides the highest BWR estimates for LOCA Categories 2 and 4. Furthermore, Panelist C's estimates increasingly diverge from the rest of the panel for each successively higher LOCA category. For LOCA Category 4, the difference between Panelist C and the minimum estimate (Panelist G) is a factor of over 1,700. Additionally, Panelist C's median estimate is higher than many of the other panelists' 95th percentile estimates. The reason for Panelist C's disparity stems from his expectation that IGSCC is nearly as likely to result in a Category 4 LOCA as it is in a Category 1 LOCA in the primary recirculation piping. While the other panelists largely agree that IGSCC of the main recirculation piping is important (Section 6.3.1), the expectation that Category 1 and 4 LOCAs are nearly as likely is not shared by the other panelists.

There is also more diversity in the 95th percentile estimates than in the medians for LOCA Categories 1 to 4. For instance, while the minimum and maximum Category 1 median estimates differ by a factor of 30, the 95th percentile estimates differ by a factor of 80. For LOCA Categories 5 and 6, there is less diversity in the 95th percentile estimates than either the median or 5th percentile estimates. This occurs because, while all panelists expect much lower frequencies for the largest piping and non-piping failures, there is simply more disagreement about how low these frequencies could be. Conversely, the 5th percentile estimates exhibit the least diversity among the percentiles for LOCA Categories 1 to 4 and the greatest diversity for Categories 5 and 6.

7-15

The PWR results also generally exhibit increasing diversity as the LOCA size increases. The minimum and maximum median Category 1 estimates differ by only a factor of 8 (Figure 7.11). This relatively good quantitative agreement occurs because nearly all panelists expect SGTR to be the dominant contributor (Section 6.3.1). However, differences between the minimum and maximum median estimates increase to between 3 and 4 orders of magnitude for LOCA Categories 2 to 4. While these differences are comparable to the BWR differences, the overall diversity associated with the PWR estimates is greater. While only Panelist C's BWR estimates differ significantly from the rest of the panelists, the PWR estimates are more uniformly spread over their range. This greater diversity reflects more disagreement about the contributing factors to PWR LOCAs of these sizes (Sections 6.3.1 and 6.3.2).

For Category 5 and Category 6 LOCAs (Figure 7.12), the differences between the minimum and maximum median estimates are about 5 and 7 orders of magnitude, respectively. This primarily reflects the low frequencies expected by Panelist J. In fact, Panelist J's 95th percentile estimate for LOCA Category 6 is less than the 5th percentile estimates for all of the other panelists. If Panelist J's estimates are ignored, the differences between the minimum and maximum median estimates for the remaining panelists decreases significantly to factors of 750 and 4,000 for LOCA Categories 5 and 6, respectively. This reduced LOCA Category 5 range also becomes consistent with the ranges for LOCA Categories 2 to 4 while the LOCA Category 6 range is not significantly larger. Panelist A's LOCA Category 6 median value also appears to be an outlier as it is greater than the 95th percentiles of the other panelists. Ignoring Panelist A's estimate decreases the range of median estimates to a factor of only 100. For LOCA Categories 5 and 6, the differences among the panelists mainly reflect the differences of opinion associated with estimating rare events (Section 6.3.6) and not from any significant disagreement about the contributing factors (Sections 6.3.1 and 6.3.2).

If Panelist J's estimates are ignored, the diversity in the PWR 5th and 95th percentile estimates is similar to the diversity in the BWR estimates. The 95th percentile estimates have the largest ranges for LOCA Categories 1 and 2, while the 5th percentile and median ranges are much smaller. However, for LOCA Categories 5 and 6, the 5th percentile and median ranges are larger than the 95th percentile ranges. The ranges for the three percentiles are similar for LOCA Categories 3 and 4. If Panelist J's estimates are included, the 5th percentile estimates exhibit the greatest diversity for LOCA Categories 2 to 6.

7.5.3 Confidence Intervals

Another measure of panel diversity is the set of statistical confidence intervals associated with the estimates of the total LOCA frequency bottom-line parameters presented in Table 7.1 and Figures 7.4 and 7.5. Panel diversity as measured by the confidence intervals is distinct from the uncertainty expressed by each panelist, which is reflected in the differences between the medians and 95th percentiles as discussed in Sections 5.1 and 7.5.1. Confidence intervals for the medians, means, and 95th percentiles as a function of threshold break diameter are presented in Figures 7.13 and 7.14 (slightly offset for visual clarity) for BWR and PWR plants, respectively. The confidence intervals have been calculated as described in Section 5.4 and are illustrated by error bars about the estimated parameters. Based on the assumed lognormal model for the individual estimates, each confidence interval has a 95% chance of covering the median of the assumed lognormal distribution for each bottom-line parameter.

Each confidence interval in Figures 7.13 and 7.14 is symmetric about its associated bottom-line parameter (i.e., symbols in Figure 7.13 and 7.14) as a consequence of the assumed lognormal distribution of the panelists' estimates (Section 5.4). From the figures, the confidence intervals generally increase with increasing break size, as would be expected given the increasing difficulty in estimating increasingly infrequent events. This trend is consistent with the observed panel diversity as discussed in Section 7.5.2.

The confidence intervals span less than one order of magnitude for the BWR and PWR Category 1 LOCA estimates. The BWR confidence interval differences then increase to a factor of approximately 10 for

LOCA Categories 2, 3, and 5, a factor of 30 for LOCA Category 4, and over two orders of magnitude for LOCA Category 6. The PWR confidence interval differences increase to a factor of approximately 20 for LOCA Categories 2 to 4, and over two orders of magnitude for LOCA Categories 5 and 6. The differences between the BWR and PWR confidence intervals and the trends with increasing break size are similar to the previously discussed trends in panel diversity (Section 7.5.2).

The group estimates in Figures 7.13 and 7.14 (i.e., the symbols) also illustrate the trends in the panelists' uncertainties discussed in Section 7.5.1. For example, the differences between the estimated medians and 95th percentiles increase with break size. Conversely, the differences between the estimated means and 95th percentiles decrease with break size. This trend implies that the skewness of the underlying distribution increases with LOCA size so that the mean becomes a higher percentile as LOCA size increases.

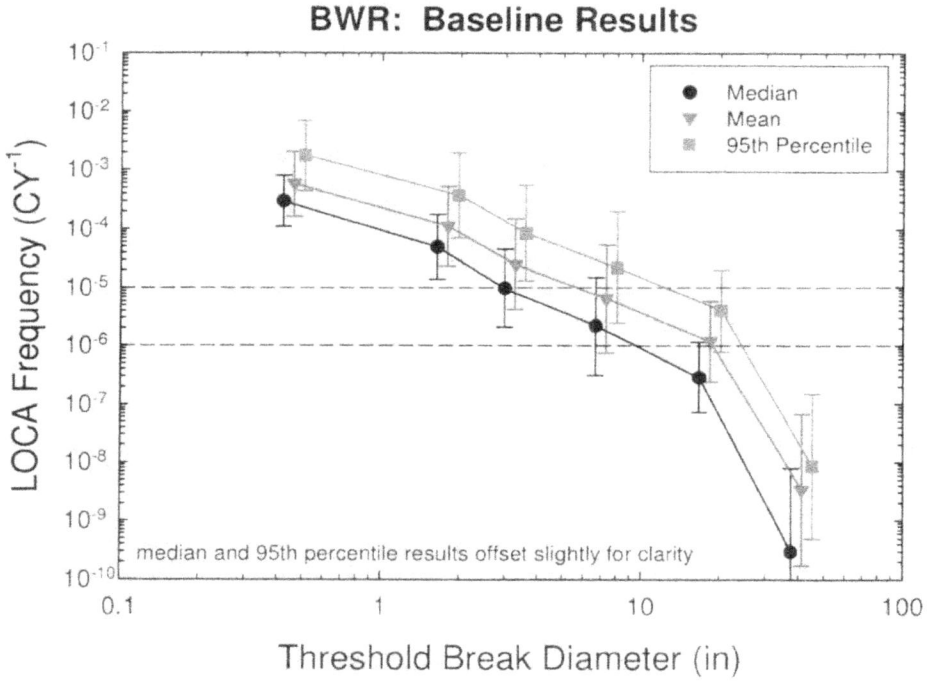

**Figure 7.13 95% Confidence Intervals for Total BWR LOCA Frequencies
at 25 Years of Plant Operations**

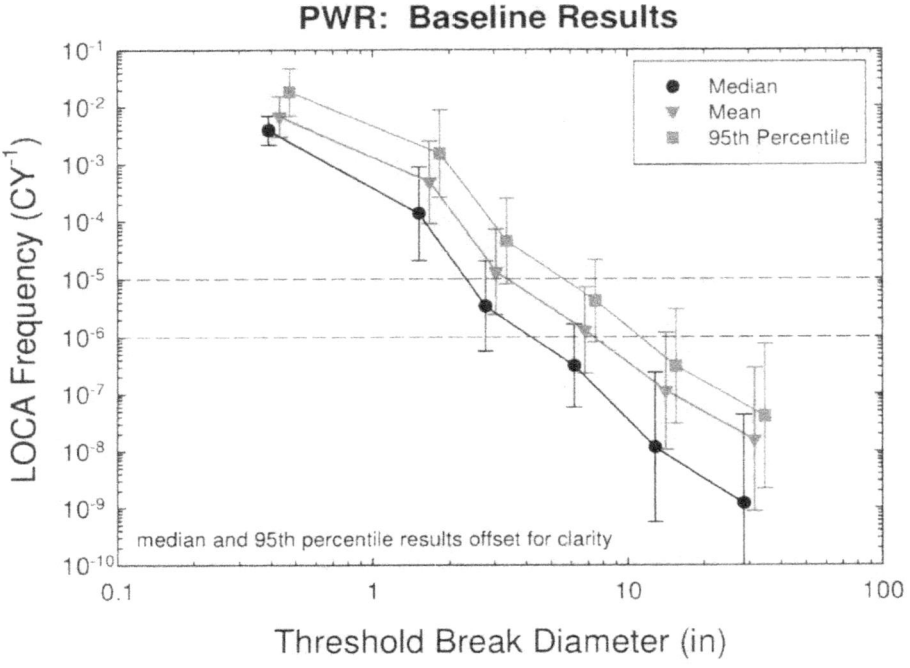

Figure 7.14 95% Confidence Intervals for Total PWR LOCA Frequencies at 25 Years of Plant Operations

7.6 Sensitivity Analyses

Several types of sensitivity analyses were conducted to examine the assumptions, structure, and techniques used to process the elicitation responses to develop the baseline LOCA frequency estimates. The elements of the baseline analysis framework were defined in Section 5.5, but they are summarized here for convenience. The framework has six key elements:

(i)   The MV, UB and LB supplied by the panelists for each elicitation question were assumed to correspond to the median, 95th percentile, and 5th percentile, respectively, of a split lognormal distribution with the mean calculated assuming the upper tail is truncated at the 99.9th percentile.

(ii)   Panelist responses are not adjusted to account for possible overconfidence in the uncertainty estimates for each elicitation response.

(iii)   Split lognormal distributions are summed by assuming perfect rank correlation among the individual terms.

(iv)   The individual estimates are the total LOCA frequency parameters (i.e., mean, median, 5th percentile, and 95th percentile) determined for each panelist.

(v)   The group estimates of the total LOCA frequency parameters were determined using the GMs of the individual estimates.

(vi)   Panel diversity is characterized with two-sided 95% confidence intervals based on an assumed lognormal model for the individual estimates.

The baseline results have been presented and discussed in Sections 7.2 to 7.5.

This section quantifies the effects of other assumptions or methods of calculation on the baseline frequency estimates. The sensitivity analyses fall into one of the following five general categories: mean determination, panelist overconfidence adjustment, correlation structure, individual response aggregation, and panel diversity. In each case, a single change in the baseline analysis procedure is made and the effect is quantified. While many sensitivity analyses could be performed, only a few representative or important sensitivity analyses were conducted for each of the five general categories. A summary of the sensitivity analyses conducted is provided in Table 7.3, along with the corresponding sections containing the approaches and results. More details on each sensitivity analysis are contained in Section 5.6.

Table 7.3 Sensitivity Analysis Summary Matrix

Analysis Category	Analysis Type	Corresponding Section
Mean Determination	Effect of Distribution Shape on Mean	7.6.1
Overconfidence Adjustment	Blanket Overconfidence Adjustment	7.6.2.1
	Targeted Overconfidence Adjustment	7.6.2.1
	Error Factor Adjustment	7.6.2.2
Correlation Structure	Perfect Rank Correlation vs. Independence	7.6.3
Response Aggregation	Measures of Group Opinion	7.6.4.1
	Aggregation Point	7.6.4.2
	Aggregation Parameters	7.6.4.3
	Mixture Distribution	7.6.4.4
Panel Diversity	Confidence Bound and Quartile Comparison	7.6.5

7.6.1 Mean Determination

Each panel member provided three responses for each elicitation question: a MV, a LB, and an UB (Section 3.8.5). These numbers are assumed to correspond to the median, 5th percentile, and 95th percentile, respectively, of the panelist's subjective uncertainty distribution (Section 5.1). However, in order to combine distributions and calculate mean values, a distribution shape must also be assumed. The multiplicative structure of the elicitation (Section 3) and the character of the elicitation responses suggest that the distributions should have a split lognormal form (Section 5.3.1).

However, to avoid the mean being dominated by an extreme upper tail of the distribution, the baseline analysis calculates the means for the individual elicitation responses by replacing the corresponding split lognormals by split lognormals truncated at the 99.9^{th} percentile (Section 5.3.3.2). The remaining 0.1 percent of each distribution is concentrated at the 99.9^{th} percentile. This truncation point was chosen for the baseline analysis to be reasonably conservative. Furthermore, because the panel members provided no information beyond the 95th percentile, any truncation point beyond the 95th percentile is consistent with their responses. Note that the truncated split lognormal is used only to calculate the means. The other parameters of the LOCA frequencies (i.e., median and the 5th and 95th percentiles) are calculated by assuming untruncated split lognormal distributions.

This sensitivity analysis evaluates the effects of other distributional shapes on the calculated means for the individual elicitation responses. There are several other plausible choices as outlined in Section 5.6.1: a split log-triangular, an untruncated split lognormal, a lognormal corresponding to the upper half of a split lognormal (upper tail lognormal), or a normalized split lognormal truncated at the 99.9^{th} percentile (the probability density function below the 99.9^{th} percentile is renormalized so that the total area remains one). The effect of the truncation point for the unnormalized split lognormal is also examined by truncating at the 95th, 96th, 97th, 98th, 99th, and 99.99^{th} percentiles instead of the 99.9^{th} percentile used in the baseline methodology.

The mean is calculated for each of these alternative distributional shapes using the assumed distributions with a nominal median of 1, a nominal LEF of 1,000, and UEFs ranging from 10 to 1,200. This UEF range encompasses the full range of actual elicitation responses. The nominal LEF was fixed at one value because it has very little effect on the mean. Any elicitation response can be modeled using these nominal parameters.

The means calculated for selected distributional shapes are plotted as a function of UEF in Figure 7.15 for UEFs up to 100 and in Figure 7.16 for larger error factors up to 1,200[3]. The results for the split lognormal, upper tail lognormal, and split log-triangular distributions are depicted in these figures because they encompass a range of plausible distributions for calculating the means. The split lognormal truncated at the 98th percentile (Truncated Split @ 98 in legends) is displayed because the calculated means are similar to those using the split log-triangular distribution. The normalized split lognormal distribution truncated at the 99.9th percentile (Normal Truncated Split @ 99.9 in legends) is used for comparison with the baseline truncated distribution, which concentrates the highest 0.1 percent of the distribution at the 99.9th percentile (Truncated Split @ 99,9 in legends). These figures do not show distributions that are truncated at less than the 98th percentile because such distributions do not appropriately estimate the means.

Some general trends are readily apparent in Figures 7.15 and 7.16. The means calculated using either a split lognormal distribution or an upper tail lognormal are virtually identical. (Their plotted curves overlap in both figures.) This verifies that the lower halves of these distributions have minimal effects on the means. This finding was also verified for larger LEFs than the nominal value of 1,000 used in Figures 7.15 and 7.16. As expected, the means of the untruncated lognormal distributions also provide UBs for the means of the truncated distributions. The effect of normalizing the truncated distribution instead of concentrating the values beyond the truncation point at the truncation point (baseline method) increases with the UEF, but it remains insignificant. The difference between these two truncation schemes is less than 35% for UEFs up to 1,200. As expected, the baseline truncation method results in slightly higher calculated means than does normalization. It is also of interest that the means calculated using the split lognormal truncated at the 98th percentile are very similar to the means calculated using the split log-triangular distribution. These two distributions provide the most appropriate LB means in this sensitivity study.

The range of plausible means for the elicitation responses can be determined from these figures by the difference between the upper tail or split lognormal and the split lognormal truncated at the 98th percentile. This range increases as the UEF increases. The calculated means differ by less than 30% for the distributions illustrated in Figure 7.15 for UEFs less than 20. The differences remain less than a factor of two for UEFs less than 50. Note that an UEF of 20 bounds almost all of the elicitation responses for LOCA Categories 1 to 3 while an UEF of 50 bounds most of the responses for Categories 4 to 6. The responses for only a few panel members have UEFs above 50, and these large error factors are almost always associated with LOCA Categories 5 and 6, which represent the most severe and rarest LOCA events. However, some UEFs are as large as 1,000 for the rarest Category 6 events.

---

[3] The plotted mean values can be used to determine the mean for any elicitation response by multiplying the MV of the elicitation response by the mean value for an assumed distributional shape corresponding to the UEF of the response (i.e., ratio of UB to MV).

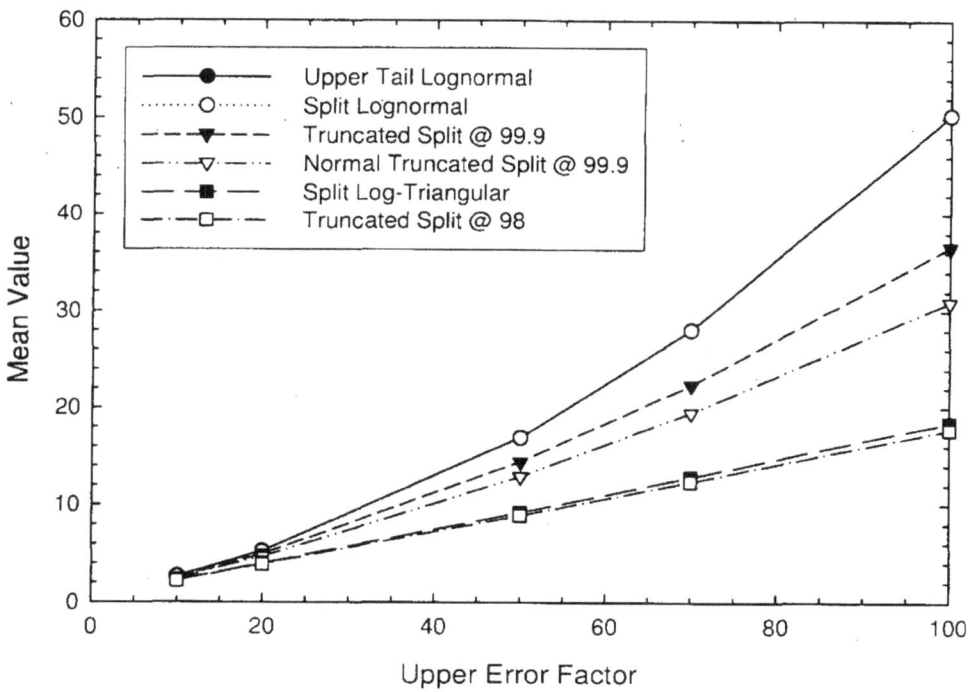

Figure 7.15 Mean Values for Selected Distributions by UEF < 100
(Median = 1 and LEF = 1000)
(Baseline Distribution is Truncated Split at 99.9%)

The effect of the distribution shape on the mean becomes significant for UEFs greater than 100 (Figure 7.16). Starting from a factor of 2.5 for an UEF of 100, the range increases to a factor of over 30 for an UEF of 1200. For example, for an UEF of 1,000, the difference between the split lognormal and split log-triangular means is a factor of 28. For this same UEF, the mean of the lognormal truncated at the 99.9[th] percentile is a factor of 5 less than the mean of the split lognormal, and is therefore slightly closer to this distribution than the split log-triangular. The choice of the truncation point is also significant for large UEFs. For example, the difference in the means between truncating at the 98th or 99.9[th] percentile when the UEF is 1000 is approximately a factor of 5.5. However, the difference when truncating at the 95th or the 99.99[th] percentile (not shown) is a factor of 30. The difference between truncating at the 95th percentile and the 99.9[th] percentile (baseline methodology) is a factor of 15 for this UEF.

Based on the sensitivity analysis, the baseline analysis uses a split lognormal truncated at the 99.9[th] percentile to calculate the means. This distributional shape was chosen as one that reflects the panelists' responses and also results in reasonably conservative values for the means. Because the mean can be a sensitive function of the tail of a distribution, the tail of the chosen distribution should not unduly influence the mean. Consequently, non-truncated split lognormal distributions should not be used because their means are overly dependent on the area beyond the 95th percentile for large error factors. However, the truncation point should be well beyond the 95th percentile, because truncating the distributions just past the 95th percentile effectively treats the UB estimates as absolute bounds. This would be counter to the instructions given to the panelists, which asked them to effectively provide 95th percentiles rather than absolute UBs (Section 3.8.5). Truncating at the 99.9[th] percentile only affects the top 0.1% percent of the distribution. Because this truncation point is well beyond the information provided by the panelists, the split lognormal distribution truncated at the 99.9[th] percentile provides a reasonable UB for the mean.

Based on the sensitivity analysis, either the split log-triangular or the split lognormal truncated at the 98th percentile provides a reasonable LB for the mean. The difference in the means between the baseline distribution and either of these two distribution shapes is only a factor of 5.5 when the UEF is 1000. For UEFs below 120, the difference is less than a factor of 2. Therefore, for the bulk of the elicitation responses, the mean is not very sensitive to the selection of an alternative distribution between these bounds. Only when the UEF is large do the differences become significant. However, the baseline distribution mean remains reasonably conservative for large error factors, which is a key reason for its selection. For simplicity and consistency, the means for all elicitation responses were calculated using this distributional shape.

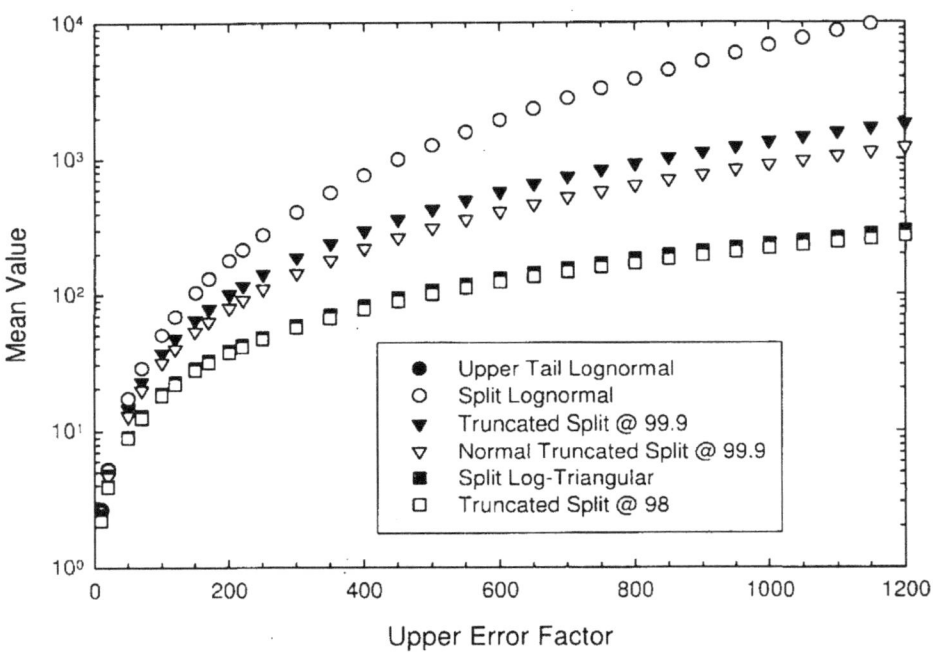

Figure 7.16 Mean Values for Selected Distributions by UEF < 1200
(Median = 1 and LEF = 1000)
(Baseline Distribution is Truncated Split at 99.9%)

7.6.2 Overconfidence Adjustment

Sensitivity analyses were conducted to investigate the effect of assuming that the panelists provided uncertainty intervals with less coverage than the nominal 90% (i.e., the difference between the 5th and 95th percentiles) requested in the elicitation responses. As noted in Section 5.6.2, nominal 90% coverage intervals provided in many applications have been typically shown to have actual coverage between 30% and 70% [7.2]. In order to correct for the typical overconfidence bias, two different types of adjustment schemes were evaluated. The first type adjusted the coverage of the intervals supplied for each individual elicitation question. This is called a coverage interval adjustment. The second type adjusted the error factors of the total LOCA frequency estimates calculated for each panelist. This is called an error-factor adjustment. Section 5.6.2 has more details on the approach of each type of adjustment. A summary of the approaches and the results are presented below.

7.6.2.1 Coverage Interval Adjustments - These overconfidence adjustments modify the coverage interval ranges associated with each panelist's elicitation responses. Each elicitation question required a MV, an UB, and a LB response. Consistent with the instructions provided to the panel, the baseline analysis assumed that the UB and LB responses correspond to the 95th and 5th percentiles, respectively, of a panelist's uncertainty distribution. As described in Section 5.6.2.1, the coverage interval adjustments associated the UB and LB responses with adjusted percentiles resulting in coverage intervals having less than the nominal 90% coverage. The MV responses were not adjusted.

This sensitivity analysis examined five different coverage interval adjustment schemes: two blanket adjustments and three targeted adjustments. The blanket adjustments adjusted all panelists by the same amount. The targeted adjustments used a two-level adjustment. Those panelists expressing larger uncertainty were adjusted by a lesser amount, while those panelists with smaller uncertainty were adjusted by a greater amount. The justification for considering a two-level adjustment is that the elicitation training and instructions provided to the panelists may have sensitized them to the expectation that uncertainties are typically underestimated in elicitations. Consequently, it is possible that those panelists expressing larger uncertainty adjusted their responses accordingly to guard against underestimation.

The various overconfidence adjustment schemes are summarized in Table 7.4. The Blanket 1 (B1) adjustment is the most severe adjustment scheme evaluated because the nominal 90% coverage intervals are reduced to 50% coverage. The Blanket 2 (B2) adjustment provides a smaller correction by reducing the coverage to 60%. The targeted adjustment schemes first used an UEF (ratio of UB to median) of 20 for LOCA Category 6 to bin the panelists responses into two groups. The most severe targeted adjustment (T1) does not adjust the coverage of the panelists having UEFs greater than 20 (more uncertain panelists) and adjusts the coverage to 50% for the panelists with UEFs less than or equal to 20 (less uncertain panelists). The second targeted adjustment (T2) adjusts the more uncertain panelists' coverage to 80% while the less uncertain panelists are adjusted to 60% coverage. The third targeted adjustment (T3) is the least severe. The less uncertain panelists are adjusted from 90% to 60% coverage while the more uncertain panelists are not adjusted.

Table 7.4 Overconfidence Adjustment Schemes

Overconfidence Adjustment	Coverage Interval 1	Panelists	Coverage Interval 2	Panelists
Blanket 1	50%	All	N/A	N/A
Blanket 2	60%	All	N/A	N/A
Targeted 1	90%	4	50%	remaining
Targeted 2	80%	4	60%	remaining
Targeted 3	90%	4	60%	remaining

Adjusted and unadjusted individual estimates of the means and 95th percentiles for PWR Category 3 LOCAs are compared using box plots (Figure 7.17) for the Blanket 1 (B1) and Targeted 1 (T1) adjustments. Several features of the B1 and T1 adjustments are apparent. Because the distribution shape is assumed to be a lognormal, these adjustments affect the means more than the 95th percentiles. This is evident by comparing the interquartile ranges and extremes of the adjusted and unadjusted means with the 95th percentiles. For both adjustments, the difference between the adjusted and unadjusted means is greater than the difference between the adjusted and unadjusted 95th percentiles. The interquartile range of the adjusted mean is also much larger than the adjusted 95th percentile interquartile range even though the unadjusted mean and 95th percentiles have similar ranges. Furthermore, the adjustments can significantly skew the distributions. While the unadjusted means are about half an order of magnitude smaller than the corresponding unadjusted 95th percentiles, the T1 means are closer to their 95th percentiles while some B1 means are even larger than their 95th percentiles. This increased skewness is not supported by the elicitation responses. The B1 and T1 adjustments for the other PWR and BWR LOCA categories exhibit similar trends to those apparent in Figure 7.17.

The blanket (B1 and B2), T1 and T2 adjustment schemes are also not supported by either the passive-system failure data for the smaller LOCA categories or the base case results (Sections 4.2 and 4.3). For instance, six of the nine individual PWR Category 3 means in Figure 7.17 are greater than 10^{-3} per calendar year and three of them are greater than 10^{-2} per calendar year for the B1 adjustment. These frequencies are substantially higher than either present operating-experience-based estimates or the associated base case frequencies. However, the affected panelists provided no rationale to support these high estimates. Other similar unsupportable results are apparent for LOCA Categories 1 and 2. The B2 and T2 adjustments are less severe, but even a modest adjustment of the coverage interval for those panelists with high uncertainty leads to results that are not supported by either operating experience or the appropriate base case frequencies.

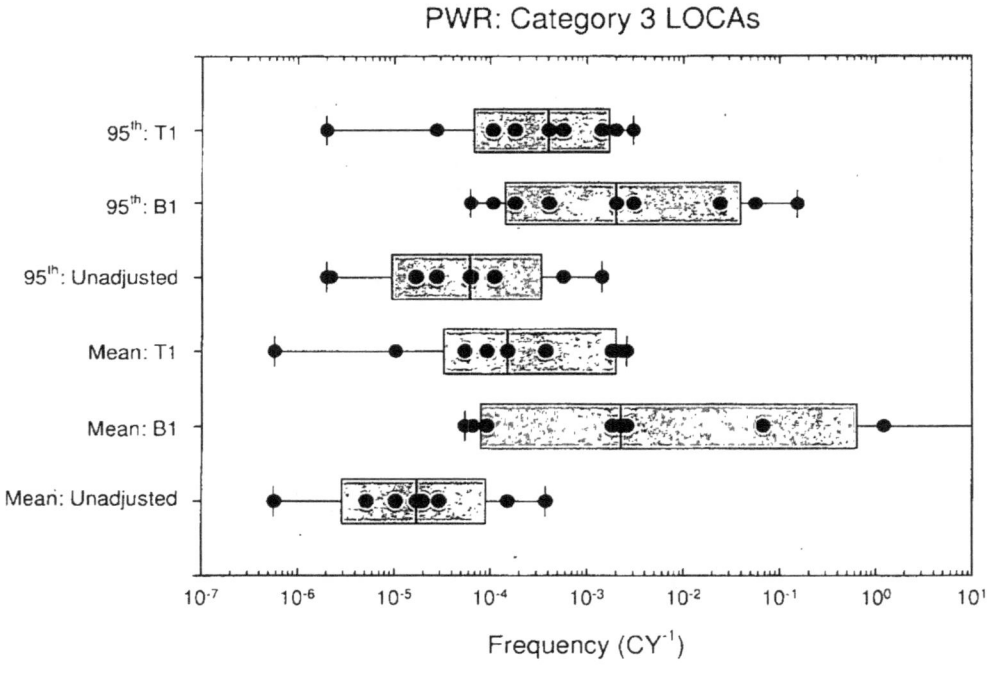

**Figure 7.17 Blanket 1 and Targeted 1 Coverage Interval Adjustments
For the Mean and 95th Percentile**

The results of the T3 coverage interval adjustment of the individual estimates are provided in Figure 7.18 for PWR Category 3 LOCAs. The T3 adjustment results in only modest increases in the unadjusted frequencies for the box plot median of the individual mean and 95th percentile estimates. Also, the interquartile ranges for the means and 95th percentiles do not significantly increase after the adjustment even though many individual estimates are increased. Surprisingly, the minimum and maximum estimates are unaffected by the adjustment. The reason is that the panelists at the extremes had more uncertainty and hence their responses were not adjusted. Because the extreme values are unchanged while many other estimates are increased, the unadjusted mean and 95th percentile interquartile regions are more symmetrically located between the minimum and maximum estimates than are the adjusted regions. Of course, the estimated medians are unchanged by the adjustment.

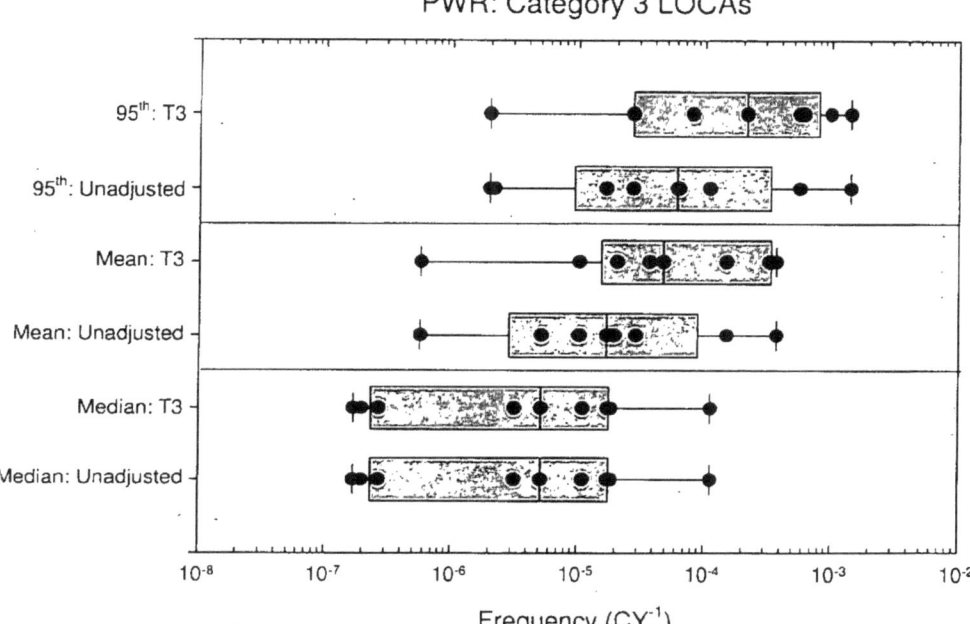

**Figure 7.18 Targeted 3 Coverage Interval Adjustments
For the Mean, Median, and 95th Percentile**

As illustrated in Figure 7.18, the least severe coverage adjustment (T3) generally appears to be supported by operating experience, the various predicted base case frequencies, and the elicitation results. With this adjustment, only the coverage intervals for the four panelists with less uncertainty (UEF < 20) are moderately adjusted from 90% to 60% (Table 7.4). Some adjustment of these responses is justified because their uncertainties are lower than the remainder of the group. While the proper adjustment magnitude cannot be rigorously determined for this scheme, the 60% coverage adjustment is within the wide range that is supported by research [7.2]. The rationale for not adjusting the remaining panelists is that their uncertainties are already large, and may not be underestimated.

The total BWR and PWR LOCA frequencies resulting from the T3 adjustment are plotted along with the associated confidence intervals in Figures 7.19 and 7.20 as a function of threshold break diameter, respectively. Table 7.5 compares the adjusted and unadjusted (baseline) results using ratios of the adjusted to the unadjusted means and 95th percentiles. Because they are not adjusted, the median frequencies do not change. In summary, the T3 overconfidence adjustment results in relatively small increases of about a half order of magnitude or less for both the BWR and PWR estimates. In fact, except for the PWR mean for LOCA Category 6, the BWR and PWR means and 95th percentiles are only increased by factors of between 2 and 4 times the unadjusted results. The mean value is generally increased as much or more than the 95th percentile estimate due to the skewness of the underlying distributions. Except for LOCA Category 1, the PWR adjustment ratios are larger than the BWR ratios. This trend occurs because more PWR than BWR panelists are adjusted (five vs. four) and because the unadjusted PWR UEFs are slightly higher than the unadjusted BWR UEFs,

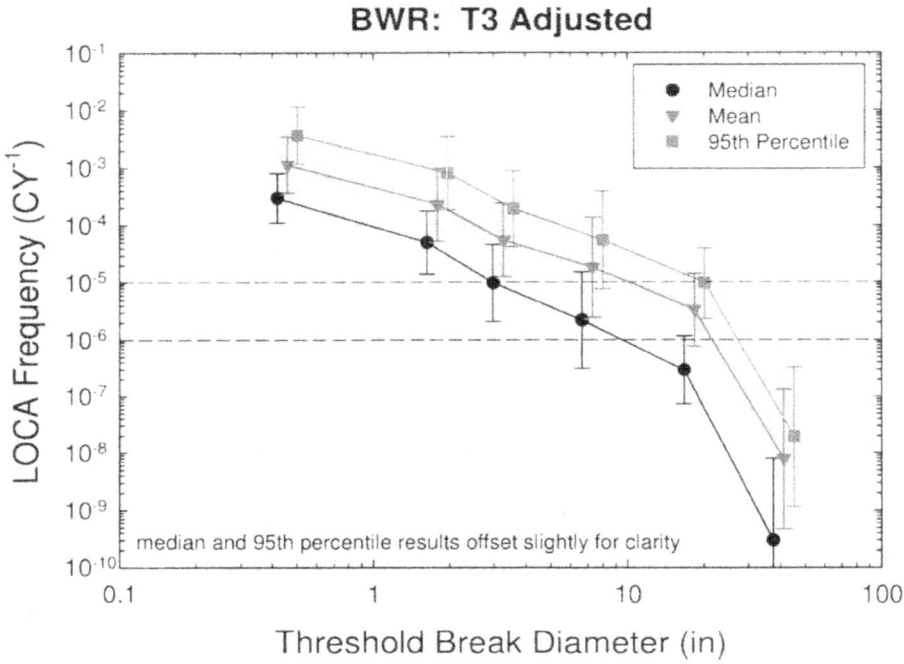

Figure 7.19 BWR LOCA Frequencies with T3 Adjustment

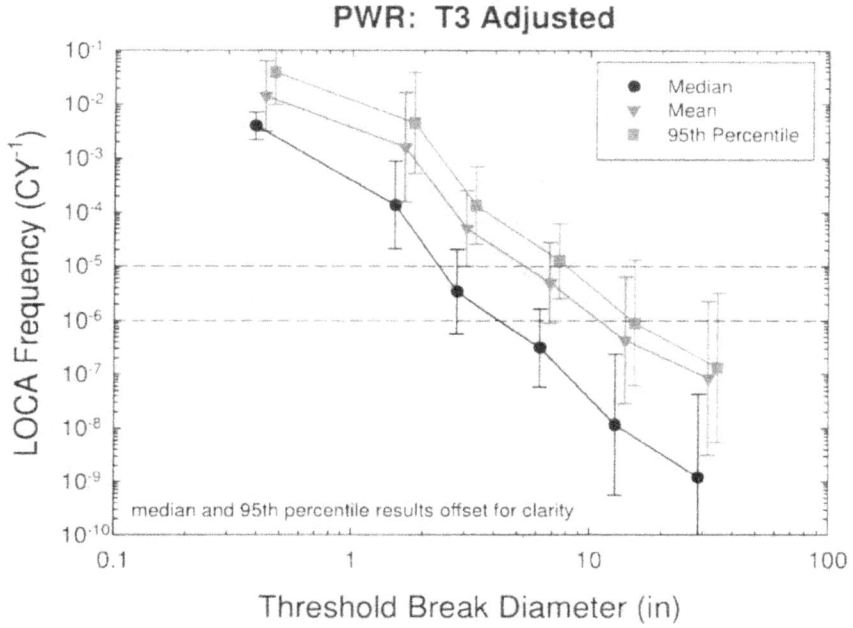

Figure 7.20 PWR LOCA Frequencies with T3 Adjustment

Table 7.5 Targeted (T3) Overconfidence Adjustment Compared to Unadjusted Results (Baseline)

	BWR: Current-Day			PWR: Current-Day	
LOCA Category	Mean Ratio	95th Percentile Ratio	LOCA Category	Mean Ratio	95th Percentile Ratio
1	1.9	2.1	1	2.1	2.1
2	2.0	2.1	2	3.3	2.9
3	2.2	2.2	3	3.8	3.0
4	2.9	2.5	4	3.8	3.0
5	2.8	2.4	5	3.8	3.0
6	2.3	2.3	6	5.3	3.3

Past studies have found that overconfidence is a typical phenomenon in elicitation exercises, which suggests that the uncertainty of the panelist responses should be adjusted at least to some degree. However, the overconfidence adjustment schemes evaluated in this section have a variety of deficiencies that limit their practical application. As noted in this section only the T3 adjustment scheme appears to be supported by operating experience, the corresponding base case frequencies, and the elicitation responses. The blanket (B1 and B2) and other targeted (T1 and T2) adjustment schemes are not consistent with this other information. Furthermore, none of the adjustment schemes uses a formal or rigorous method to determine the appropriate adjustment level, and the choice of panelist responses to adjust in the targeted adjustment schemes is also somewhat arbitrary.

Another deficiency of the adjustment schemes considered is that the adjustment level for each panelist is the same regardless of the LOCA category and the magnitude of his uncertainty relative to the other panelists. For instance, uncertainty is expected to increase with LOCA size because greater extrapolation from the passive-system failure data is required. Therefore, it may be appropriate to have a larger adjustment for larger LOCAs. Also, it may be appropriate to adjust a panelist's responses to a greater extent if he is very overconfident. This situation can be inferred from the panelist's results if his uncertainty is much less than the other panelists.

7.6.2.2 Error-Factor Adjustment – The error-factor adjustment scheme was developed to address several of the deficiencies associated with the coverage interval adjustment schemes (Section 7.6.2.1). As described in Section 5.6.2.2, this scheme adjusts the one-sided error factors associated with each panelist's 5th and 95th LOCA frequency estimates instead of the coverage intervals associated with the individual elicitation responses. Using the unadjusted median and the adjusted error factors, an adjusted mean and 5th and 95th percentiles are calculated for each panelist.

To implement the error-factor adjustment scheme, the GM of all the individual unadjusted error factors is first determined as a group estimate of the error factor for each plant type, LOCA category, and operating time period evaluated. Next, error factors less than the GM for a given estimate are adjusted up to the GM. Error factors above the GM are not adjusted. This scheme leads to a variable adjustment as a function of plant type, LOCA category, and operating time. Furthermore, the magnitude of the adjustment of the error factors less than the GM increases with the difference between the unadjusted error factor and the GM.

The GMs of the unadjusted error factors for the current-day estimates are tabulated in Table 7.6. As before, the UEF is the ratio of the 95th percentile to the median while the LEF is the ratio of the median to the 5th percentile. The LEF GM tends to be larger than the UEF GM, except for the larger category LOCAs, i.e., BWR LOCA Category 6 and PWR LOCA Categories 5 and 6. This reflects the panel's opinion that there is more uncertainty associated with how low these frequencies may be compared to how large they may be. The LEF GMs are similar for the BWR and PWR plant types. However, the PWR UEF GMs are somewhat larger than the BWR UEF GMs, reflecting more uncertainty about the effects of PWSCC (Section 6.3.2).

Table 7.6 Geometric Means of the Panelists' Error Factors
Current-day Estimates

LOCA Category	BWR Plants		PWR Plants	
	LEF	UEF	LEF	UEF
1	7	6	4	5
2	12	7	13	11
3	12	9	12	13
4	17	10	15	13
5	23	14	20	26
6	27	29	25	33

The median, mean, 5th, and 95th percentile BWR and PWR LOCA frequency values adjusted using the error-factor scheme are tabulated in Table 7.7 for both the current-day and end-of-plant-license periods. The current-day median, mean, and 95th percentile estimates and their associated 97.5% upper confidence bounds (Section 5.4.2) are tabulated in Table 7.8 and plotted in Figures 7.21 and 7.22, for BWR and PWR plants, respectively. Table 7.9 tabulates the ratios of the adjusted error-factor mean and 95th percentiles along with their associated 97.5% upper confidence bounds compared to the baseline results. Figure 7.23 illustrates the differences between the unadjusted means and means adjusted using the error-factor scheme for the current-day PWR estimates. It is important to note in these results, that the medians are identical to those indicated in Table 7.1, because they are not adjusted by the error-factor scheme.

As seen in Table 7.9 and Figure 7.23, the adjusted error-factor and baseline results are not markedly different for either BWR or PWR plants. The BWR adjustment increases (Table 7.9) both the mean and 95th percentile estimates by less than a factor of 1.5 for LOCA Categories 1 to 5 and approximately a factor of 2 for LOCA Category 6. Similarly, the PWR adjustment increases the mean and 95th percentile estimates by no more than a factor of 2 for all LOCA categories. The 95th percentile adjustments are just slightly larger than the mean estimates. Also, except for the largest LOCAs, the adjustments to upper confidence bounds are comparable to the corresponding adjustments to the mean and 95th percentiles. More importantly, the differences between the adjusted and baseline estimates do increase with increasing LOCA category as the panelists' uncertainties increase.

Table 7.7 Total BWR and PWR LOCA Frequencies
(After Overconfidence Adjustment using Error-Factor Scheme)

Plant Type	LOCA Size (gpm)	Eff. Break Size (inch)	Current-Day Estimate (per cal. year) (25 years fleet average operation)				End-of-Plant-License Estimate (per cal. year) (40 years fleet average operation)			
			5th Per.	Median	Mean	95th Per.	5th Per.	Median	Mean	95th Per.
BWR	>100	½	3.3E-05	3.0E-04	6.5E-04	2.3E-03	2.8E-05	2.6E-04	6.2E-04	2.2E-03
	>1,500	1 7/8	3.0E-06	5.0E-05	1.3E-04	4.8E-04	2.5E-06	4.5E-05	1.2E-04	4.8E-04
	>5,000	3 ¼	6.0E-07	9.7E-06	2.9E-05	1.1E-04	5.4E-07	9.8E-06	3.2E-05	1.3E-04
	>25K	7	8.6E-08	2.2E-06	7.3E-06	2.9E-05	7.8E-08	2.3E-06	9.4E-06	3.7E-05
	>100K	18	7.7E-09	2.9E-07	1.5E-06	5.9E-06	6.8E-09	3.1E-07	2.1E-06	7.9E-06
	>500K	41	6.3E-12	2.9E-10	6.3E-09	1.8E-08	7.5E-12	4.0E-10	1.0E-08	2.8E-08
PWR	>100	½	6.9E-04	3.9E-03	7.3E-03	2.3E-02	4.0E-04	2.6E-03	5.2E-03	1.8E-02
	>1,500	1 5/8	7.6E-06	1.4E-04	6.4E-04	2.4E-03	8.3E-06	1.6E-04	7.8E-04	2.9E-03
	>5,000	3	2.1E-07	3.4E-06	1.6E-05	6.1E-05	4.8E-07	7.6E-06	3.6E-05	1.4E-04
	>25K	7	1.4E-08	3.1E-07	1.6E-06	6.1E-06	2.8E-08	6.6E-07	3.6E-06	1.4E-05
	>100K	14	4.1E-10	1.2E-08	2.0E-07	5.8E-07	1.0E-09	2.8E-08	4.8E-07	1.4E-06
	>500K	31	3.5E-11	1.2E-09	2.9E-08	8.1E-08	8.7E-11	2.9E-09	7.5E-08	2.1E-07

Table 7.8 Current-Day Total BWR and PWR LOCA Frequencies and Upper Confidence Bounds
(After Overconfidence Adjustment using Error-Factor Scheme)

Plant Type	LOCA Size (gpm)	Eff. Break Size (inch)	Median	Median – 97.5% conf.	Mean	Mean – 97.5% conf.	95th Per.	95th Per. – 97.5% conf.
BWR	>100	½	3.0E-04	8.0E-04	6.5E-04	2.1E-03	2.3E-03	7.9E-03
	>1,500	1 7/8	5.0E-05	1.8E-04	1.3E-04	5.3E-04	4.8E-04	2.1E-03
	>5,000	3 ¼	9.7E-06	4.6E-05	2.9E-05	1.6E-04	1.1E-04	6.4E-04
	>25K	7	2.2E-06	1.5E-05	7.3E-06	6.0E-05	2.9E-05	2.4E-04
	>100K	18	2.9E-07	1.2E-06	1.5E-06	6.9E-06	5.9E-06	2.6E-05
	>500K	41	2.9E-10	8.0E-09	6.3E-09	1.4E-07	1.8E-08	3.9E-07
PWR	>100	½	3.9E-03	7.1E-03	7.3E-03	1.5E-02	2.3E-02	5.2E-02
	>1,500	1 5/8	1.4E-04	8.9E-04	6.4E-04	3.0E-03	2.4E-03	1.2E-02
	>5,000	3	3.4E-06	2.0E-05	1.6E-05	8.9E-05	6.1E-05	3.4E-04
	>25K	7	3.1E-07	1.7E-06	1.6E-06	1.0E-05	6.1E-06	3.6E-05
	>100K	14	1.2E-08	2.4E-07	2.0E-07	2.6E-06	5.8E-07	7.9E-06
	>500K	31	1.2E-09	4.3E-08	2.9E-08	6.7E-07	8.1E-08	2.0E-06

Table 7.9 Ratio of Adjusted Error Factor to Baseline Results

	LOCA Category	Mean	Mean: 97.5% confidence	95th Percentile	95th percentile, 97.5% confidence
BWR	1	1.2	1.1	1.3	1.1
	2	1.2	1.0	1.3	1.1
	3	1.2	1.1	1.3	1.2
	4	1.2	1.1	1.3	1.2
	5	1.3	1.2	1.5	1.3
	6	1.9	2.2	2.2	2.7
PWR	1	1.1	1.0	1.3	1.1
	2	1.4	1.2	1.6	1.4
	3	1.3	1.3	1.4	1.4
	4	1.3	1.5	1.5	1.7
	5	1.8	2.3	2.0	2.7
	6	1.9	2.4	2.0	2.7

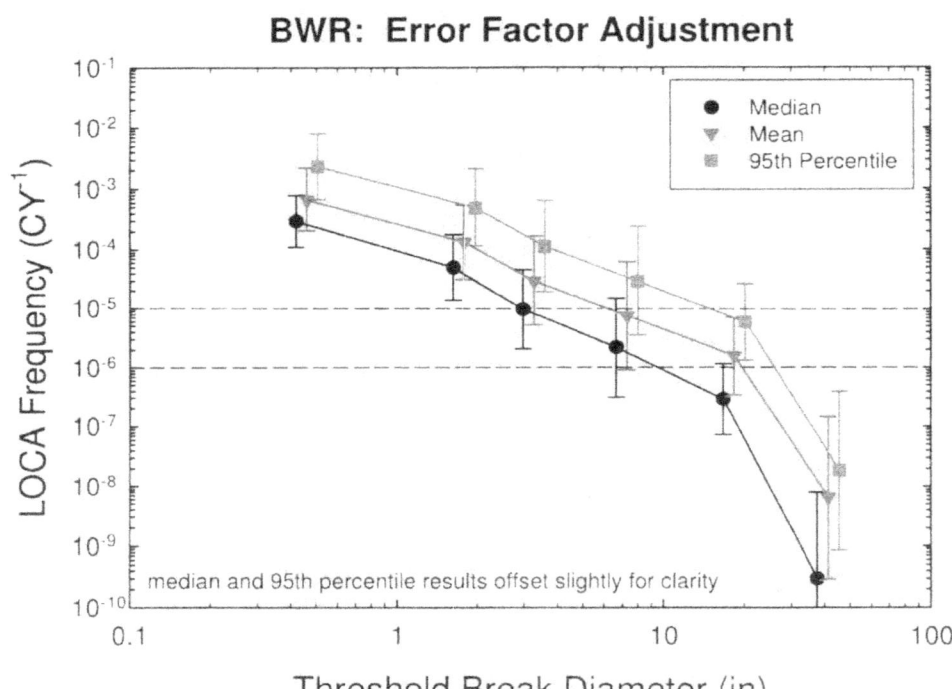

Figure 7.21 BWR LOCA Frequencies with Error-Factor Adjustment

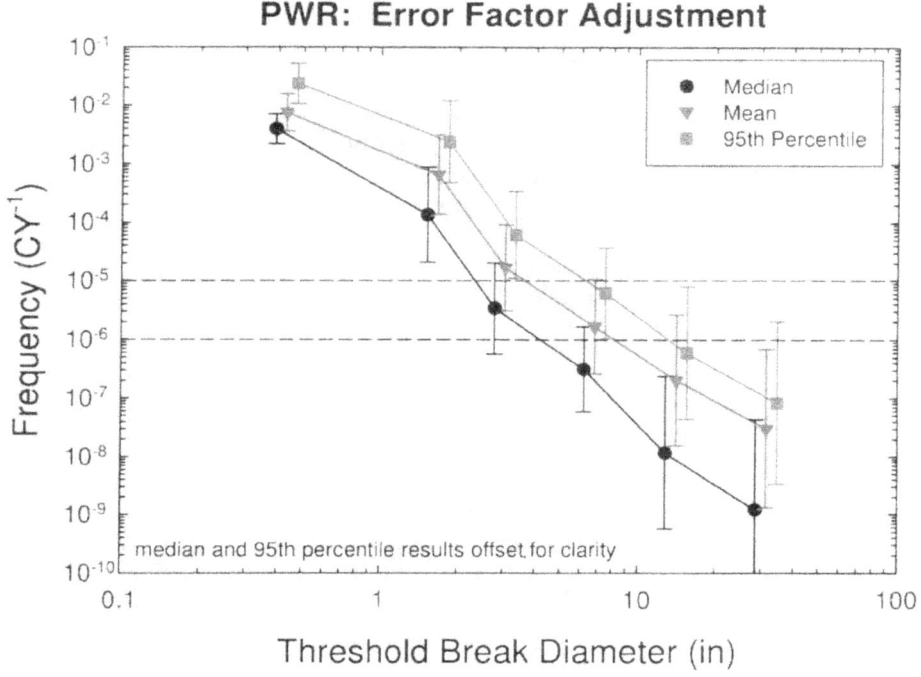

Figure 7.22 PWR LOCA Frequencies with Error-Factor Adjustment

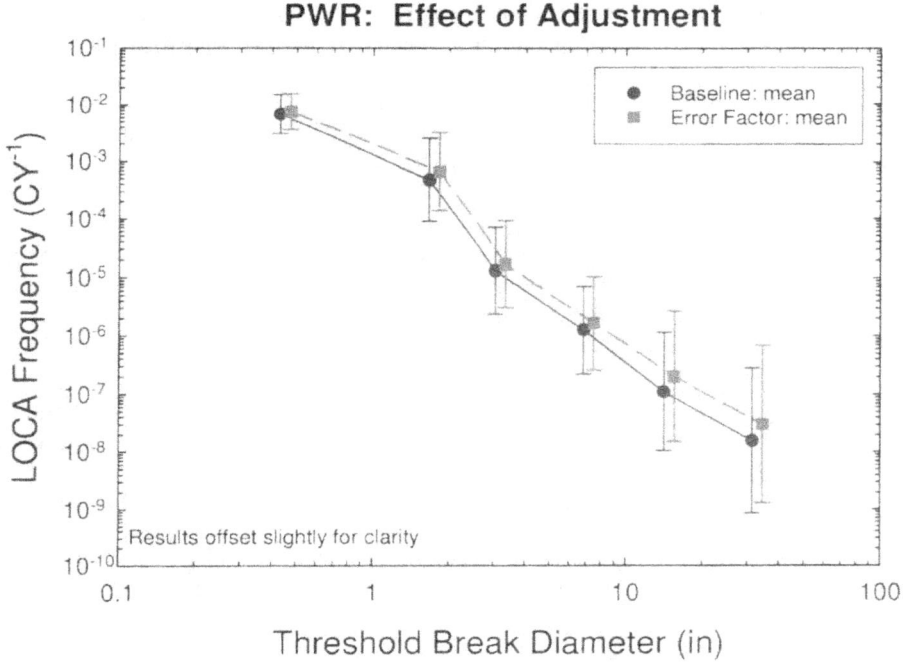

Figure 7.23 Comparison of Baseline and Error-Factor Adjustment Results

The results from the error-factor adjustment and the targeted (T3) adjustment (Section 7.6.2.1) are also reasonably comparable. The mean and 95th percentile estimates from these two adjustment schemes differ by less than a factor of 5 for all LOCA categories and plant types. The differences are not obviously correlated with LOCA category. However, the error-factor adjustments are consistently smaller than the T3 adjustments. The error-factor adjustment is less because it modifies the nominal 90% coverage intervals of the bottom-line estimates so that they effectively remain greater than 60% (the value for the T3 adjustment) for the adjusted panelists. It might be possible to determine a coverage interval for the adjusted panelists which would lead to a more accurate comparison of the differences between the T3 and error-factor schemes. However, a true comparison requires that the effective coverage intervals for individual elicitation responses be determined, and this is not possible using the error-factor adjustment scheme.

The error-factor adjustment scheme is preferable to the coverage interval adjustment schemes evaluated in Section 7.6.2.1 because the error-factor adjustment varies as a function of the difference between the individual and group (GM) estimates for each parameter. Furthermore, the error-factor scheme requires no arbitrary determination of the panelists to adjust and the adjustment level, as is the case for the coverage interval schemes. Finally, the error-factor adjustment increases with LOCA size, as does group uncertainty. This reflects the expectation that panelist overconfidence will increase with increasing LOCA size as the extrapolation from the passive-system failure data increases. For these reasons, the error-factor scheme is used to adjust for overconfidence to calculate the summary estimates for this study.

7.6.3 Correlation Structure

The percentiles of the total BWR and PWR LOCA frequency distributions are a function of the correlation structure of the distributions associated with the various LOCA-sensitive piping systems and non-piping subcomponents and which are summed to obtain the total frequency distributions (Sections 5.3.4 and 5.6.3). While it is plausible to assume that the components of the sums are positively correlated, the correlation structure is otherwise unknown. However, the correlation structure can be bounded by assuming that the piping systems and non-piping subcomponents are either independent (zero correlation) or have perfect rank correlation (maximum achievable correlation). The baseline analysis (Section 5.5) assumes perfect rank correlation to calculate percentiles during any step which requires the summation of distributions. This sensitivity analysis examines the effect of assuming independence when summing distributions.

Monte Carlo simulation was used to evaluate the differences between the perfect rank correlation (PRC) and independent correlation (IC) structures. Because of the large number of distributions that represent the individual elicitation responses, Monte Carlo trials were used to calculate the bottom line estimates only for selected panelist responses. Ten simulation trials were selected to span several important variables as listed in Table 7.10. A number of criteria were considered in selecting the trials. First, a representative sample of the panelists and the range of LOCA size, location, plant type, and time period combinations[4] were chosen. Second, the number of distributions representing contributing piping components (or non-piping subcomponents) which are summed to develop the bottom-line estimates was varied. Third and most important, the Monte Carlo trials were selected to span the range of distinguishing characteristics of the distributions representing the elicitation responses. These distinguishing characteristics include the general shape [symmetric (S in Table 7.10) or asymmetric (U in Table 7.10)] and the relative magnitudes (small, moderate, large) of the larger of the LEF or UEF in Table 7.10). The range of distinguishing characteristics reflects the variability of the elicitation responses.

---

[4] For example, BWR-1 Piping @ 25 years refers to a Category 1 LOCA in a BWR piping system after 25 years of operation.

Table 7.10 Summary of Monte Carlo Trials

Trial Number	Number of Summed Distributions	Panelist	LOCA Size, Location. Plant Type and Time Period	Distinguishing Characteristics
1	12	A	BWR-1 Piping @ 25 yrs	S, small LEF/UEF
2	12	A	BWR-2 Piping @ 25 yrs	U, moderate LEF
3	2	C	PWR-6 Piping @ 25 yrs	U, large UEF
4	4	C	BWR-3 Piping @ 25 yrs	S, moderate LEF/UEF
5	14	G	PWR-5 Non-Piping @ 60 yrs	S, large LEF/UEF
6	5	C	BWR-3 Non-Piping @ 25 yrs	S, large LEF/UEF
7	8	J	PWR-5 Non-Piping @ 25 yrs	U, large UEF
8	7	I	PWR-4 Piping @ 25 yrs	S, moderate LEF/UEF
9	4	E	BWR-4 Non-Piping @ 25 yrs	U, large LEF
10	9	B	PWR-3 Non-Piping @ 25 yrs	U, moderate UEF

Each of the individual piping component (or non-piping subcomponent) distributions is the product of two independent elicitation response distributions: (1) the anchoring or base case frequencies (Blocks 1.2 or 1.9 in Figure 5.1) and (2) the corresponding adjustment ratios (Blocks 1.3 or 1.10 in Figure 5.1). These distributions were simulated in the Monte Carlo trials by split lognormal distributions truncated at the 99.9[th] percentile (Section 5.3.3). The products of the distributions were determined by generating 10,000 independent pairs of the anchoring frequency and adjustment ratio distributions, multiplying the paired values and sorting them into ascending order. The distribution of the sum, S, of any number, N (column 2 in Table 7.10), of such products was then determined using a PRC structure in the following manner. The rank 1 values of the N product distributions were summed to obtain the rank 1 value of S, the rank 2 values of the N product distributions were summed to obtain the rank 2 value of S, and so on. The resulting 10,000 ordered values of S determine the distribution of S under the assumption of a perfect rank correlation structure. The required percentiles (median, 5th and 95th) were then determined directly from the distribution for S.

The results obtained by this Monte Carlo procedure are somewhat different than those obtained by the baseline calculation methodology described in Section 5.3.5 because the baseline calculation methodology uses an approximation to the assumed truncated split lognormal distributions to calculate the percentiles. Figure 7.24 compares the 5th, 50th, and 95th percentiles calculated using the baseline calculation methodology with the Monte Carlo simulations using the assumed PRC structure. This figure plots the ratios of the Monte Carlo results to the baseline results for each Monte Carlo trial in Table 7.10. The 5th and 95th baseline percentiles differ by less than 10% from the Monte Carlo percentiles for all the trials. The medians also differ by less than 10%, except for trials 2 and 9 where the baseline medians are approximately 30% and 20% higher, respectively, than the Monte Carlo medians.

It is somewhat surprising that the two largest differences are for the medians, because the differences between the medians in the other eight trials are generally smaller than the differences for the 5th and 95th percentiles. However, it is interesting to note that the distributions in trials 2 and 9 are grossly asymmetric and the LEF is at least a factor of 3 larger than the UEF. As expected, the means determined by the Monte Carlo and baseline methods are approximately equal (not illustrated in Figure 7.24) since they assume identical distributions for calculating the mean.

The conclusion is that the differences between the baseline calculation methodology and the Monte Carlo procedure are generally insignificant for a PRC structure. The largest differences are expected for the median estimates for distributions which are asymmetric and have LEFs at least 3 times larger than the UEFs. In these cases, the approximate median estimates determined by the baseline method will be marginally conservative, up to 30%. However, there are relatively few elicitation responses having these characteristics. Therefore, any differences in the individual bottom line estimates will be further reduced upon aggregating the individual responses to determine the group estimates.

As previously discussed, Monte Carlo simulation was used to evaluate the differences between the PRC and IC structures using the previously selected distributions (Table 7.10). This approach was necessary because no closed-form formulas exist for summing distributions using an IC structure so that individual estimates (Block 1.18 in Figure 5.1) calculated using an IC and PRC (i.e., baseline method) structure could be directly compared.

Therefore, summations of the distributions in Table 7.10 were determined using an IC structure as follows. First, as described previously for the PRC simulation, the products of the independent anchoring frequency and adjustment ratio distributions for the selected piping components (or non-piping subcomponents) were determined using Monte Carlo simulation. Next, the product distributions for each term in the sum, S, were independently sampled and then added for the number of required distributions, N, to determine the total piping (or non piping) distribution. This calculation was then repeated to obtain 10,000 samples for S. These 10,000 samples were then ordered to determine the IC cumulative distribution function of S for comparison with the PRC cumulative distribution function of S.

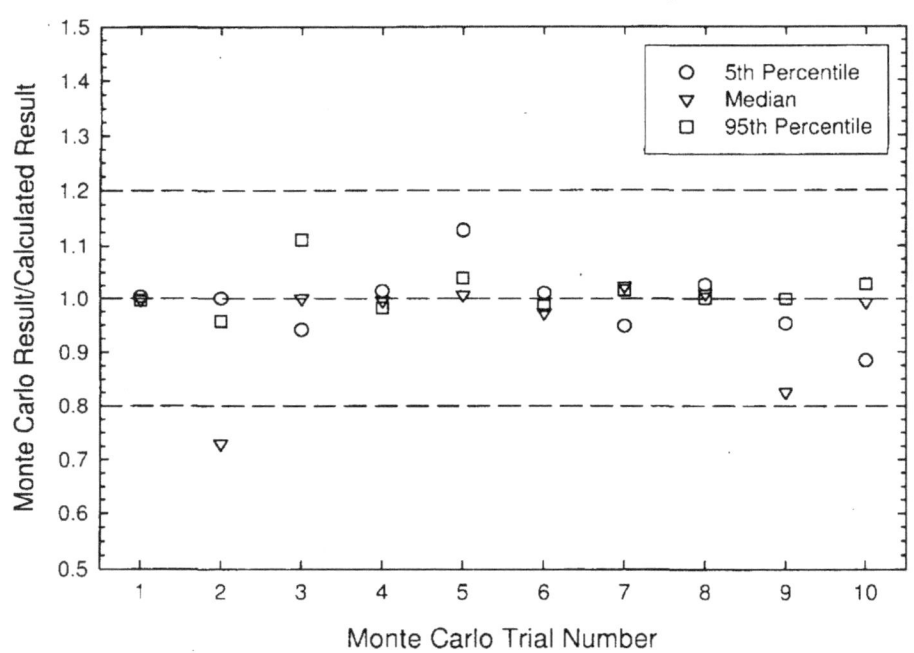

Figure 7.24 Ratio of Monte Carlo to Baseline Results Assuming a Perfect Rank Correlation Structure

Figure 7.25 plots the ratios of the median, 5th and 95th percentile estimates determined by assuming a PRC structure to those calculated assuming an IC structure as a function of the trial number in Table 7.10. The mean is not illustrated because it is unaffected by the correlation structure. For these 10 trials, the median and 5th percentile estimates based on the independent correlation structure are consistently larger than those based on the PRC structure. However, the PRC estimates of the 95th percentile are usually larger than the independent estimates with the exception of trials 3 and 7. Therefore, based on these results, the

PRC structure generally results in increased differences between the 5th and 95th percentiles compared with the IC structure. This is not surprising because the PRC structure leads to maximum correlation (Section 5.3.4.1) between the terms in the sum and this broadens the distribution of the sum, S.

It is also apparent from Figure 7.25 that differences resulting from the correlation structure decrease as the percentiles of S, increase. Differences in the 5th percentile can be nearly two orders of magnitude. The largest differences in the 5th percentile are greater than one order of magnitude and occur for trials 5, 7, and 9. The distributions in these trials all have relatively large LEFs. Differences in the medians are less than a factor of two for all the Monte Carlo trials except 3, 5, and 7, which have distributions with the largest error factors. The largest difference in the median estimates is more than an order of magnitude and occurs for trial 7, which is determined by summing several (8) distributions with large UEFs of $\approx$ 300. Differences in the 95th percentiles are generally less than a factor of two. The largest differences occur for trials 3 and 7, which consist of asymmetric distributions characterized by large error factors (up to 1,000).

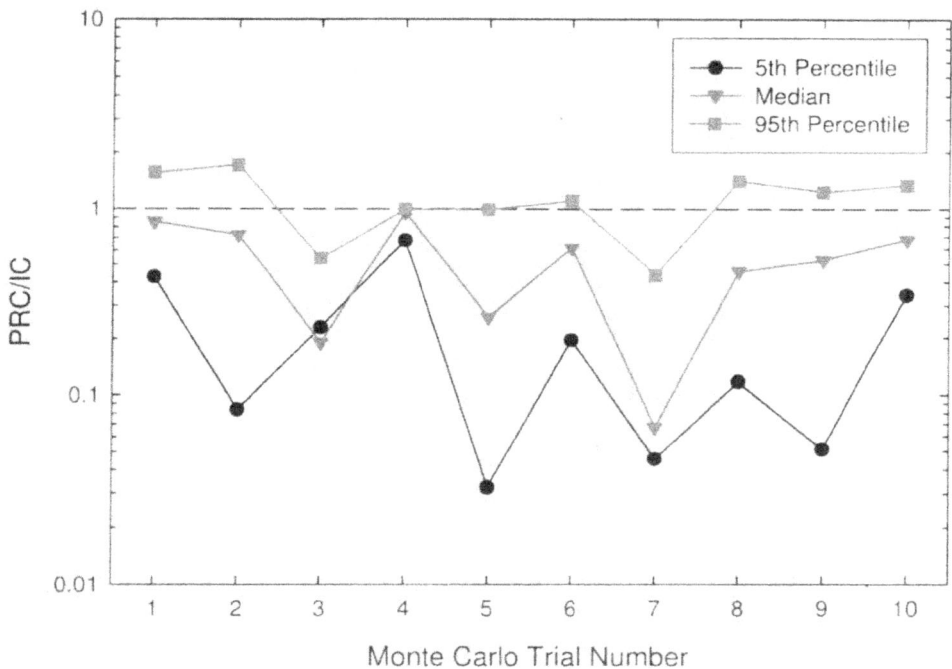

Figure 7.25 Ratio of PRC to IC Structure Results

In summary, the correlation structure affects the lower percentiles more significantly than the 95th percentile. The PRC structure leads to significant decreases in the medians and especially the 5th percentiles. Differences between the correlation structures increase as both the error factor and the number of distributions summed increase. This point should be considered when interpreting the baseline results for the median or 5th percentile. The 95th percentiles are not particularly sensitive to the correlation structure with the PRC structure leading to higher 95th percentile estimates except for extremely asymmetric distributions with large UEFs (> 100). Of course, neither correlation structure affects the means of the summed distributions.

It should be emphasized that the distributions chosen for the Monte Carlo trials spanned the full range of characteristics of the elicitation responses. Consequently, the largest differences exhibited in Figure 7.25 are not representative of the majority of the elicitation distributions. Therefore, when the individual results are aggregated to form group estimates, any differences between the lower percentiles due to the correlation structure will be reduced.

The assumption of a PRC structure for the baseline calculation method is justified by the observation that the elicitation responses tend to be strongly correlated. Most panelists noted that degradation mechanisms that significantly contribute to the LOCA frequencies affect many systems similarly. That is, if conditions lead to higher failure probabilities in one piping system due to a specific degradation mechanism, the failure probabilities of other susceptible systems will also increase. Furthermore, the choice of correlation structure has little effect on the estimated means and 95th percentiles of the total LOCA frequencies (Block 1.18 in Figure 5.1), which are the most important parameters for most applications.

7.6.4 Aggregation

Several analyses were performed to examine the sensitivity of the baseline results to other possible aggregation schemes for combining the panelists' individual estimates to form group estimates. Sensitivity studies evaluated the effect of using different measures of group opinion other than the GM (Section 5.6.4.1), the effect of aggregating panelist responses at different stages of the analysis (Section 5.6.4.2), and the sensitivity of the results to the method of calculating parameters using the lognormal parameterization (Section 5.6.4.3). Aggregation by averaging the panelists' distributions to create a mixture distribution was also evaluated (Section 5.6.4.4). More details on each of these analyses are available in Section 5.6.4.

7.6.4.1 Measures of Group Opinion - The objective of the elicitation (Section 2) is to determine a group estimate of LOCA frequency. The technique used to determine group opinion can have a significant effect on the group estimates. Several different aggregation schemes were used to determine group estimates from the individual total LOCA frequency estimates: GM, TGM, median, and AM. The baseline LOCA frequency estimates use the GM (Section 5.5).

There are no significant differences between the median, GM, and TGM group estimates of the mean PWR LOCA frequency estimates at 25 years of plant operation (Figure 7.26). In general, the differences among these three measures are typically less than a factor of 2 for all the bottom line parameters (i.e., mean, median, 5th, and 95th percentiles). There was some initial concern that the very low estimates provided by Panelist J for the PWR Category 5 and 6 LOCA frequencies might significantly affect the GM estimates. However, from Figure 7.26, this is not the case.

Differences among these measures are also slightly less than a factor of 2 for the BWR LOCA frequency estimates (Figure 7.27) except for LOCA Category 6, where the median group estimate is approximately a factor of five higher than the GM estimate. In this particular case, there were two panelist's responses for the mean and 95th percentile estimates that were approximately 2 orders of magnitude less than the remaining estimates. These low individual estimates cause the GM group estimates to fall below the median.

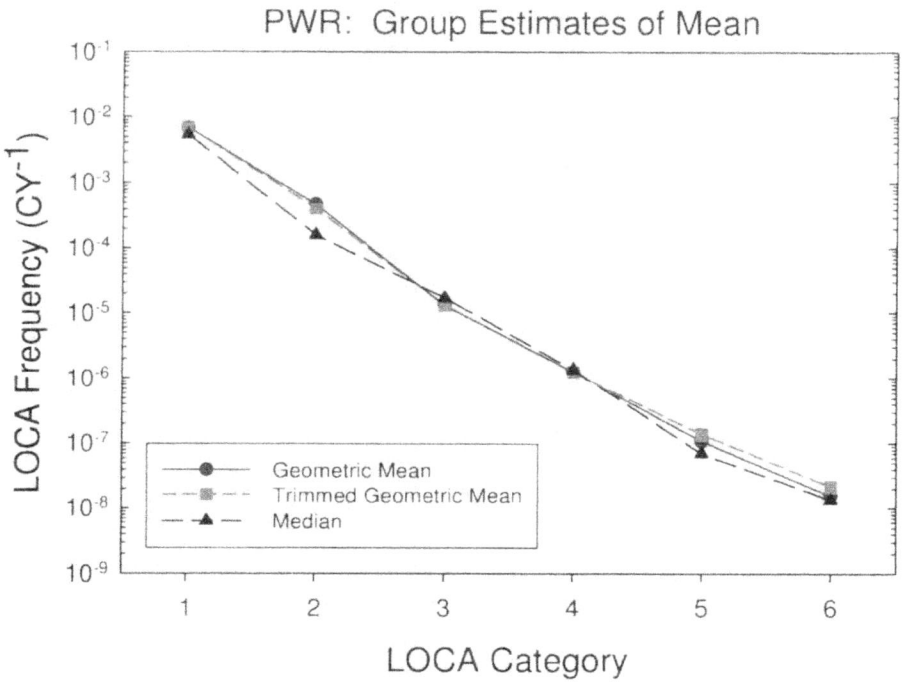

Figure 7.26 Group Estimates of Mean PWR LOCA Frequencies at 25 Years

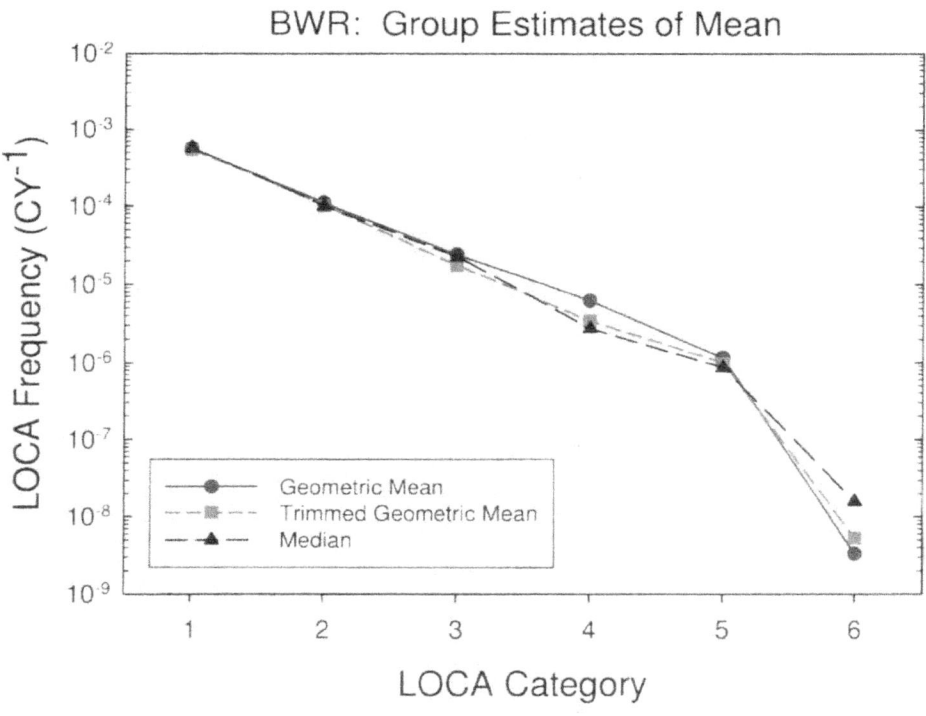

Figure 7.27 Group Estimates of Mean BWR LOCA Frequencies at 25 Years

The AM and GM group estimates of the BWR and PWR LOCA mean frequencies at 25 years are plotted in Figures 7.28 and 7.29, respectively. Because the AM of any data set is always larger than its GM, the AM LOCA frequency estimates will always be larger than the GM estimates. However, these figures indicate that the AM mean LOCA frequency estimates are significantly higher than the GM estimates. The difference between the estimates can be as large as two orders of magnitude for BWRs and one and a half orders of magnitude for PWRs.

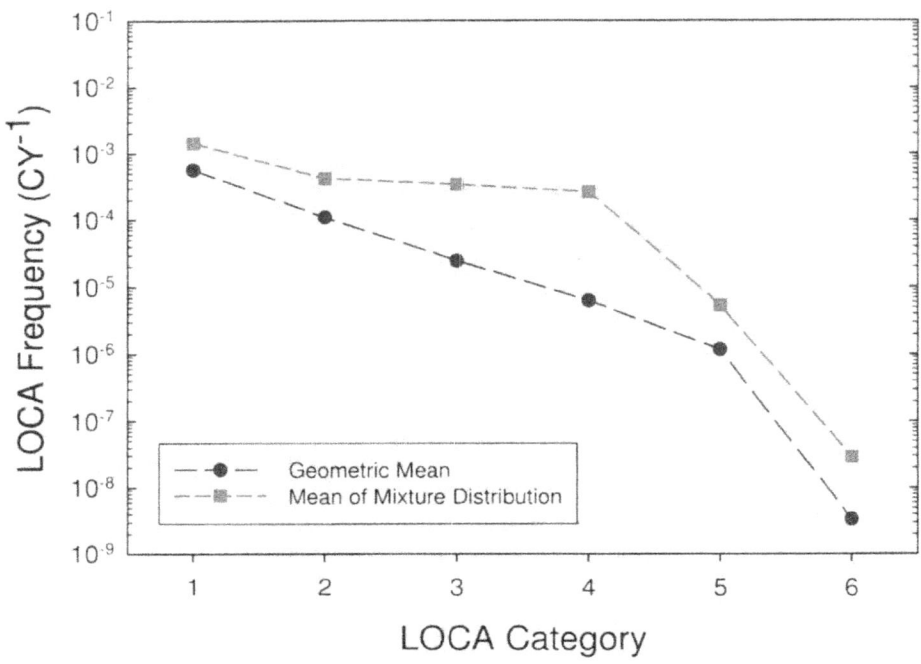

Figure 7.28 Arithmetic-Mean and Geometric-Mean Group Estimates of Mean BWR LOCA Frequencies at 25 Years

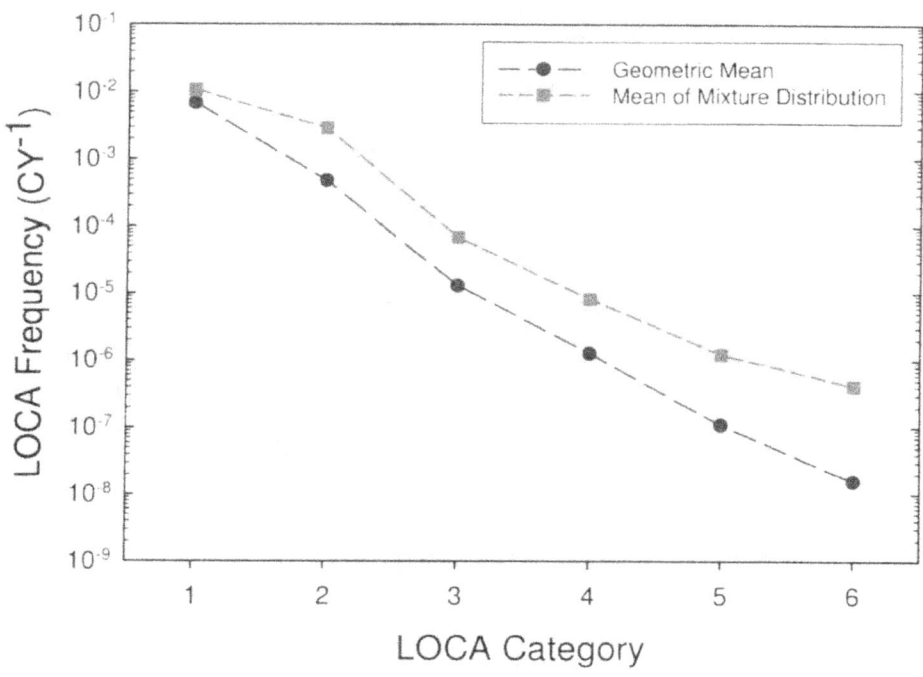

**Figure 7.29 Arithmetic-Mean and Geometric-Mean Group Estimates of
Mean PWR LOCA Frequencies at 25 Years**

The LOCA frequencies determined using the AM aggregation method at 25 and 40 years of plant operations are provided in Table 7.11. These values are the baseline values determined without using any overconfidence adjustment. The ratios of the AM to GM aggregated LOCA frequencies for the baseline (unadjusted) mean and 95th percentile estimates are provided in Table 7.12 by dividing the LOCA frequencies in Table 7.11 by those in Table 7.1. The largest ratios are greater than a factor of 10 and occur for the BWR Category 3 and 4 and the PWR Category 5 and 6 LOCA estimates. In each of these instances, there are one or two individual estimates that are significantly higher than the others. The AM aggregated group estimates and therefore the ratios are dominated by these high values.

Specifically, a single panelist, C, expects the Category 2 to 4 LOCA frequencies for BWR plants to be essentially constant due to IGSCC concerns in large recirculation system lines and the expectation that failure of these lines will propagate directly to a larger LOCA (Section 7.5.2). The expected frequency of these failures is approximately two orders of magnitude higher than other estimates for Category 4 LOCAs. The AM estimates largely reflect this single panelist's expectation for the Category 3 and 4 LOCAs. The same panelist also has large uncertainty about the Category 5 and 6 PWR LOCA estimates. While this panelist's median values are consistent with the remainder of the group, the large uncertainty leads to means that are significantly higher than most other panelists' means. Also, Panelist A's Category 5 and 6 LOCA median estimates are significantly higher than the remaining panelists' medians (Section 7.5.2), which leads to comparatively high mean estimates. These two panelists' estimates cause the increases in the PWR Category 5 and 6 AM estimates relative to the GM estimates.

Table 7.11 Total BWR and PWR LOCA Frequencies
(Using Arithmetic-Mean Aggregation)

Plant Type	LOCA Size (gpm)	Eff. Break Size (inch)	Current-Day Estimate (per cal. year) (25 years fleet average operation)				End-of-Plant-License Estimate (per cal. year) (40 years fleet average operation)			
			5th Per.	Median	Mean	95th Per.	5th Per.	Median	Mean	95th Per.
BWR	>100	½	5.8E-05	5.1E-04	1.4E-03	5.0E-03	4.7E-05	4.3E-04	1.3E-03	4.9E-03
	>1,500	1 7/8	7.3E-06	1.3E-04	4.1E-04	1.6E-03	5.4E-06	8.6E-05	2.9E-04	1.1E-03
	>5,000	3 ¼	3.1E-06	7.0E-05	3.3E-04	1.3E-03	1.8E-06	3.6E-05	1.6E-04	6.4E-04
	>25K	7	1.6E-06	5.2E-05	2.6E-04	1.0E-03	8.9E-07	2.7E-05	1.4E-04	5.4E-04
	>100K	18	3.5E-08	7.6E-07	5.2E-06	1.9E-05	4.4E-08	9.4E-07	7.1E-06	2.5E-05
	>500K	41	4.5E-10	5.0E-09	2.8E-08	4.8E-08	3.7E-10	4.7E-09	5.3E-08	7.2E-08
PWR	>100	½	1.1E-03	4.8E-03	1.0E-02	3.3E-02	7.5E-04	4.2E-03	9.3E-03	3.2E-02
	>1,500	1 5/8	5.7E-05	7.0E-04	2.8E-03	1.0E-02	4.3E-05	7.6E-04	3.2E-03	1.2E-02
	>5,000	3	1.8E-06	1.9E-05	6.7E-05	2.5E-04	1.6E-06	2.2E-05	1.3E-04	4.5E-04
	>25K	7	1.3E-07	1.3E-06	7.9E-06	2.1E-05	1.9E-07	2.0E-06	2.2E-05	6.1E-05
	>100K	14	1.6E-08	2.6E-07	1.2E-06	1.9E-06	3.1E-08	3.3E-07	2.3E-06	3.1E-06
	>500K	31	6.2E-09	1.5E-07	4.0E-07	7.8E-07	6.3E-09	1.5E-07	6.9E-07	9.9E-07

Table 7.12 Ratio of AM to GM (Baseline) Group Estimates

BWR: Current-Day			PWR: Current-Day		
LOCA Category	Mean Ratio	95th Percentile Ratio	LOCA Category	Mean Ratio	95th Percentile Ratio
1	2.5	2.9	1	1.5	1.8
2	3.8	4.2	2	5.9	6.9
3	14	16	3	5.2	5.7
4	41	47	4	6.3	5.1
5	4.5	4.9	5	11	6.5
6	8.6	5.6	6	26	20

Table 7.13 provides the 25 and 40-year LOCA frequency estimates determined using AM aggregation for individual estimates that have been adjusted using the error-factor scheme (Section 7.6.2.2). These results can be directly compared with those in Table 7.7 to evaluate the differences between AM and GM aggregation for the adjusted individual estimates. The ratios between the AM and GM aggregated results exhibit trends similar to those in Table 7.12, although they are slightly less because the error-factor adjustment decreases the diversity of the individual estimates. Greater diversity generally increases the difference between AM and GM aggregation.

Table 7.13 Total BWR and PWR LOCA Frequencies
(Using Arithmetic-Mean Aggregation with Error-Factor Overconfidence Adjustment)

Plant Type	LOCA Size (gpm)	Eff. Break Size (inch)	Current-Day Estimate (per cal. year) (25 years fleet average operation)				End-of-Plant-License Estimate (per cal. year) (40 years fleet average operation)			
			5th Per.	Median	Mean	95th Per.	5th Per.	Median	Mean	95th Per.
BWR	>100	½	5.0E-05	5.1E-04	1.4E-03	5.4E-03	3.8E-05	4.3E-04	1.3E-03	5.2E-03
	>1,500	1 7/8	6.4E-06	1.3E-04	4.1E-04	1.6E-03	4.6E-06	8.6E-05	2.9E-04	1.2E-03
	>5,000	3 ¼	3.0E-06	7.0E-05	3.4E-04	1.3E-03	1.6E-06	3.6E-05	1.6E-04	6.5E-04
	>25K	7	1.6E-06	5.2E-05	2.6E-04	1.0E-03	8.3E-07	2.7E-05	1.4E-04	5.5E-04
	>100K	18	2.5E-08	7.6E-07	5.5E-06	2.1E-05	3.0E-08	9.4E-07	7.5E-06	2.7E-05
	>500K	41	1.7E-10	5.0E-09	5.6E-08	1.6E-07	1.5E-10	4.7E-09	8.4E-08	1.9E-07
PWR	>100	½	8.1E-04	4.8E-03	1.0E-02	3.6E-02	5.2E-04	4.2E-03	9.2E-03	3.3E-02
	>1,500	1 5/8	4.2E-05	7.0E-04	3.0E-03	1.2E-02	3.9E-05	7.6E-04	3.5E-03	1.3E-02
	>5,000	3	1.3E-06	1.9E-05	7.3E-05	2.9E-04	1.4E-06	2.2E-05	1.4E-04	5.1E-04
	>25K	7	6.9E-08	1.3E-06	9.4E-06	3.0E-05	1.0E-07	2.0E-06	2.1E-05	6.8E-05
	>100K	14	9.9E-09	2.6E-07	2.4E-06	7.2E-06	1.4E-08	3.3E-07	4.0E-06	1.0E-05
	>500K	31	5.9E-09	1.5E-07	1.5E-06	5.2E-06	6.1E-09	1.5E-07	1.9E-06	5.5E-06

7.6.4.2 Aggregation Point - There are three possible aggregation points in the elicitation structure (Section 5.6.4.2). One is after the total LOCA frequencies have been calculated for each panelist (Block 1.18 in Figure 5.1). Another is after each panelist's piping and non-piping LOCA frequencies have been calculated (Blocks 1.7 and 1.17, respectively, in Figure 5.1). The last is after each panelist's piping system and non-piping sub-component frequencies have been calculated (Blocks 1.4 and 1.11, respectively, in Figure 5.1) and is the earliest possible aggregation point in the elicitation structure. However, this last alternative is problematic because the panelists' elicitation responses did not all have the same basic structure and the same level of completeness. For instance, each panelist may have used a slightly different definition of the piping system or non-piping sub-component boundaries. These basic differences make it difficult to ensure consistent LOCA frequency estimates among the panelists' responses at this elicitation step. Also, as discussed previously (Section 3), the elicitation questions only required each panelist to assess the dominant LOCA contributors. Additional assumptions would be required to impute missing or incomplete responses for each panelist's less-important LOCA contributing piping systems or non-piping sub-components in order to implement the aggregation. Because of these issues, aggregation at this earliest point was not performed in this study.

The current elicitation responses are suitable for aggregation at the other two points. The baseline aggregation point (Section 5.5) is after the total LOCA frequencies have been calculated for each panelist. As previously discussed (Section 5.5), the group estimate for the baseline analysis is calculated using the GM of the individual estimates. This aggregation scheme is referred to as individual aggregation within the rest of this section. The aggregation of the piping and non-piping contributions is accomplished by first calculating the group estimates separately for the piping and non-piping LOCA frequency parameters using the GM. As with the baseline approach, the group estimates of the parameters are assumed to be lognormally distributed and the piping and non-piping group estimates are then summed using the procedure described in Section 5.3.5. This aggregation scheme is referred to as component aggregation.

The differences resulting from the chosen aggregation point are plotted in Figure 7.30 for the PWR current-day median, mean and 95th percentile LOCA frequency estimates for both individual and component aggregation. The differences are small, usually less than a factor of two for all the bottom-line parameters, although the differences do increase as the LOCA size increases. The differences in the BWR results are similar, but they are not a function of the LOCA size. For both plant types, the baseline (individual) aggregation results lead to slightly higher frequencies. Therefore, while the results are not particularly sensitive to the aggregation point, the baseline methodology results in slightly more conservative estimates.

Individual aggregation is also the better choice based on the elicitation structure. This aggregation scheme is not affected by panelist differences in the piping and non-piping boundary definitions which could affect whether contributions (e.g., CRDM failures) were counted as piping or non-piping by a panelist. Individual aggregation also ensures that the relative piping and non-piping contributions and their associated uncertainties reflect the correlations expressed by the panelists. For example, a panelist might expect that piping is the dominant contributor to the total LOCA frequency estimates, but could be much less confident (large uncertainty) about the non-piping contribution. For this panelist, the total LOCA frequency estimates would be dominated by the piping contribution and its associated uncertainty. Thus, the individual aggregation results are dominated by those contributions judged to be most important by this panelist.

In contrast, with component aggregation, the more uncertain, but less important non-piping contribution for this panelist is explicitly considered in the non-piping aggregation. Because these contributions have a direct effect on the combined results, their actual values become important with component aggregation, while individual aggregation more accurately reflects the opinion of this panelist that the non-piping contribution should be negligible. In short, individual aggregation after the total LOCA frequencies have been calculated (baseline methodology) ensures that the results are more consistent with the panelists' opinions and that only the most important contributors identified by the panelist dominate the total LOCA frequency estimates.

7.6.4.3. Aggregation Parameters - This sensitivity analysis compares the bottom-line mean estimates determined from direct aggregation with estimates calculated from the other aggregated percentiles. In the baseline methodology, as discussed in Section 5.3, the mean estimates are calculated directly from the elicitation responses for each panelist by summing the means of the piping system (Block 1.4 in Figure 5.1) and non-piping sub-component (Block 1.11 in Figure 5.1) frequencies to determine the individual total LOCA frequencies (Block 1.18 in Figure 5.1). The group estimate for the mean is then determined by aggregation. The baseline methodology also aggregates the other percentiles (i.e., median, 5th, and 95th) directly from the panelists' responses (Section 5.3). An alternative approach is to calculate the bottom-line mean estimates only after the aggregated percentile estimates have been determined by assuming a distributional relationship for the percentile estimates.

This alternative scheme is called the median and error factor (MEF) approach. In this approach, the 5th and 95th percentile estimates are calculated using the GM to aggregate the median and error factors instead of directly aggregating the percentiles as in the baseline methodology. The bottom-line mean estimates are then calculated from the group estimates of the median and the 5th and 95th percentiles by assuming that the underlying distribution is a split lognormal which is truncated at the 99.9^{th} percentile. This is the same distributional form assumed for calculating the mean estimates from the individual elicitation responses in the baseline methodology (Section 5.3).

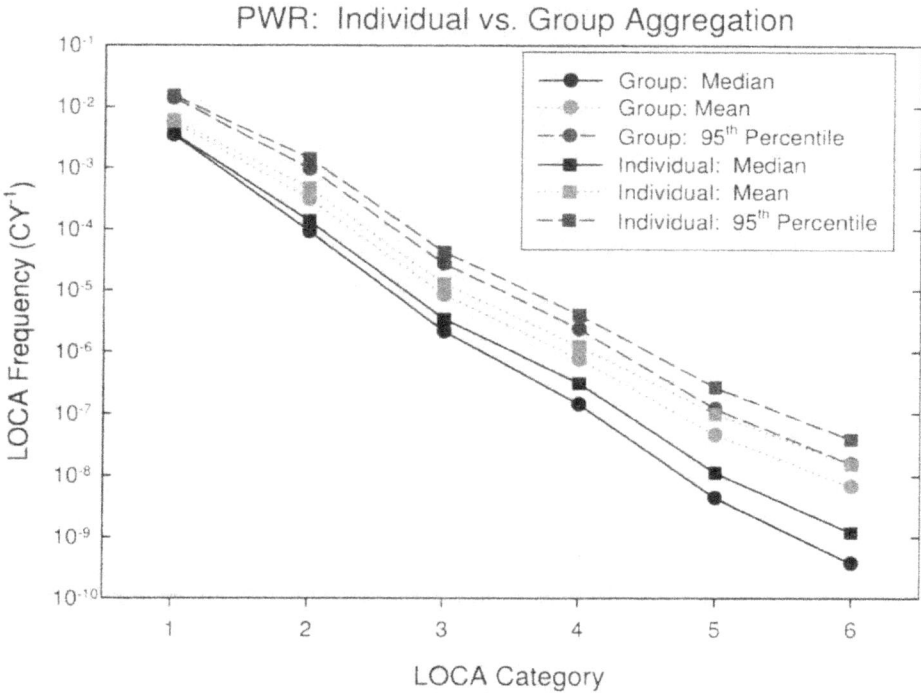

Figure 7.30 Effect of Aggregation Point on PWR LOCA Frequency Estimates at 25 Years of Operation

The 5th and 95th percentile estimates determined by the MEF approach are identical to the estimates determined by direct aggregation since all the individual estimates can be calculated either directly from the median and the 5th and 95th percentiles (baseline methodology) or from the median and the UEFs and LEFs (MEF approach). In fact, the MEF approach was used to calculate the percentiles in order to validate the baseline calculations. However, the mean can be determined at several points in the analysis and differences do result from this choice. Another possible calculation point is after the individual bottom-line percentile estimates are determined. However, it is postulated that, compared with other calculation points, the MEF approach will maximize differences in the mean estimates from the baseline approach because the characteristics of the aggregated distributions are most likely to differ from the individual elicitation response distributions due to the intermediate processing steps (Section 5). Hence, it is postulated that the means determined after group aggregation (MEF approach) will result in the largest differences from the means determined by summing the means from the individual responses (baseline method).

Representative results from this sensitivity analysis are plotted in Figure 7.31 for the current-day PWR MEF and baseline means and 95th percentiles. As expected, the 95th percentile estimates (as well as the medians and 5th percentiles) from both approaches are identical. In general, the mean estimates from both approaches are also consistent. Differences in the means do increase with increasing LOCA size, but the maximum differences are less than 35% for the PWR results (Figure 7.31) and less than 40% for the BWR results (not shown). Therefore, the approach used to determine the aggregated means has little influence on the mean estimates. However, the baseline approach of directly aggregating the mean estimates from the individual responses always results in slightly higher frequencies.

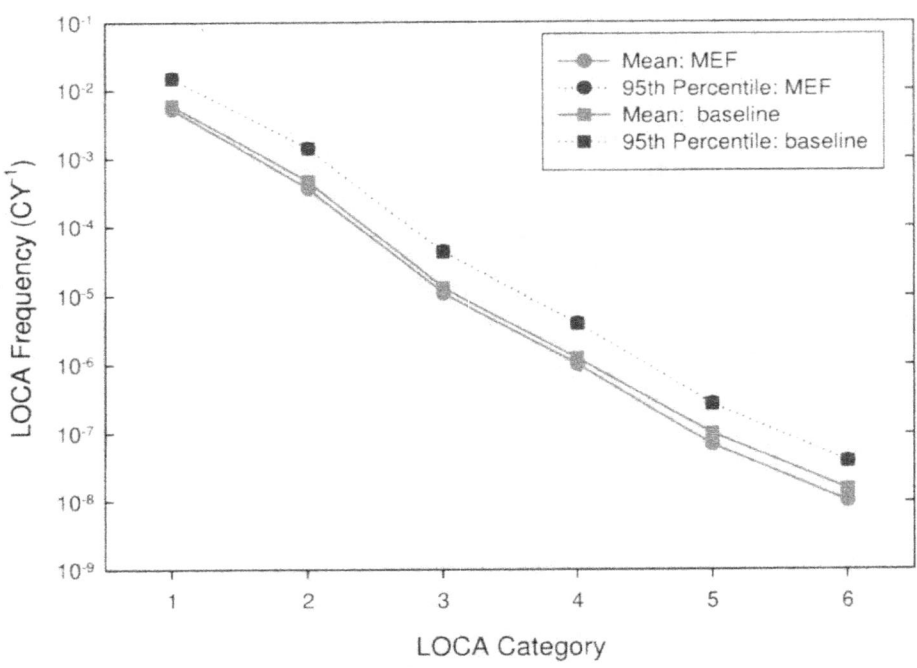

**Figure 7.31 Comparison of Baseline and MEF Aggregation Parameters for
PWR LOCA Frequencies at 25 Years**

7.6.4.4 Mixture-Distribution Aggregation - An alternative method of aggregating the individual
bottom-line estimates is to use a mixture distribution (Section 5.6.4.4). The mixture distribution used is
the arithmetic average of the panelists' LOCA frequency distributions for each LOCA category and plant-
type (i.e., BWR or PWR). This aggregation methodology assumes that the expert panel is a random
sample from the population of all experts (Section 5.6.4.4). The mixture-distribution approach does not
develop group estimates of the four bottom-line parameters as does the baseline methodology, but rather
estimates the full LOCA frequency distribution. More information on the mixture-distribution approach
is provided in Section 5.6.4.4.

LOCA frequency distributions were developed using mixture-distribution aggregation for both the
current-day, unadjusted estimates (baseline) and the current-day estimates adjusted for overconfidence
using the error-factor scheme (Sections 5.6.2.2 and 7.6.2.2). The following procedure was used. For each
panelist, a split lognormal distribution was fit to the three estimated bottom-line percentile estimates
(median, 5th, and 95th percentiles) calculated as described in Section 5.3. The distribution below the
median was determined using the LEF and the distribution above the median was determined using the
UEF.

The cumulative distribution function (CDF) of the mixture distribution for each plant type and LOCA
category is simply the average of the corresponding split lognormal CDFs of all panelists who provided
bottom-line estimates. Each mixture CDF was calculated as follows. First the frequency range for each
mixture CDF was determined by setting its lower end equal to the 0.1th percentile of the panelist CDF
with the smallest 5th percentile and setting its upper end equal to the 99.9th percentile of the panelist CDF
with the largest 95th percentile. Next, this frequency range was divided into 1,000 equal logarithmic
intervals and the individual panelist CDFs were calculated and averaged at each of the 1,000 points in the

frequency range. This procedure yielded a CDF evaluated at 1,000 points for each plant type and LOCA category.

The 5th, 50th, and 95th percentiles were then determined from each mixture CDF. Because the mean of a mixture distribution is equal to the average of the means of the contributing panelist distributions (Section 5.6.4.4), the mean of each mixture distribution was set equal to the AM of the panelists' mean estimates from Section 7.6.4.1. The means determined in this way are not precisely the same as the means determined directly from the 1,000-point CDF, but the two means are generally within 10 percent of each other. Table 7.14 provides the mixture-distribution percentiles and means for the current-day, unadjusted estimates. The means, medians and 95th percentiles for the mixture distributions are plotted in Figures 7.32 and 7.33 for BWR and PWR LOCA frequencies, respectively. Although confidence bounds have not been determined for these parameters, they could be calculated from the mixture CDF using a bootstrap technique [7.3].

Table 7.14 Current-day Total BWR and PWR LOCA Frequencies
(Using Mixture Distribution for Aggregation)

Plant Type	LOCA Size (gpm)	Eff. Break Size (inch)	5th Percentile	Median	Mean	95th Percentile
BWR	>100	½	2.6E-05	2.4E-04	1.4E-03	4.7E-03
	>1,500	1 7/8	2.2E-06	3.8E-05	4.1E-04	1.6E-03
	>5,000	3 ¼	3.8E-07	6.6E-06	3.3E-04	8.3E-04
	>25K	7	4.8E-08	1.3E-06	2.6E-04	6.5E-04
	>100K	18	4.3E-09	2.5E-07	5.2E-06	1.3E-05
	>500K	41	5.3E-14	7.4E-10	2.8E-08	5.5E-08
PWR	>100	½	6.1E-04	3.9E-03	1.0E-02	3.3E-02
	>1,500	1 5/8	6.8E-07	1.3E-04	2.8E-03	9.1E-03
	>5,000	3	5.2E-08	3.9E-06	6.7E-05	2.4E-04
	>25K	7	4.0E-09	3.4E-07	7.9E-06	1.4E-05
	>100K	14	1.7E-12	1.7E-08	1.2E-06	2.6E-06
	>500K	31	3.0E-14	1.6E-09	4.0E-07	1.7E-06

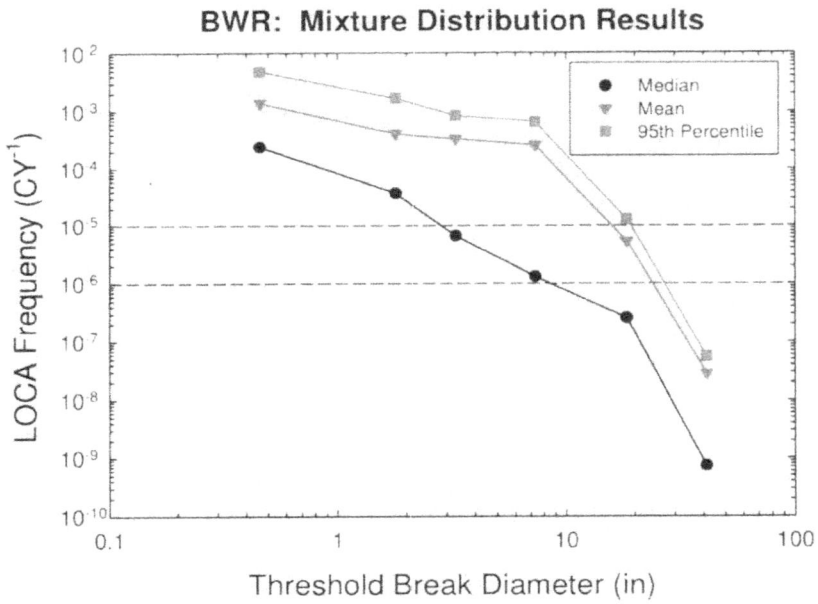

Figure 7.32 BWR LOCA Frequencies Determined by Mixture-Distribution Aggregation: Current-day Estimates

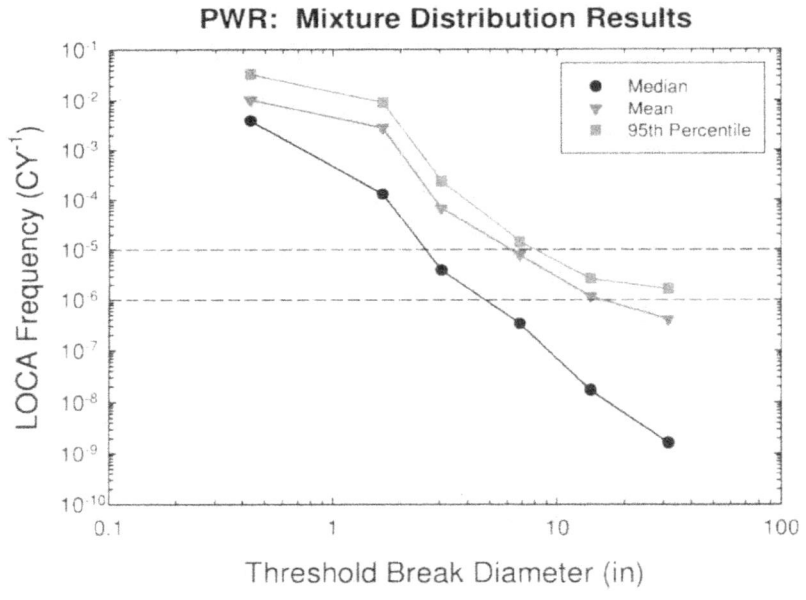

Figure 7.33 PWR LOCA Frequencies Determined by Mixture-Distribution Aggregation: Current-day Estimates

Table 7.15 compares the mixture-distribution and GM (baseline methodology) aggregation techniques by providing the ratios of the median, mean, and 95th percentile estimates. The median estimates determined from the two aggregation techniques are relatively consistent, usually within a factor of 2. Because the AM and mixture mean are identical, the ratios of the means in Table 7.15 are identical to those contained in Table 7.12. The 95th percentile ratios in Table 7.15 are also similar to the 95th percentile ratios in Table 7.12 and usually there is less than a factor of 2 difference between the two ratios.

Table 7.15 Ratios of Mixture-Distribution to Geometric-Mean Aggregated Results (not adjusted for overconfidence)

LOCA Cat.	BWR: Current-Day			PWR: Current-Day		
	Median	Mean	95th	Median	Mean	95th
1	0.8	2.5	2.7	1.0	1.5	1.8
2	0.8	3.8	4.4	1.0	5.9	6.0
3	0.7	14	10	1.2	5.2	5.4
4	0.6	42	30	1.1	6.3	3.5
5	0.9	4.5	3.3	1.5	11	8.8
6	2.5	8.6	6.4	1.4	26	42

The reasons for the discrepancies between the two aggregation schemes for the mean and 95th percentile estimates as shown in Table 7.15 is the same as for the discrepancies between the AM and GM aggregation schemes as shown in Table 7.12. When differences between the mixture-distribution and GM aggregated parameters are large, it is an indication that one or even two panelists' estimates are significantly greater than the other estimates for that category. This is particularly true for BWR LOCA Categories 3 and 4 and PWR LOCA Categories 5 and 6 (Section 7.6.4.3). Smaller differences are indicative of more uniformity in the panelists' estimates. This also explains why the AM aggregated 95th percentile estimates are similar to the mixture-distribution 95th percentiles. In this study, because of the diversity of the estimates, the largest panelist estimate often dominates the AM and mixture-distribution aggregation schemes.

The similarity of the parameter estimates aggregated by the AM scheme to the mixture-distribution parameter estimates decreases as the percentiles decrease. This occurs because the AM aggregation is most greatly influenced by the highest panelist estimates for each parameter while the mixture distributions become more influenced by the lowest panelist estimates as the percentiles decrease. In the extreme, the 5th percentile of the mixture distribution (not shown) can be significantly lower than either the GM or AM 5th percentile estimate if one or two panelist estimates are significantly lower than the other panelist estimates.

The mixture-distribution aggregation scheme will always result in higher mean and 95th percentile estimates and lower 5th percentile estimates than the GM aggregation scheme. Consequently, the mixture-distribution scheme exhibits larger differences between the 5th and 95th percentiles. However, the mean and 95th percentile estimates are often dominated by the maximum panelist estimate while the 5th percentile is often dominated by the minimum panelist estimate. This characteristic implies that the extreme panelist estimates will often dominate the mean, 5th, and 95th percentile group estimates.

The mixture-distribution percentiles and means for the current-day estimates adjusted using the error-factor scheme are provided in Table 7.16. The trends observed by comparing mixture-distribution aggregation to both the GM and AM aggregation schemes (Section 7.6.4.3) for the adjusted estimates are similar to those discussed above for the unadjusted estimates. Ratios of the mixture-distribution median, mean and 95th percentiles to the GM parameters are typically within 30 percent of the values in Table 7.15 for the unadjusted results.

Table 7.16 Current-day Total BWR and PWR LOCA Frequencies
(Using Mixture-Distribution Aggregation with Error-Factor Overconfidence Adjustment)

Plant Type	LOCA Size (gpm)	Eff. Break Size (inch)	5th Percentile	Median	Mean	95th Percentile
BWR	>100	½	2.0E-05	2.7E-04	1.4E-03	5.3E-03
	>1,500	1 7/8	1.5E-06	4.1E-05	4.1E-04	1.6E-03
	>5,000	3 ¼	2.7E-07	7.3E-06	3.4E-04	8.4E-04
	>25K	7	3.2E-08	1.4E-06	2.6E-04	6.5E-04
	>100K	18	2.9E-09	2.7E-07	5.5E-06	1.5E-05
	>500K	41	5.2E-14	6.4E-10	5.6E-08	1.4E-07
PWR	>100	½	4.8E-04	3.9E-03	1.0E-02	3.5E-02
	>1,500	1 5/8	6.7E-07	1.5E-04	3.0E-03	1.1E-02
	>5,000	3	3.8E-08	4.0E-06	7.3E-05	2.8E-04
	>25K	7	3.4E-09	3.1E-07	9.4E-06	2.5E-05
	>100K	14	1.7E-12	1.7E-08	2.4E-06	5.7E-06
	>500K	31	3.0E-14	1.7E-09	1.5E-06	2.5E-06

7.6.5 Panel Diversity

An important objective of this elicitation is to measure panel diversity (Section 2). In this study, panel diversity refers to the range of individual LOCA frequency estimates provided by the panelists. Panel diversity has been evaluated in this study by two methods. The baseline method calculates two-sided 95% statistical confidence intervals assuming that the individual LOCA frequency estimates of the mean and 95th percentile are lognormally distributed (Section 5.4.2). Under this assumption, the group estimate is the GM and a measure of panel diversity is a confidence interval centered on the group estimate (Section 7.5.3).

This sensitivity analysis uses a second method (quartile method) to measure panelist diversity. The quartile method does not rely on the assumption of an underlying distributional structure for the individual estimates. Instead, it uses a *quartile interval* corresponding to the interquartile range of the individual estimates. The quartile interval is defined by the lower (25th percentile) and upper (75th percentile) quartiles of the individual estimates. With the quartile method, the median (or 50th percentile estimate) is a natural measure of central group opinion. As discussed in Section 7.6.4.1, the GM and median estimates of group opinion produce similar results in this study. Thus, a quartile interval is a measure of group diversity about the median while, analogously, a confidence interval is a measure of group diversity about the GM.

The current-day, GM aggregated (i.e., baseline method) and median group estimates of the mean and 95th percentile, along with their corresponding diversity measures, are plotted in Figures 7.34 and 7.35 for BWRs and PWRs, respectively. The BWR confidence and quartile intervals are quite similar for LOCA Categories 1, 2, and 5 for both the mean and 95th percentile estimates. The confidence intervals tend to be slightly wider than the quartile intervals, but this reflects the fact that the confidence intervals have 90 percent coverage while the quartile intervals cover only the middle 50 percent of the estimates. However, the upper confidence bounds are approximately a factor of 5 higher than the upper quartiles for both the means and 95th percentiles for BWR LOCA Categories 3 and 4. As before, these differences reflect the high estimates by a single panelist for these LOCA categories, which affect the confidence intervals but do not affect the quartile intervals. For LOCA Category 6, the upper confidence bounds for the mean and 95th percentiles are similar to the upper quartile, but the lower confidence bounds are much lower than the lower quartiles. These differences reflect the very low estimates by a single panelist.

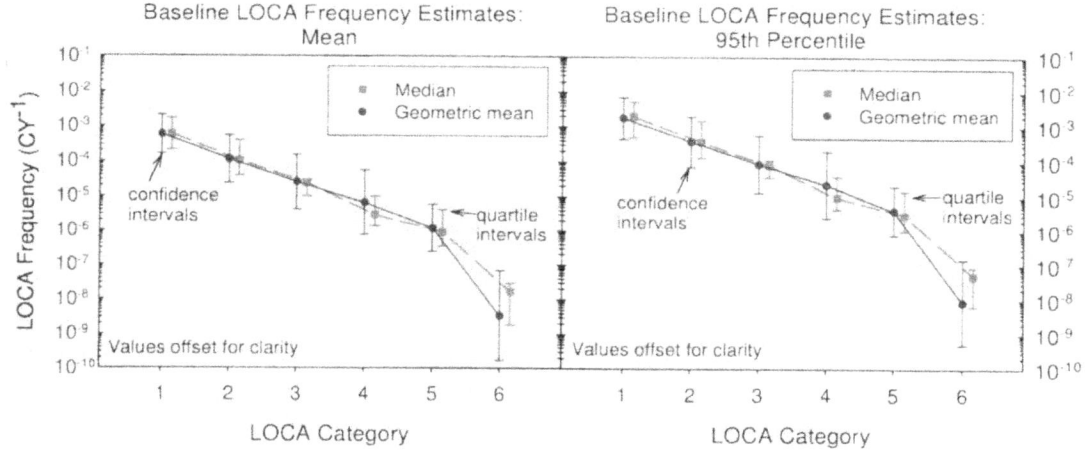

Figure 7.34 Measures of Panel Diversity for BWR LOCA Frequencies at 25 Years

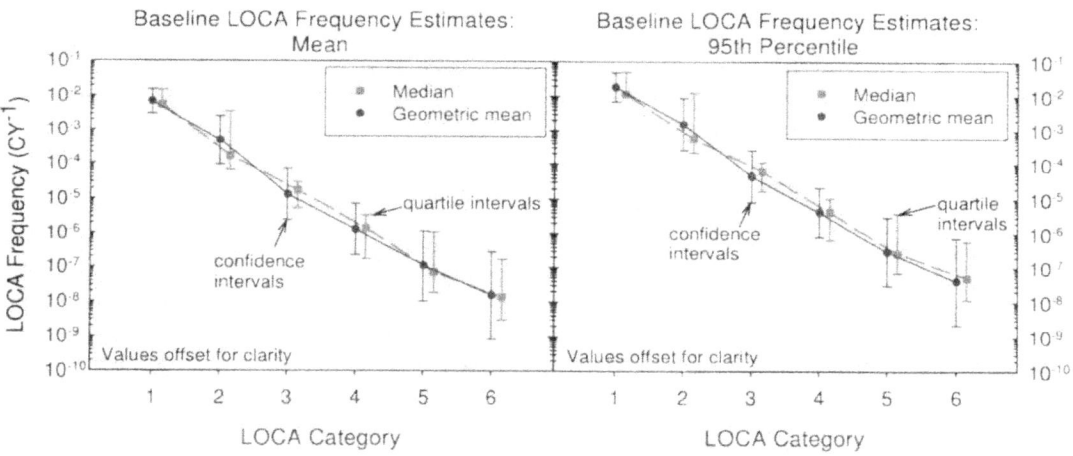

Figure 7.35 Measures of Panel Diversity for PWR LOCA Frequencies at 25 Years

The PWR mean and 95th percentile diversity measures (Figure 7.35) exhibit many trends similar to the BWR trends. Once again, the confidence intervals are typically wider than the quartile intervals. The differences between the confidence and quartile intervals also tend to increase slightly with LOCA size. The only exception to these trends is for LOCA Category 2, where the quartile intervals are wider than the confidence intervals. This occurs because two panelists expect relatively high LOCA frequencies, due to PWSCC concerns, compared with the rest of the group (Section 6.3.2). These two high estimates skew the upper quartiles with respect to the medians. The upper quartiles for LOCA Categories 5 and 6 are similarly skewed with respect to the median by two high estimates, and they are similar to the upper confidences bounds. However, the quartile intervals are not wider than the confidence intervals in these cases.

In summary, with only a few exceptions, the confidence intervals are wider than the quartile intervals. Therefore, the baseline method and use of confidence bounds provides a more conservative measure of panel diversity than the quartile method. Of course, the quartile intervals could be widened to cover a larger percentage of the panelists' estimates. However, if their coverage was increased to the 90 percent coverage of the confidence intervals, their spreads would be strongly influenced by the extreme panelists' estimates.

7.7 Summary Estimates

The baseline LOCA frequency estimates in Table 7.1 were calculated using the analysis framework described in Section 5.5. The individual and group *summary* estimates of the means, medians and 5th and 95th percentiles of the LOCA frequency distributions were determined using the following summary analysis framework:

(i)   The MV, UB, and LB supplied by the panelists for each elicitation question were assumed to correspond to the median, 95th percentile and 5th percentile, respectively, of a split lognormal distribution, with the mean calculated assuming that the upper tail is truncated at the 99.9th percentile.

(ii)   Only those panelists whose uncertainty ranges were relatively small were adjusted using an error-factor adjustment scheme to account for possible overconfidence (Section 7.6.2.2).

(iii)   Split lognormal distributions were summed by assuming perfect rank correlation among the individual terms.

(iv)   The individual estimates are the total LOCA frequency parameters (i.e., mean, median, 5th percentile, and 95th percentile) determined for each panelist.

(v)   The group estimates of the total LOCA frequency parameters were determined using the GMs of the individual estimates.

(vi)   Panel diversity was characterized with a two-sided 95% confidence interval based on an assumed lognormal model for the individual estimates.

Based on the sensitivity analyses, the resultant individual and group estimates are consistent with the elicitation objectives and structure and are a reasonable reflection of the panelists' quantitative judgments. In particular, the group estimates are not dominated by extreme results, either on the high or low end. The only difference between the baseline estimates discussed up to this point and the summary estimates is that the summary estimates are adjusted for overconfidence using the error-factor adjustment scheme in (ii) above, while the baseline estimates are determined without any such adjustment. For the reasons given in Sections 5.6.2 and 7.6.2, the adjusted individual LOCA frequency estimates using (ii) above result in improved group LOCA frequency estimates.

The summary LOCA frequency estimates for the current-day and end-of-plant-license period are provided in Table 7.17 for both BWR and PWR plant types. Table 7.17 is identical to Table 7.7 and is reproduced here for convenience. The current-day median, mean, and 95th percentile summary estimates are plotted in Figures 7.36 and 7.37 for BWRs and PWRs, respectively. Figures 7.36 and 7.37 are identical to Figures 7.21 and 7.22 and are also reproduced here for convenience. As for the baseline estimates, a measure of the panelists' uncertainty in the summary estimates is given by the difference between the median and 5th or 95th percentile in Table 7.17 and Figures 7.36 and 7.37. Also, the 95% confidence intervals plotted in these figures provide a measure of the diversity among the panelists.

7-50

Table 7.17 Total BWR and PWR LOCA Frequencies
(After Overconfidence Adjustment using Error-Factor Scheme)
(reproduced from Table 7.7)

Plant Type	LOCA Size (gpm)	Eff. Break Size (inch)	Current-Day Estimate (per cal. year) (25 years fleet average operation)				End-of-Plant-License Estimate (per cal. year) (40 years fleet average operation)			
			5th Per.	Median	Mean	95th Per.	5th Per.	Median	Mean	95th Per.
BWR	>100	½	3.3E-05	3.0E-04	6.5E-04	2.3E-03	2.8E-05	2.6E-04	6.2E-04	2.2E-03
	>1,500	1 7/8	3.0E-06	5.0E-05	1.3E-04	4.8E-04	2.5E-06	4.5E-05	1.2E-04	4.8E-04
	>5,000	3 ¼	6.0E-07	9.7E-06	2.9E-05	1.1E-04	5.4E-07	9.8E-06	3.2E-05	1.3E-04
	>25K	7	8.6E-08	2.2E-06	7.3E-06	2.9E-05	7.8E-08	2.3E-06	9.4E-06	3.7E-05
	>100K	18	7.7E-09	2.9E-07	1.5E-06	5.9E-06	6.8E-09	3.1E-07	2.1E-06	7.9E-06
	>500K	41	6.3E-12	2.9E-10	6.3E-09	1.8E-08	7.5E-12	4.0E-10	1.0E-08	2.8E-08
PWR	>100	½	6.9E-04	3.9E-03	7.3E-03	2.3E-02	4.0E-04	2.6E-03	5.2E-03	1.8E-02
	>1,500	1 5/8	7.6E-06	1.4E-04	6.4E-04	2.4E-03	8.3E-06	1.6E-04	7.8E-04	2.9E-03
	>5,000	3	2.1E-07	3.4E-06	1.6E-05	6.1E-05	4.8E-07	7.6E-06	3.6E-05	1.4E-04
	>25K	7	1.4E-08	3.1E-07	1.6E-06	6.1E-06	2.8E-08	6.6E-07	3.6E-06	1.4E-05
	>100K	14	4.1E-10	1.2E-08	2.0E-07	5.8E-07	1.0E-09	2.8E-08	4.8E-07	1.4E-06
	>500K	31	3.5E-11	1.2E-09	2.9E-08	8.1E-08	8.7E-11	2.9E-09	7.5E-08	2.1E-07

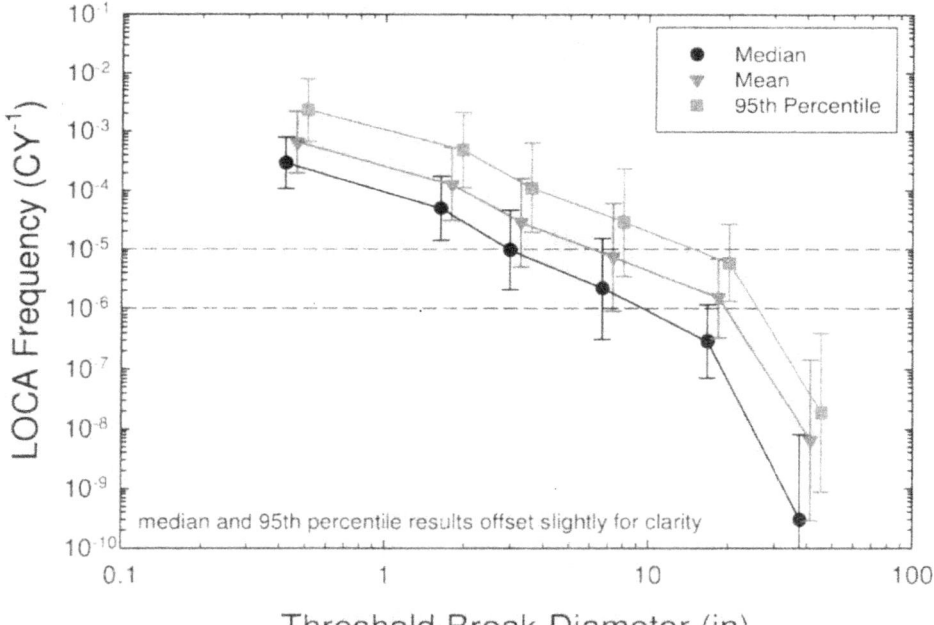

Figure 7.36 BWR LOCA Frequencies with Error-Factor Adjustment
(reproduced from Figure 7.21)

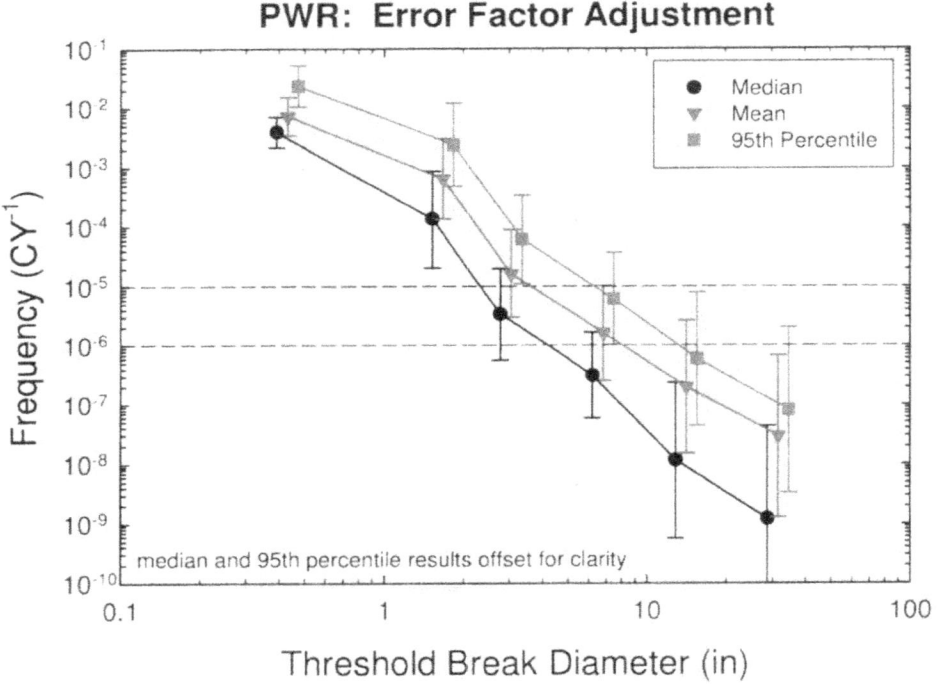

**Figure 7.37 PWR LOCA Frequencies with Error-Factor Adjustment
(reproduced from Figure 7.22)**

7.8 Steam Generator Tube Rupture and Small Break (Category 1) LOCA Frequencies

Steam generator tube ruptures (SGTRs) are usually analyzed separately from other passive system failures in PRAs because they have occurred relatively frequently and the event trees associated with SGTRs are modeled separately from all other passive system LOCA failures. Therefore, it is useful to identify the SGTR frequency contributions to the total PWR LOCA frequency estimates provided in Table 7.17. The mean frequencies associated with the Category 1 (i.e., flow rate > 100 gpm [380 lpm]) PWR LOCA frequencies of other passive system failures can then be simply determined by subtraction. However, the other frequency parameters cannot be so determined.

As explained in Section 4, a generic Category 1 SGTR frequency of 3.5E-03 per calendar year was determined based on operating experience. This estimate was provided to the panelists for use as a base case. The panelists then assessed the applicability of this base case for estimating current and future SGTR frequencies. Category 1 LOCA failures are possible with a single tube rupture. Other higher category LOCAs require multiple tube ruptures due to either a primary or secondary side pressure transient or another common cause failure mechanism. The assessment of multiple tube ruptures was specifically defined as a failure scenario and panelists were asked to assess the relative likelihood of multiple tube rupture if they believed it to be a significantly contributing factor.

The portions of the total LOCA frequencies (Table 7.17) attributed by the panelists to SGTR are provided in Table 7.18 for the current-day and end-of-plant-license periods. These SGTR frequencies have been aggregated from the individual estimates using the GM and the error-factor overconfidence adjustment used for the summary estimates discussed in Section 7.7. The number of panelists (# of Est.) who made SGTR frequency estimates for each LOCA category is also provided in Table 7.18. Only group estimates for LOCA Categories 1 and 2 are provided because the majority of panelists judged that SGTR contributions are insignificant for LOCA Categories 3 and higher.

Table 7.18 PWR Steam Generator Tube Rupture Frequencies (Geometric Mean Values After Overconfidence Adjustment)

LOCA Size (gpm)	# of Est.	Eff. Break Size (inch)	Current-Day Estimate (per cal. year) (25 year fleet average operation)				Next 15 Year Estimate (per cal. year) (End-of-plant license)			
			5th	Median	Mean	95th	5th	Median	Mean	95th
>100	9	½	5.0E-04	2.6E-03	3.7E-03	1.0E-2	2.6E-04	1.5E-03	2.1E-03	6.0E-03
>1,500	7	1 5/8	3.4E-07	4.8E-06	1.5E-05	5.4E-05	1.5E-07	2.3E-06	6.5E-06	2.3E-05

All the panelists who provided PWR elicitation responses identified SGTR as a major contributor to Category 1 PWR LOCAs. In fact, the current-day mean and median SGTR frequencies from Table 7.18 are 51 and 67 percent, respectively, of the total Category 1 LOCA frequencies from Table 7.17. Similar results hold for the end-of-plant-license period. Note that the mean current-day Category 1 SGTR frequency of 3.7E-03 per calendar year is nearly identical to the base case and operating-experience estimated frequency of 3.5E-03 per calendar year (Section 4.4.1). Clearly, the panelists as a group did not believe that these frequencies needed any significant adjustment. As noted in Section 4.4.1, this frequency estimate is also consistent with the prior estimate from NUREG/CR-5750. The SGTR 95th percentile of 1.0E-02 per calendar year is also less than the mean steam generator leak frequency (Section 4.4.1) of 1.6E-02 per calendar year. This result is not surprising, because the leak frequency can reasonably be expected to bound the rupture frequency.

From Table 7.18, seven panelists provided SGTR estimates for LOCA Category 2, and the group estimates reported are based only on these seven estimates. The fact that two of the panelists – C and H - did not provide estimates means that they believe that SGTR contributions to LOCA Category 2 are relatively insignificant. However, ignoring the responses of these two panelists might bias the results in Table 7.18. One way to evaluate this possible bias is to use bounding estimates. After the formal elicitation was completed, Panelist C provided an UB estimate on the conditional rupture probability of a second steam generator tube given that one has already failed. This UB estimate is somewhat conservative because it is likely that more than two tubes would need to severely rupture for the flow rate to reach 1,500 gpm (5,700 lpm).

A lower estimate was also obtained from Panelist C's responses. Panelist C only provided estimates for 80 percent of the steam generator failure modes that contribute to Category 2 PWR non-piping LOCAs. Therefore, it was assumed that the remaining 20 percent of these contributions consist of multiple SGTRs. This is a reasonable assumption because Panelist C provided direct estimates for the frequency contributions of all the other steam generator tube components that were evaluated in the elicitation (Appendix B).

There is approximately one order of magnitude difference between these UBs and LBs. Aggregating Panelist C's UB estimate with the other seven estimates increases the Category 2 value in Table 7.18 by less than 40 percent. Aggregating the LB estimate decreases the value in Table 7.18 by only 4 percent. Hence, Panelist C's influence on the Category 2 SGTR frequencies is minimal.

Similarly, bounding values were determined from Panelist H's responses. A LB value was determined by assuming that the SGTR Category 2 frequency is equivalent to the RPV shell failure estimation. This was the smallest Category 2 PWR non-piping estimate that panelist H provided. The UB estimate was determined by assuming that the 20 percent of the Category 2 PWR non-piping frequency contributions that were not directly evaluated by this panelist is all attributed to SGTRs. Panelist H's dominant contribution is from CRDM failures so the implication of this assumption is that multiple SGTRs are about 4 times less likely than CRDM failures. After the elicitation, Panelist H indicated that these are suitable bounding assumptions. Unfortunately, there is a large disparity between these UBs and LBs (approximately 5 orders of magnitude) and there is no additional information in this panelist's response that can be used to reduce this discrepancy. However, aggregating panelist H's bounding estimates with the other seven estimates modifies the values in Table 7.18 by a factor of approximately ±2.

From this sensitivity analysis, it can be seen that excluding these two panelists' results has relatively little effect on the aggregated estimates. Therefore, the Category 2 estimates in Table 7.18 are considered to be appropriate group estimates for multiple steam generator tube failures.

The SGTR contributions to the Category 2 LOCA frequencies are less than 3 percent of the mean total LOCA frequencies from Table 7.17. Thus, these contributions are much less significant than the contributions of SGTRs to the Category 1 frequencies. Also, the results in Table 7.18 predict slight decreases in both the SGTR Category 1 and 2 frequencies over the next 15 years. These expected decreases are due to continued steam generator replacement and improved programs to mitigate steam generator tube degradation.

The PWR LOCA frequencies of passive system failures other than SGTRs can also be estimated using the following procedure. First the LOCA frequencies without SGTR contributions are calculated for each panelist. Using the previous assumption that the non-piping subcomponent frequencies have perfect rank correlation (Section 5.3.4.1), the percentiles (i.e., 5th, 50th, and 95th) are calculated by subtracting the SGTR percentiles from the corresponding total LOCA frequency percentiles. Also, the means are calculated by subtracting the SGTR means from the total LOCA frequency means. Because the mean is linear, it is not necessary to assume a perfect rank correlation structure for this calculation. The individual estimates are then adjusted for overconfidence using the error-factor scheme as described in Sections 5.6.2.2 and 7.6.2.2. Finally, group estimates are calculated using the GM to aggregate the individual estimates as discussed in Section 5.4.

Note that the non-SGTR contributions cannot be estimated by directly subtracting the group SGTR estimates (Table 7.18) from the total LOCA frequency estimates (Table 7.17) because the overconfidence adjustment and aggregation are both non-linear. The procedure described above ensures that the individual estimates are consistent with the original elicitation responses before overconfidence adjustment and aggregation are performed. This procedure is also consistent with the procedure used to calculate the total LOCA and SGTR estimates. This estimate is treated as a LB for this analysis.

This procedure was used to calculate both Category 1 and Category 2 LOCA frequency estimates without SGTR contributions. No additional assumptions are required for calculating the Category 1 estimates without SGTR contributions since all the panelists provided Category 1 SGTR estimates. As discussed above, all panelists except C and H also provided Category 2 SGTR estimates. Using a bounding sensitivity analysis, the Category 2 SGTR contributions for Panelists C and H are relatively insignificant. Specifically, if the UB Category 2 SGTR estimates discussed above for Panelists C and H are used, then the differences between the means of their individual Category 2 estimates with and without SGTR contributions are less than 5 and 20 percent, respectively. Therefore, it can be conservatively assumed that the total LOCA estimates for Panelists C and H are equivalent to their non-SGTR LOCA estimates.

The panelists also judged that the SGTR failure contributions to Categories 3 – 6 are insignificant. Therefore, it was not necessary to adjust the estimates for these larger LOCAs to remove SGTR contributions. The PWR total LOCA frequencies after excluding SGTR contributions are provided in Table 7.19. The Category 3 – 6 estimates in Table 7.19 are identical to the PWR results in Table 7.17. The results in Table 7.19 are appropriate to use for PRA applications that separately consider SGTRs.

Table 7.19 Total PWR LOCA Frequencies without SGTR Contributions
(LOCA Categories 3 – 6 Reproduced from Table 7.17)

Plant Type	LOCA Size (gpm)	Eff. Break Size (inch)	Current-Day Estimate (per cal. year) (25 years fleet average operation)				End-of-Plant-License Estimate (per cal. year) (40 years fleet average operation)			
			5th Per.	Median	Mean	95th Per.	5th Per.	Median	Mean	95th Per.
PWR	>100	½	6.8E-05	6.3E-04	1.9E-03	7.1E-03	7.0E-05	7.2E-04	2.1E-03	7.9E-03
	>1,500	1 5/8	5.0E-06	8.9E-05	4.2E-04	1.6E-03	6.1E-06	1.2E-04	5.8E-04	2.2E-03
	>5,000	3	2.1E-07	3.4E-06	1.6E-05	6.1E-05	4.8E-07	7.6E-06	3.6E-05	1.4E-04
	>25K	7	1.4E-08	3.1E-07	1.6E-06	6.1E-06	2.8E-08	6.6E-07	3.6E-06	1.4E-05
	>100K	14	4.1E-10	1.2E-08	2.0E-07	5.8E-07	1.0E-09	2.8E-08	4.8E-07	1.4E-06
	>500K	31	3.5E-11	1.2E-09	2.9E-08	8.1E-08	8.7E-11	2.9E-09	7.5E-08	2.1E-07

7.9 Comparison with Prior Estimates

In this section, selected results from this study are compared with LOCA frequency estimates developed in prior studies. As noted in Section 1.2, LOCA frequency estimates were previously developed in WASH-1400 [7.4], NUREG-1150 [7.5], and NUREG/CR-5750 [7.6]. Furthermore, estimates were developed for each plant type for the individual plant examinations (IPEs) [7.7] using an early 1990's piping precursor database as documented in EPRI TR-100380 [7.8]. Prior estimates were also obtained during the pilot elicitation in this exercise as discussed in Section 3.1 [7.9]. As indicated earlier (Section 1), the LOCA frequency estimates from these earlier studies were typically based, at least in part, on the analysis of operating experience. This elicitation also provided the relevant operating experience to the panelists and used it as the basis for many of the base case conditions (Section 3.5).

Prior studies differed in their analysis and interpretation of operating experience. It is important to understand the differences between these prior studies and this elicitation before making comparisons. First, these prior estimates defined the range of small break LOCAs as 100 to 1,500 gpm (380 to 5,700 lpm) and the range of medium break LOCAs as 1,500 to 5,000 gpm (5,700 to 19,000 lpm). The current study defines each LOCA category using flow rate thresholds (Section 3.7). To facilitate comparison with prior estimates, the summary estimates of the mean LOCA frequencies (Table 7.17) are recalculated as interval-based estimates. Interval-based estimates for the mean LOCA frequencies were determined by simply subtracting the estimates for adjacent cumulative LOCA categories. This technique cannot strictly be used for determining other percentiles of the LOCA frequency distribution. However, because the differences between the results for adjacent LOCA categories in Table 7.17 are much smaller than the uncertainties, the differences between the cumulative threshold and interval-based estimates are not statistically significant. Therefore, the percentiles from Table 7.17 could be directly compared with prior estimates if desired.

Another important distinction between the current elicitation and prior studies is that the prior studies generally used the WASH-1400 and NUREG/CR-5750 pipe break to leak rate correlation which distinguishes between medium and large break LOCAs at an approximately 6-inch (152 mm) effective pipe break diameter (Section 3.7). In this study, Category 3 (i.e., flow rate > 5000 gpm [19,000 lpm]) and Category 4 (i.e., flow rate > 25,000 gpm [95,000 lpm]) LOCAs are correlated with generic minimum effective break diameters of approximately 3 and 7-inches (76 to 178 mm), respectively. Because the break size definitions are not consistent, LB LOCA frequencies from prior studies are compared with both Category 3 and 4 LOCA estimates from this study.

Table 7.20 compares the summary mean frequency estimates of this study with NUREG/CR-5750 estimates, which is the most recent generic study. In this table, as previously discussed, the elicitation means at 25 years are calculated from Table 7.17 after converting the cumulative frequency estimates to interval ranges consistent with the NUREG/CR-5750 LOCA definitions. The NUREG/CR-5750 ratio entries in this table are the ratios of the NUREG/CR-5750 estimates to the summary elicitation estimates. Note that the NUREG/CR-5750 SB LOCA PWR estimates include SGTR frequency contributions to make these estimates comparable with the elicitation estimates, which include these events. Furthermore, as indicated in Sections 4.4.1 and 7.8, the mean SGTR frequency estimates from this study, NUREG/CR-5750, and operating experience are all consistent.

Table 7.20 Comparison of Current Elicitation and Selected Prior Mean Estimates

Plant Type	Summary Elicitation Estimates			NUREG/CR-5750 Estimates			Pilot Elicitation (60 years)	
	LOCA Intervals	Mean Freq. (25 yrs)	Mean Freq. (60 yrs)	LOCA Size	Mean Freq.	Ratio	Mean Freq.	Ratio
BWR	1-2	5.2E-04	1.1E-03	SB	4.0E-04	0.76	1.5E-03	1.36
	2-4	1.2E-04	2.0E-04	MB	3.0E-05	0.25	9.1E-05	0.47
	>3	2.9E-05	6.1E-05	LB	2.0E-05	0.70	5.2E-05	0.85
	>4	7.3E-06	1.7E-05	LB	2.0E-05	2.74	5.2E-05	2.98
PWR	1-2	6.6E-03	1.1E-02	SB	7.4E-03	1.12	1.5E-03	0.13
	2-4	6.4E-04	1.9E-03	MB	3.0E-05	0.05	6.1E-05	0.03
	>3	1.6E-05	1.5E-04	LB	4.0E-06	0.25	7.2E-06	0.05
	>4	1.6E-06	1.9E-05	LB	4.0E-06	2.49	7.2E-06	0.38

From Table 7.20, the current mean elicitation estimates are generally consistent with the NUREG/CR-5750 estimates [7.6], as the differences between the two studies are generally less than a factor of 4. The one exception is for the current-day mean PWR MB LOCA estimate which is about twenty times higher than the corresponding NUREG/CR-5750 estimate (Table 7.20 and Figure 7.38). The increase is largely due to current PWR PWSCC concerns (Section 6.3.2) in both piping and non-piping (CRDM) components. However, the expectation that aging will affect intermediate piping to a greater extent than small or large diameter piping (Section 6.3.3) also contributes to increases in both the current BWR and PWR MB LOCA estimates (Table 7.17).

The NUREG/CR-5750 BWR and PWR LB LOCA estimates in Table 7.20 are less than a factor of 4 lower than the Category 3 elicitation estimates, but are less than a factor of 3 higher than the Category 4 estimates. For LB LOCAs, the current elicitation Category 4 estimates are more comparable with the NUREG/CR-5750 estimates because the effective break sizes are closer (i.e., 7 inches (180 mm) for the elicitation and 6 inches (150 mm) for NUREG/CR-5750). The Category 3 LOCA frequency estimates correspond to break sizes greater than 3 inches, which is one-half of the threshold break size for LB LOCAs as defined in NUREG/CR-5750. The generally good agreement between the NUREG/CR-5750 and current elicitation estimates provided in Table 7.20 is a bit surprising given the markedly different methodologies used to arrive at these results.

Figures 7.38 and 7.39 compare the elicitation estimates of the median, mean, 5th and 95th percentiles with estimates from prior studies for PWR MB LOCAs and BWR LB LOCAs, respectively. These LOCA categories were chosen because their differences between the elicitation and NUREG/CR-5750 estimates are the largest. For PWR MB LOCAs (Figure 7.38), the current elicitation yields mean and 95th percentile estimates, while higher than in NUREG/CR-5750, are very consistent with estimates from WASH-1400 [7.4], NUREG-1150 [7.5], the IPE submittals [7.7], and EPRI TR-100380 [7.8]. From Figure 7.39, the BWR LB LOCA mean and 95th percentile estimates are approximately one to two orders of magnitude less than all the prior estimates. Once again, these comparisons are consistent with the elicitation rationale provided in Section 6 that intermediate piping is most affected by aging degradation while the failure of larger diameter piping is less likely than estimated in prior studies.

While these total LOCA frequency estimates are relatively consistent, it is also of interest to compare the PWR SB LOCA estimates after excluding the large SGTR contribution. As previously discussed in Sections 4.4.1 and 7.8, the SGTR estimates from this study, NUREG/CR-5750, and operating experience are all consistent. Therefore, the differences in the total LOCA frequency estimates are largely due to the failure frequencies of other passive systems. From Table 7.19, the mean PWR LOCA Category 1 and 2 estimates without SGTR contributions are 1.9E-03 and 4.2E-04 per calendar year, respectively. Therefore, the interval-based estimate that corresponds to the historical SB LOCA definition (i.e., breaks between 100 and 1,500 gpm [380 and 5,700 lpm]) is 1.5E-03 per calendar year. This estimate is half an order of magnitude higher than the mean SB LOCA NUREG/CR-5750 estimate of 4.0E-04 per calendar year.

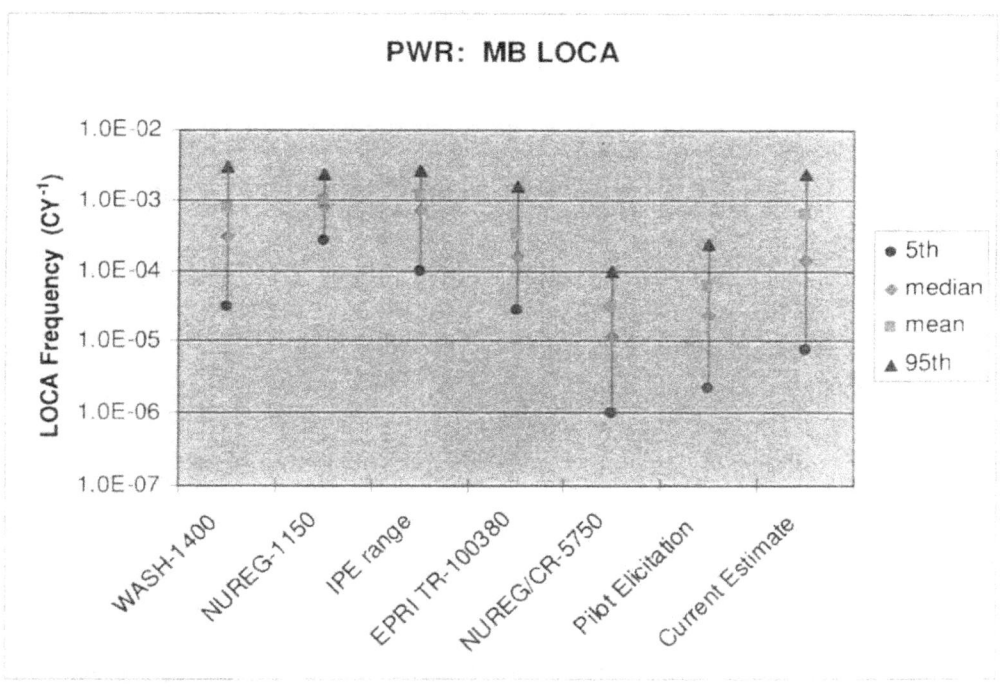

Figure 7.38 Comparison of Elicitation Estimates for PWR Medium Break LOCAs at 25 Years with Estimates from Prior Studies

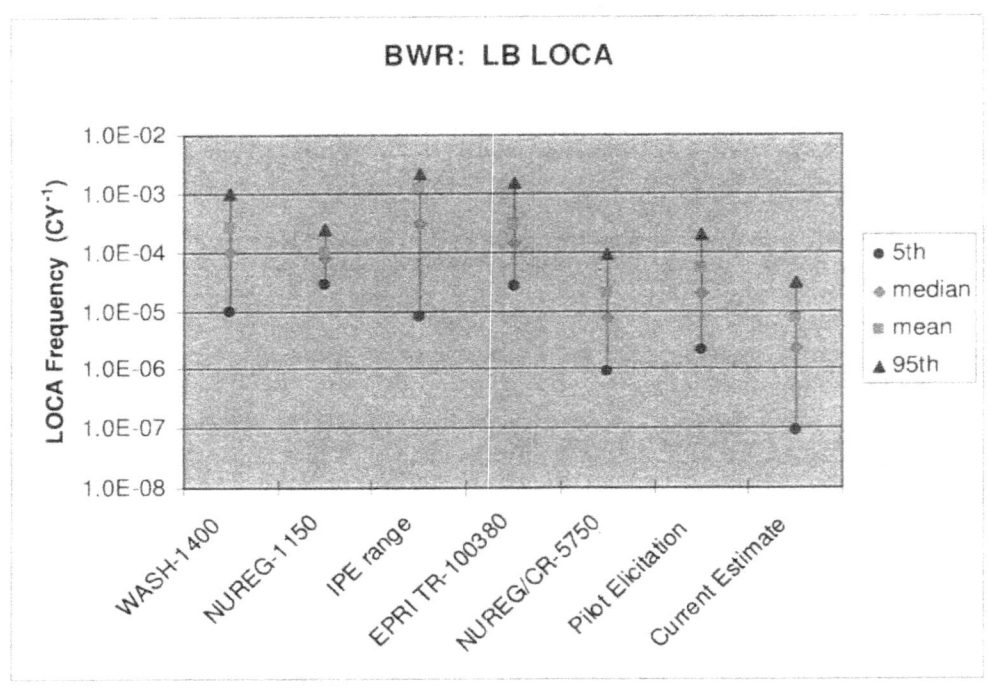

**Figure 7.39 Comparison of Elicitation Estimates for BWR Large Break LOCAs
at 25 Years with Estimates from Prior Studies**

The NUREG/CR-5750 estimate translates to an expected PWR SB LOCA every 36 years for the current fleet of 69 operating PWRs and the elicitation frequency translates to an expected PWR SB LOCA every 11 years. This increase reflects the panelists' opinion that the current-day PWR SB LOCA frequency for components other than steam generator tubes is higher than historical averages, primarily due to current PWSCC concerns (Section 6.3.2).

Many PRA applications currently utilize NUREG/CR-5750 initiating event frequencies. Based on the discussion in this section, the current-day, summary estimates may be used to update the NUREG/CR-5750 estimates using the results in Table 7.17. Estimates for flow rates greater than 100 gpm (380 lpm), i.e., effective break sizes greater than ½ inch (12.7 mm), can be in lieu of total passive system PWR SB LOCA. The current-day SGTR estimates in Table 7.18 can be used in applications that analyze these ruptures separately from all other PWR passive system failures. For these applications, the mean current-day PWR estimates for breaks greater than 100 gpm in Table 7.19 should be also used for the frequency associated with all other passive system failures. The elicitation estimates for flow rates greater than 1,500 gpm (5,700 lpm) (i.e., breaks greater than 1 7/8 inches (47.6 mm) for BWRs and 1 5/8 inches (41.3 mm) for PWRs) can be used instead of NUREG/CR-5750 MB LOCA frequencies. As indicated above, the Category 4 LOCA estimates (flow rates greater than 25,000 gpm (95,000 lpm)) should be used instead of NUREG/CR-5750 LB LOCA estimates because this break size is most similar to the NUREG/CR-5750 LB LOCA break size definition.

Table 7.20 also includes the internal NRC pilot elicitation results, which were developed in July 2002 [7.9]. As discussed in Section 3.1, the pilot elicitation used the NUREG/CR-5750 frequencies as the basis for determining the impact of aging on passive system LOCA frequencies over the next 35 years. The pilot elicitation panelists expect a moderate increase in these frequencies over the next 35 years.

Consequently, the pilot elicitation estimates are between factors of 2 to 4 times higher than the NUREG/CR-5750 estimates for all BWR and PWR LOCA sizes, with the exception of the PWR SB LOCA estimates (Table 7.20). A comparable increase is apparent in the PWR SB LOCA estimates (Table 7.20) if the SGTR frequency of 7E-03 per calendar year is excluded from the NUREG/CR-5750 estimates.

Table 7.20 also compares the summary mean elicitation estimates with the pilot elicitation estimates 35 years into the future (i.e., at 60 years of average plant life) using the ratios of the pilot elicitation estimates to the summary estimates. The comparison with the 60-year estimates was made because this was the only time period evaluated in the pilot elicitation (Section 3.1). More details on the 60-year elicitation results can be found in Appendix L.

From Table 7.20, the current elicitation BWR SB and LB (Category 4) estimates are approximately one and one-half and three times lower, respectively, than the pilot elicitation estimates. Conversely, the current elicitation BWR MB estimates are twice as high as the pilot elicitation estimates. These differences are not striking when considering that the estimates reflect the next 35 year time period, but they are supported by the panelists' rationale. The study elicitation panelists appeared to provide more credit for IGSCC mitigation measures than did the pilot elicitation panelists, which may have led to the lower SB and LB LOCA estimates. The effectiveness of IGSCC mitigation is offset by the elicitation panelists' concern for PWSCC cracking in BWR CRDM nozzles. This concern elevated the current elicitation BWR MB LOCA estimates slightly higher than the pilot elicitation estimates

For PWRs, the differences between the current and pilot elicitation estimates are larger, and the current elicitation estimates are consistently higher than the pilot elicitation estimates. The largest difference is for the PWR MB LOCAs, where the current elicitation estimates are 30 times higher than the pilot estimates. The PWR SB and LB LOCA current elicitation estimates are approximately 5 and 2 times larger after SGTR contributions are eliminated from the current elicitation estimates. One important reason for the increased MB and SB estimates is that the current elicitation panel made a more comprehensive assessment of the severity of CRDM and PWSCC cracking in PWR plants. Therefore, this difference is consistent with the panelists' expectation that this mechanism will more severely impact intermediate and small piping failure frequencies.

7.10 Comparison with Operating Experience Data

There have been very few failures of primary pressure boundary systems that have led to LOCAs. Steam generator tube rupture is the most common failure, and a few other smaller diameter pipes have also failed. These events can be classified as either very small (VS), or SB LOCAs depending on the break size. No MB or LB LOCAs have occurred. Because of the paucity of actual failures, estimating LOCA frequencies by considering only the number of actual failures becomes less accurate as the break size increases.

Operating experience-based estimates can also be obtained using failure precursor information (e.g., leaks and part-through-wall cracks). This information is much more prevalent than actual failures, and failure precursors have been documented on a broad range of primary pressure boundary systems. Loss-of-coolant accident frequencies can then be estimated if the CFP given a precursor is known. The challenge is that the CFP depends on many factors whose effects are difficult to quantify, such as the piping system, degradation mechanism, precursor type, and precursor size. Therefore, using the CFP and failure precursor information to estimate VS and SB LOCA frequencies is likely to be less accurate than directly estimating these frequencies from the failure events. Also, as the piping size increases, the uncertainty of the CFP increases. This effect should be accounted for when estimating MB and LB LOCA frequencies.

Another consideration is that operating experience reflects only historical performance. New degradation mechanisms (e.g., PWSCC) can result in systematic increases in generic frequencies while wholesale mitigation programs (e.g., IGSCC mitigation) can result in systematic decreases in LOCA frequencies over time. Because LOCAs are so rare, the effects that these changes have on the LOCA frequencies are typically not reflected in operating experience until some time (often years) has passed. Therefore, using operating experience to predict current or future performance is not necessarily conservative. These considerations, along with other rationales discussed in Section 1, explain why elicitation is preferable to exclusively relying on operating experience for estimating current and future LOCA frequencies, even for small break LOCAs.

However, operating experience is still valuable as a basis for identifying trends, assessing historical performance, and estimating failure precursor frequencies (Section 1). The elicitation procedure in this study made extensive use of operating experience as the basis for many of the piping and non-piping base case estimates and sensitivity analyses that the panelists used to anchor their elicitation responses (Section 3.5). Piping and non-piping failure and failure precursor databases were also provided to the panelists so that they could analyze plant systems and failure mechanisms that were not specifically evaluated in a base case. Most panelists used operating experience as the basis for their current and future LOCA frequency estimates and made modifications to account for the current state of knowledge.

Therefore, it is of considerable value to compare the elicitation results with operating experience. Differences between operating experience and the elicitation results reflect the modifications that the panelists have deemed necessary to accurately account for current and future changes that may alter the average failure frequencies associated with past commercial nuclear plant performance. This section only compares the Category 1, or SB LOCA, frequencies with operating-experience estimates developed directly from failure events because, as discussed above, these estimates require the least extrapolation or interpretation of passive-system failure data and should therefore be the most accurate.

As of December 2006, US BWRs had accumulated 1,023 reactor-years of operating experience and no Category 1 LOCAs have occurred. When no events have occurred, a 50 percent confidence bound is analogous to a mean frequency estimate. It is assumed that the number of Category 1 LOCAs in N reactor years has a Poisson distribution with mean λ. With no events, a 50 percent confidence bound on λ is 0.70 [7.10]. The estimated frequency is then 0.70/N per reactor year. Accordingly, an operating experience-based estimate of the frequency of a BWR Category 1 LOCA is (0.70/1023) = 6.8E-04 per reactor year. Multiplying this value by the conversion factor of 0.8 reactor-yr /calendar-yr (Section 4) results in an estimated BWR SB LOCA frequency of 5.5E-04 per calendar year based on historical plant performance. From Table 7.17, the corresponding elicitation estimate of the current-day mean frequency is 6.5E-04 per calendar year.

Similarly, for US PWRs, there have been 9 total events (8 steam generator tube failures and 1 piping failure[5]) as of December 2006 resulting in leak rates that were greater than 100 gpm (380 lpm). This is the threshold leak rate defined for SB LOCAs in NUREG-1150 and NUREG/CR-5750 and Category 1 LOCAs in this report. Up to that point in time, US PWRs had operated for 1,986 reactor-years. Therefore, an operating experience-based estimate for the frequency of a PWR Category 1 LOCA (including the SGTR data) is (9/1986) = 4.5E-03 per reactor year, or 3.6E-03 per calendar year. From Table 7.17, the corresponding estimate of the current-day mean frequency is 7.3E-03 per calendar year.

---

[5] The one Category 1 event involving a US PWR plant was an instrument line fitting failure in November 1991 at Oconee Unit 3 which resulted in an observed maximum flow rate of about 130 gpm (490 lpm) (LER 50-287/1991-008 and IAEA IRS Report 1238). During this event approximately 87,000 gallons (330,000 liters) of reactor coolant was discharged into containment.

From the analysis above, the BWR elicitation estimate is less than 20 percent higher than the operating-experience estimate which, given the uncertainty of these estimates, is not statistically significant. There is a larger difference between the PWR estimates. The PWR elicitation estimate is about 100 percent (or a factor of two) higher than the operating-experience-based estimate. Additional insight into this difference can be gained by separately considering SGTR and other passive systems failure frequencies.

Based on the 8 SGTR events listed above, the estimated SGTR Category 1 LOCA frequency is (8/1986) = 4.0E-03 per reactor year, or 3.2E-03 per calendar year. From Table 7.18, the comparable current-day elicitation estimate is 3.7E-03 per calendar year. In light of the uncertainties, these two estimates are statistically equivalent. That is, the panelists' believe that the average, historical SGTR frequency is applicable to present conditions. This outcome is also supported by the panelists' qualitative rationale (Section 6.3.1). Note that the base case SGTR frequency in Section 4 is 3.5E-03 per calendar year and was determined similarly to the above operating-experience-based estimate The two values are slightly different due to the additional years of operating experience that have accumulated since the elicitation was conducted.

Similarly, the operating-experience-based estimate of the non-SGTR Category 1 LOCA frequency for the one documented event[5] is (1/1986) = 5.0E-04 per reactor year, or 4.0E-04 per calendar year. From Table 7.19, the current-day Category 1 elicitation estimate for non-SGTR events is 1.9E-03 per calendar year. This value is 5 times the value obtained from operating experience, and encompasses nearly all of the difference between the total PWR Category 1 estimates. This discrepancy also raises the possibility that the elicitation estimate may not be consistent with operating experience.

The importance of this discrepancy between the two frequency estimates can be assessed by first determining if it is statistically significant. To assess statistical significance, it is necessary to consider the uncertainty of the estimates. Based on the one observed non-SGTR piping failure, an upper 95 percent confidence bound for the failure frequency is (4.7/1986) = 2.4E-03 per reactor year or 1.9E-03 per calendar year [7.10]. Because this bound is equivalent to the elicitation estimate of 1.9E-03 per calendar year, it follows that the difference between the UB operating experience and elicitation frequency estimates is not statistically significant.

The factor of 5 discrepancy between the operating-experience and elicitation frequency estimates is based on one piping failure in US PWRs up to December 2006 that apparently exceeded the Category 1 leak rate threshold. However, one other piping system break was identified in NUREG/CR-5750 (Catawba Unit 1 in 1986, LER 413/86-031), with a leak rate of approximately 80 gpm (300 lpm). This leak rate is just below the Category 1 threshold of 100 gpm (380 lpm). Considering this event, a conservative assumption is that there have been two rather than one Category 1 LOCAs in the 1,986 reactor-years of experience. The resulting operating-experience-based non-SGTR Category 1 LOCA frequency is 8.0E-04 per calendar year with an upper 95 percent confidence bound of (0.8)(6.3)/(1986) = 2.5E-03 per calendar year. This bound is greater than the elicitation estimate of 1.9E-03 per calendar year. The conclusion is that the difference between the elicitation and operating-experience-based estimates of the non-SGTR, PWR Category 1 LOCA frequency, whether based on one or two observed events, is not statistically significant.

Even though the difference between the elicitation and operating-experience estimates of the non-SGTR PWR Category 1 LOCA frequency is not statistically significant, it must be emphasized that this does not imply that the two estimation procedures are in fact estimating the same frequency. Because the operating-experience-based estimate is an historical average based on many years of operation, a difference will exist if the panelists believe that the current failure frequency differs from the historical average. In fact, the increased elicitation estimate is supported by the panelists' qualitative and quantitative responses. The panelists indicated that medium and, to a lesser extent, small LOCAs in

PWRs are most dramatically impacted by PWSCC in relatively small diameter passive system component (e.g., CRDMs, instrument nozzles, etc.) (Section 6.3.2). This impact was previously reflected in the 20-fold increase in the MB PWR LOCA elicitation estimates compared to the NUREG/CR-5750 estimates (Section 7.9), and explains the smaller, five-fold increase in non-SGTR Category 1 failure frequencies.

7.11 References

7.1 Chokshi, N.C., Shaukat, S.K., Hiser A.L., DeGrassi, G., Wilkowski, G., Olson, R., and Johnson, J.J., "Seismic Considerations For the Transition Break Size," NUREG-1903, U.S. Nuclear Regulatory Commission, February 2008.

7.2 Meyer, M.A. and Booker, J.M., "Eliciting and Analyzing Expert Judgment: A Practical Guide," NUREG/CR-5424, U.S. Nuclear Regulatory Commission, January 1990.

7.3 Efron, B. and Tibshirani, R.J., "An Introduction to the Bootstrap," Chapman & Hall, 1993.

7.4 "Reactor Safety Study: An Assessment of Accident Risks in U.S. Commercial Nuclear Power Plants," WASH-1400, U.S. Nuclear Regulatory Commission, October 1975.

7.5 "Severe Accident Risks: An Assessment for Five U.S. Nuclear Power Plants," NUREG-1150, U.S. Nuclear Regulatory Commission, December 1990.

7.6 Poloski, J.P., Marksberry, D.G., Atwood, C.L., and Galyean, W.J., "Rates of Initiating Events at U.S. Nuclear Power Plants: 1987-1995," NUREG/CR-5750, February 1999.

7.7 "Individual Plant Examination Program: Perspective on Reactor Safety and Plant Performance," NUREG-1560, Vol. 2, Parts 2 – 5, December 1997.

7.8 Bush S., "Pipe Failures in U.S. Commercial Nuclear Power Plants," EPRI TR-100380, 1992.

7.9 Memorandum from A.C. Thadani to S.J. Collins, Transmittal of Technical Work to Support Possible Rulemaking on a Risk-Informed Alternative to 10 CFR 50.46/GDC 35, dated July 31, 2002.

7.10 Lurie, D. and Moore, R.H., "Applying Statistics," NUREG-1475, U.S. Nuclear Regulatory Commission, February 1994.

8. BASIS FOR FUTURE LOCA FREQUENCY UPDATES

The LOCA frequencies developed in this report will be periodically reevaluated to determine if they need to be updated based on information and knowledge gained subsequent to the expert elicitation. There are several areas of work ongoing or planned for the near future that will support/augment this effort. Those activities include:

    (1) updating the pipe failure database as part of the OPDE project,

    (2) developing a new PFM computer code for predicting the frequencies of various size LOCA events, and

    (3) expanding on the efforts of this program through a new international cooperative research program called MERIT (Maximizing Enhancements in Risk Informed Technology).

The OECD Pipe Failure Data Exchange (OPDE) project is a collaboration between twelve OECD member countries to capture international nuclear power plant piping event information in a central database. Because of the rarity of piping failures and precursor events, it is crucial to have both an accurate count of these events and a large pool of relevant reactor experience to draw from in order to develop meaningful piping failure frequency estimates. Data captured in the OPDE database allows for examination of LOCA precursor event occurrences versus time directly from operating experience. The database is striving for complete international data starting from 1998. Information from events dating back to the early 1970s will be included for some countries, including the United States. The OPDE database will form part of the technical database used to periodically update LOCA frequency estimates.

As a means of verifying/augmenting the results from this elicitation effort, additional probabilistic analyses will be conducted using a probabilistic LOCA code (PRO-LOCA) that is currently under development. This new code includes many deterministic aspects that have been developed in various NRC piping programs but not included in any of the currently available PFM codes, such as PRAISE, SRRA, or PRODIGAL. The PRO-LOCA code includes improved crack initiation/crack growth models, additional weld residual stress solutions, improved fracture models, improved leak-rate models with newly developed crack morphology parameters, and new methods for addressing multiple crack initiation sites and crack coalescence. While the code currently estimates frequencies through direct Monte-Carlo simulation, new more efficient processing routines (such as Importance Sampling) are currently under development. Once the development work on this code has been completed and the necessary benchmarking and quality assurance checks completed, the code will be exercised for a series of test cases as a means of validating the results from this elicitation effort. The final objective will be to combine this code with the OPDE database, to either directly calculate LOCA frequencies or use as a basis for conducting a subsequent elicitation to update LOCA frequency estimates contained in this report.

Additionally, a new international cooperative research program (MERIT) is being developed as a means of further developing the tools for making these types of probabilistic-based risk-informed decisions. The two main outcomes of the MERIT program as related to the determination of LOCA frequencies are the continued development of the previously mentioned probabilistic LOCA code (PRO-LOCA) and the further assessment of weld residual stresses and their impact on SCC. As part the continued development of the PRO-LOCA code, work within the MERIT program will focus on LOCA frequency contributions from non-piping component degradation. Also, an attempt will be made to reach an international consensus as to a set of standardized procedures for making these types of assessments. With regards to the further assessment of weld residual stresses and their impact on SCC, some of the key activities to be undertaken as part of the MERIT program are: (1) developing new solutions for additional bimetal weld locations and weld procedures, (2) examining the effect of weld repairs on the residual stress state, (3) identifying mitigation techniques/strategies for the re-qualification of leak-before-break (LBB) for bimetal welds susceptible to PWSCC, and (4) optimizing weld repair techniques to minimize weld residual stresses.

9. SUMMARY OF RESULTS AND CONCLUSIONS

An expert elicitation process has been used to consolidate operating experience and insights from PFM studies with knowledge of plant design, operation, and material performance to estimate LOCA frequencies. The approach allowed the panelists to decompose complex issues which impact LOCA frequencies into simpler, more fundamental issues which are easier to assess. The panelists were provided with frequency estimates for precursor LOCA events associated with degradation in piping and non-piping components. Additionally, selected panelists developed frequencies for several simplified base case conditions. Based on their knowledge of passive system component failure, the panelists extrapolated from this information to respond to a lengthy questionnaire structured to develop LOCA frequency contributions for potential piping and non-piping passive systems failures. The panelists' responses were processed to calculate LOCA frequency estimates that reflect individual panelist uncertainty and panel diversity.

The panelists also provided qualitative insights into a number of topics including evaluation of base case predictive methodologies; the effect of safety culture on LOCA frequencies; important degradation mechanisms; piping and non-piping LOCA frequency contributors; the effects of component size and operating time on LOCA frequencies; the influence of mitigation and maintenance; and uncertainty in making estimations, to name a few. The following insights were generally shared by most, if not all, of the panelists. Most panelists believe that operating-experience-based analyses provide the most accurate basis for estimating current-day (i.e., 25 year) frequencies of the simplified base case conditions. However, PFM studies also offer valuable insights into the possible future effect of degradation mechanisms. The use of sensitivity analysis, using either approach, is also valuable for identifying important contributing factors. Many panelists believe that combining information from both approaches is the optimal way to estimate LOCA frequencies, even for the simplified base case conditions.

There are several competing factors that affect the general safety culture of the industry and regulatory bodies. However, the panel members generally expressed the belief that the future safety culture will not differ dramatically from the current culture. In fact, most panelists expect a small improvement due mainly to continued nuclear experience and technology advancements. Because of this general expectation, the facilitation team elected not to adjust the LOCA frequency estimates provided by the panelists to account for safety culture effects. This decision was endorsed by the elicitation panelists. The panelists also generally expressed the view that utility and regulatory safety cultures are highly correlated. Many panelists also believe that safety culture can significantly affect LOCA frequencies at specific plants, either positively or negatively. However, there is also an expectation that timely regulatory actions using existing enforcement measures would diminish both the possibility and impact of safety culture deficiencies, and LOCA frequency increases, at particular plants. Many panelists believe that the possible effect of a deleterious safety culture motivates the NRC to remain vigilant in addressing individual plant deficiencies.

Many participants also believe that the number of precursor events (e.g., cracks and leaks) is a good barometer of the LOCA susceptibility to the associated degradation mechanism. Welds were almost universally recognized as important because of the high residual stress, the preferential attack of many degradations mechanisms, and the high likelihood of defects in these locations. Nozzle, elbows, and tees were also thought to be important locations. The panelists generally identified thermal fatigue, SCC, and mechanical fatigue as the degradation mechanisms which most significantly contribute to LOCA frequencies in PWR plants. These mechanisms and FAC are important LOCA frequency contributors for BWR plants. The panel consensus is that the susceptibility of BWR piping systems to IGSCC is greatly reduced compared to the past. However, this mechanism remains an important LOCA contributor in many of the panelists' estimates. The panelists were concerned that PWSCC is an important contributor to current-day PWR LOCA frequency estimates. However, most panelists expect that this mechanism will be mitigated within the next 15 years.

The panelists generally believe that complete rupture of a smaller pipe or non-piping component is more likely than an equivalent size opening in a larger pipe because of the increased susceptibility to fabrication or service cracking. Additionally, smaller bore piping, compared to larger bore piping, is more likely to have fabrication flaws, has lower inspection quality and quantity, and is more susceptible to external failure mechanisms arising from human error. This is a primary reason why the largest contributors in each LOCA category tend to be the smallest pipes which can lead to that size LOCA. There is substantially more disagreement about the most likely contributing systems for the intermediate-size (i.e., Category 3, 4) LOCAs because of the sheer number of possible contributing systems. However, many panelists thought that aging may have the greatest effect on intermediate diameter (6 to 14-inch) piping systems.

Generally, non-piping LOCA frequency assessment is believed to be more challenging than the piping assessments. There are multiple non-piping components to consider, each associated with different operating requirements, designs, materials, and inspection considerations. Furthermore, there is a dearth of precursor data available for the non-piping components compared with piping. Nevertheless, the panelists believe that non-piping components contribute significantly to Category 1 and 2 LOCA frequencies. For PWRs, non-piping components are the dominant contributor to Category 1 and 2 LOCA frequencies due to SGTRs and CRDM cracking. Postulated non-piping failures contribute much less significantly to Category 3, 4, and 5 LOCA frequency estimates. Non-piping contributions are more important for LOCA Category 6 because there is little PWR and no BWR piping which can result in this size flow rate at failure. Non-piping contributions are more significant in PWR plants because of the increased number of LOCA-sensitive components and past and current experience with degradation in these components.

Frequency estimates are not expected to change dramatically over the next 15 years for any size LOCA, or even over the next 35 years for smaller BWR (< Category 6) and PWR (< Category 3) LOCA categories. However, larger, (i.e., order of magnitude) frequency increases for the larger LOCA categories beyond the next 15 years are expected. There are several competing factors which will affect these future trends. Decreases in LOCA frequencies are expected to be driven by improvements in mitigation strategies and inspection techniques, material replacement, and continued reduction in fabrication-related problems. Increases in LOCA frequencies are primarily expected to be driven by continued aging, which will also be exacerbated by improper maintenance and other human errors.

However, while aging will continue, the panelists' consensus is that mitigation procedures are in place, or will be implemented in a timely manner, to alleviate significant increases in future LOCA frequencies due to known degradation mechanisms. Therefore, the significant increases expected beyond the next 15 years are largely due to uncertainty about the future and the concern that new degradation mechanisms could arise in the operating fleet after this time.

While the panelists generally agreed on the important technical issues and factors contributing to LOCAs, the quantitative estimates were much more difficult to assess because of the underlying scientific uncertainty and the lack of truly relevant LOCA data. The panelists generally expressed greater uncertainty in their predictions as the LOCA size increased. As noted, most panelists also believe that uncertainty increases with future operating time. Both trends a consequence of the greater extrapolation required of passive-system failures as the LOCA size and forecasting period increase. Panelist uncertainty was generally similar for BWR and PWR plants. Although, there are also significant differences among the panelists' estimates, panel diversity as measured by the quartiles or confidence bounds is generally much less than each panelist's uncertainties. Furthermore, panel diversity does not vary significantly with either LOCA size or plant type.

Baseline LOCA frequency estimates for the 5th percentile, median, mean and 95th percentile (i.e., bottom-line parameters) were calculated from each panelist's elicitation responses. These individual responses were then aggregated into a group estimate using the GM of each of these bottom-line parameter estimates. Group variability was measured by calculating 95 percent confidence bounds for each bottom-line estimate under the assumption that the individual estimates are lognormally distributed. The group estimates of the bottom-line parameters were calculated for each of the six LOCA size categories and for each of the three distinct time-periods: current-day (25 years fleet average), end-of–plant-license (40 years fleet average), and end-of-plant-license-renewal (60 years fleet average). Because of the predicted stability in these estimates over the near-term, it is recommended that the current-day (25 year) results be used to estimate the LOCA frequencies over the next 15 years of fleet operation.

While it is acknowledged that operating-experience-based estimates do not necessarily reflect current conditions, it is informative to compare such estimates with the elicitation frequencies for the smallest LOCAs (Category 1). Because no larger LOCAs have occurred, this comparison requires the least extrapolation of passive-system failure data. The BWR and PWR Category 1 LOCA frequencies (including the SGTR frequency for PWRs) were estimated up through December 2006. For BWR plants, the average SB LOCA frequency based solely on the number of reported events is 5.5E-04 per calendar year. The mean elicitation BWR SB LOCA estimate is 6.5E-04 per calendar year. Thus, the BWR elicitation estimate is less than 20 percent higher than the operating-experience-based estimate which, given the uncertainty of these estimates, is not statistically significant.

For PWR plants, the average SB LOCA frequency is 3.6E-03 per calendar year. The mean elicitation PWR SB LOCA estimate is 7.3E-03 per calendar year. The PWR elicitation estimate is about 80 percent higher than the operating-experience-based estimate. Additional insight into this difference can be gained by partitioning the PWR passive-system failure data into frequencies for SGTRs and for all other passive-system SB LOCAs. Based on reported failures, the average SGTR LOCA frequency is 3.2E-03 per calendar year. This result is almost identical to the current-day elicitation estimate of 3.7E-03 per calendar year. Based on operating experience, the average frequency of all other Category 1 PWR passive-system failures is 4.0E-04 per calendar year. The corresponding elicitation estimate is 1.9E-03 per calendar year. While this value is 5 times greater than the operating-experience-based estimate, this difference is explained by the elicitation panelists' assessment of the effect of PWSCC on small diameter component failures.

There are several prior studies that also estimated LOCA frequencies, including the most recent study documented in NUREG/CR-5750. Some care is needed when comparing the elicitation LOCA frequency estimates with these earlier studies because the LOCA categories are defined differently. Specifically, the current-day, LOCA Category 1 and 2 estimates in Table 7.17 are comparable to total system SB, MB, and LB LOCA frequencies, respectively, reported in NUREG/CR-5750. Additionally, current-day SGTR frequencies (Table 7.18) and PWR LOCA frequencies for all other passive system failures (i.e., frequencies for breaks greater than 100 gpm [380 lpm] in Table 7.19) are comparable to NUREG/CR-5750 SGTR and PWR SB LOCA frequencies when these failure modes are analyzed separately. The NUREG/CR-5750 LB LOCA frequency estimates are best compared to the elicitation LOCA Category 4 frequency estimates because the pipe break sizes are most similar.

After accounting for these differences, the elicitation LOCA frequency estimates are generally much lower than the WASH-1400 estimates and more consistent with the NUREG/CR-5750 estimates. The SB LOCA PWR elicitation estimates after subtracting the SGTR frequencies are approximately 3 times greater than the NUREG/CR-5750 estimates, due to the aforementioned PWSCC concerns. However, the total BWR and PWR SB LOCA frequency estimates are similar once the SGTR frequencies are added to the NUREG/CR-5750 PWR results. The elicitation MB LOCA estimates are higher than the NUREG/CR-5750 estimates by factors of approximately 4 and 20 for BWR and PWR plant types, respectively. These differences are partly due to concerns about PWSCC in piping and non-piping (e.g., CRDM) components as well as general aging concerns with piping in this size range. The NUREG/CR-5750 LB LOCA frequency estimates are slightly higher (less than a factor of 3) than the current elicitation results for both PWR and BWR plants. The generally good agreement between the NUREG/CR-5750 and current elicitation estimates is somewhat surprising given the markedly different methodologies used to arrive at these results.

A number of sensitivity studies were also conducted to examine the robustness of the quantitative results to the underlying analysis procedure. Sensitivity analyses investigated the effect of distribution shape on the means, and the effects of correlation structure, panel diversity measure, panelist overconfidence and aggregation measure on the estimated parameters. The mean calculation in this study used a split lognormal distribution truncated at the 99.9^{th} percentile to obtain reasonably conservative values compared with other possible choices. The choice of distribution shape is not significant for distributions with error factor less than 100, which represents the bulk of the elicitation responses. Split log-triangular and the baseline truncated split lognormal distribution are the most credible bounds for characterizing the elicitation responses. The difference in the means calculated for these distributions is less than a factor of 6 for distributions with error factors up to 1,000.

The correlation structure used in this study assumed maximal correlation, which is a reasonable characterization of the elicitation structure and the panelists' responses. An independent correlation structure provides an alternative bounding assumption. Based on selected Monte Carlo simulations, the maximal correlation structure provides 95th percentile estimates that are either conservative or not significantly different from those obtained using a bounding independent correlation structure. However, the independent correlation structure leads to median and 5th percentile estimates that are as much as one or two orders of magnitude higher than estimates calculated assuming maximal correlation. These differences increase with the error factor of the underlying distribution and the number of distributions which contribute to the total LOCA frequency estimates for each panelist. Because the relatively large sensitivity to the correlation structure is manifested for only some of the panelists, the sensitivity of the median and 5th percentile estimates to the correlation structure is less for the aggregated results. The means, of course, are unaffected by the choice of the correlation structure.

The assumption of a maximal correlation structure in this study is justified by the observation that the elicitation responses tend to be strongly correlated. Most panelists noted that degradation mechanisms that significantly contribute to the LOCA frequencies affect many systems similarly. That is, if conditions lead to higher failure probabilities in one piping system due to a specific degradation mechanism, the failure probabilities of other susceptible systems will also increase. This assumption is also justified since the correlation structure is not important for determining the means and 95th percentiles of the total LOCA frequency estimates. These are the most important parameters for most applications.

The analysis procedure in this study used confidence intervals for the group estimates as a measure of panel diversity. The 95% confidence bounds associated with each LOCA frequency parameter (i.e., the median, mean, 95th percentile) were calculated based on an assumed lognormal structure for the panelist's responses. The confidence bounds are based on the variance of all the panelists' estimates. An alternative approach used the interquartile ranges of the individual estimates to measure panel diversity. The interquartile ranges, by definition, encompass approximately one-half of the individual estimates. With their nominal 95 percent coverage, the confidence intervals are a more conservative measure of diversity than the interquartile ranges of the individual results. That is, the 95% confidence intervals based on the 2.5% and 97.5% confidence bounds cover at least half of all the individual estimates for each parameter.

The initial estimates discussed in Sections 7.2 – 7.5, and used as the basis for comparison of the sensitivity analyses in Section 7.6, did not adjust the individual estimates for panelist overconfidence. These results are referred to throughout the report as the *baseline* estimates. Sensitivity analyses were conducted to investigate the effects on the baseline estimates of adjusting individual estimates to account for panelists' overconfidence in their elicitation responses. The blanket overconfidence adjustments considered are difficult to justify and can result in large, unsupportable increases in the frequency estimates The blanket and more severe targeted adjustments generally lead to greatly skewed estimates (e.g., the mean is much greater than the 95th percentile) that are not consistent with either the elicitation structure or the panelists' responses. Additionally, the blanket adjustments can lead to high mean frequency estimates that are not supported by the operating experience. The targeted adjustment of only those panelists whose estimates express a relatively large certainty can be supported by the post-adjusted results. However, all these adjustment schemes require arbitrary assumptions about which panelists to adjust and the appropriate level of adjustment.

The error-factor adjustment scheme used in this study is less subject to these drawbacks. This adjustment scheme provides a variable overconfidence adjustment as a function of LOCA size, time period, and plant type. The adjustment is greatest for those panelists who expressed the least amount of uncertainty compared to the group average. This adjustment leads to a relatively small increase in the baseline LOCA frequency estimates and the results are comparable to the least severe targeted adjustment. The estimates calculated using the error-factor adjustment scheme are referred to throughout the report as the *summary* estimates. Because there is a known tendency for people, including experts, to be overconfident when making subjective judgments, these *summary* estimates are deemed to result in improved LOCA frequencies compared to the baseline estimates.

The largest sensitivity is associated with the method used to aggregate the individual panelist estimates to obtain group estimates. The baseline and summary estimates were developed using GM aggregation. In this study, the GM aggregation produces frequency estimates that approximate the median of the panelists' estimates and can therefore be interpreted as consensus-type results. Hence, the summary estimates in Table 7.17 are believed to be a reasonable representation of the expert panel's current state of knowledge regarding LOCA frequencies.

A sensitivity study evaluated the effect of using alternative aggregation methods to calculate the group estimates. Specifically, mixture-distribution and AM aggregation were evaluated. For the panelists' responses in this study, these alternative aggregation methods can lead to significantly higher mean and 95th percentile estimates than those obtained using GM aggregation. For example, while the mixture-distribution increases compared to the baseline estimates are less than a factor of 6 for approximately half of the six LOCA categories, the increases for the other LOCA categories vary from factors of 9 to 43. The LOCA categories exhibiting the greatest discrepancies all have a common attribute - there are one or two individual frequency estimates which are much greater than the other estimates. Alternative LOCA frequency estimates that are higher than the *summary* estimates (Table 7.17) can be derived by using either the summary estimates with 95% confidence bounds (Tables 7.8), the AM aggregated results (Table 7.13), or the mixture-distribution results (Table 7.16). These estimates also incorporate the same overconfidence adjustment as the summary estimates.

Because alternative aggregation methods can lead to significantly different results, a particular set of LOCA frequency estimates is not generically recommended for all risk-informed applications. The purposes and context of the application must be considered when determining the appropriateness of any set of elicitation results. In particular, during the selection of the BWR and PWR transition break sizes for the proposed 10 CFR50.46a rulemaking, the NRC staff considered the totality of the results from the sensitivity studies, rather than only the summary frequency estimates from this study. The NRC anticipates that a similar approach will be used in selecting appropriate replacement frequencies for the estimates provided in NUREG/CR-5750 and for other applications that require frequencies for break sizes other than those in NUREG/CR-5750. While the lack of clear application guidance places an additional burden on the users of the study results, those users are in the best position to judge which study results are most appropriate to consider for their particular applications.

NRC FORM 335
(9-2004)
NRCMD 3.7

U.S. NUCLEAR REGULATORY COMMISSION

BIBLIOGRAPHIC DATA SHEET

(See instructions on the reverse)

1. REPORT NUMBER
(Assigned by NRC, Add Vol., Supp., Rev., and Addendum Numbers, if any.)

NUREG-1829, Vol. 1

2. TITLE AND SUBTITLE	3. DATE REPORT PUBLISHED	
Estimating Loss-of-Coolant Accident (LOCA) Frequencies Through the Elicitation Process Main Report	MONTH	YEAR
	April	2008
	4. FIN OR GRANT NUMBER	
	N6360	

5. AUTHOR(S)	6. TYPE OF REPORT
Robert Tregoning and Lee Abramson, US NRC Paul Scott, Battelle	Technical
	7. PERIOD COVERED *(Inclusive Dates)*
	N/A

8. PERFORMING ORGANIZATION - NAME AND ADDRESS *(If NRC, provide Division, Office or Region, U.S. Nuclear Regulatory Commission, and mailing address; if contractor, provide name and mailing address.)*

Division of Engineering
Office of Regulatory Research
U.S. Nuclear Regulatory Commission
Washington, DC 20555-0001

Battelle-Columbus
505 King Avenue
Columbus, OH 43201

9. SPONSORING ORGANIZATION - NAME AND ADDRESS *(If NRC, type "Same as above"; if contractor, provide NRC Division, Office or Region, U.S. Nuclear Regulatory Commission, and mailing address.)*

Same as Above

10. SUPPLEMENTARY NOTES

A. Csontos, NRC Project Manager

11. ABSTRACT *(200 words or less)*

The NRC is developing a risk-informed revision of the design-basis pipe break size requirements in 10 CFR 50.46, Appendix K to Part 50, and GDC 35 which requires estimates of loss-of-coolant-accident (LOCA) frequencies as a function of break size. Separate BWR and PWR piping and non-piping passive system LOCA frequency estimates were developed as a function of effective break size and operating time through the end of license extension. The estimates were based on an expert elicitation process which consolidated service history data and insights from probabilistic fracture mechanics studies with knowledge of plant design, operation, and material performance.

The elicitation required each member of an expert panel to qualitatively and quantitatively assess important LOCA contributing factors and quantify their uncertainty. The quantitative responses were combined to develop BWR and PWR total LOCA frequency estimates for each contributing panelist. The individual estimates were then aggregated to obtain group estimates, along with measures of panel diversity. Sensitivity studies were conducted to examine the effects of distribution shape, correlation structure, panelist overconfidence, measures of panel diversity, and aggregation method. The group estimates are most sensitive to the method used to aggregate the individual estimates.

12. KEY WORDS/DESCRIPTORS *(List words or phrases that will assist researchers in locating the report.)*	13. AVAILABILITY STATEMENT
piping risk-informed emergency core cooling system (ECCS) loss-of-coolant accident (LOCA) break frequencies design-basis break size LOCA frequency estimates expert elicitation aging	unlimited
	14. SECURITY CLASSIFICATION
	(This Page) unclassified
	(This Report) unclassified
	15. NUMBER OF PAGES
	16. PRICE

www.ingramcontent.com/pod-product-compliance
Lightning Source LLC
Chambersburg PA
CBHW080240180526
45167CB00006B/2355